Linear Algebra Thoroughly Explained

Milan Vujičić

Linear Algebra Thoroughly Explained

Springer

Author
Milan Vujičić
(1931–2005)

Editor
Jeffrey Sanderson
Emeritus Professor,
School of Mathematics & Statistics,
University of St Andrews,
St Andrews,
Scotland

ISBN: 978-3-642-09410-1   e-ISBN: 978-3-540-74639-3

Cover Design: eStudio Calamar S.L.

Printed on acid-free paper

9 8 7 6 5 4 3 2 1

springer.com

Foreword

There are a zillion books on linear algebra, yet this one finds its own unique place among them. It starts as an introduction at undergraduate level, covers the essential results at postgraduate level and reaches the full power of the linear algebraic methods needed by researchers, especially those in various fields of contemporary theoretical physics, applied mathematics and engineering. At first sight, the title of the book may seem somewhat pretentious but it faithfully reflects its objective and, indeed, its achievements.

Milan Vujičić started his scientific carrier in theoretical nuclear physics in which he relied heavily in his research problems on linear algebraic and group theoretic methods. Subsequently, he moved to the field of group theory itself and its applications in various topics in physics. In particular, he achieved, together with Fedor Herbut, important results in the foundations of and distant correlations in quantum mechanics, where his understanding and skill in linear algebra was precedent. He was known as an acute and learned mathematical physicist.

At first Vujičić taught group theory at graduate level. However, his teaching career blossomed when he moved to the Physics Faculty of the University of Belgrade, and it continued, even after retirement, at the University of Malta, where he taught linear algebra at the most basic level to teaching diploma students. He continuously interested himself in the problems of teaching, and with worthy results. Indeed, his didactic works were outstanding and he was frequently singled out by students, in their teaching evaluation questionnaires, as a superb teacher of mathematical physics.

This book is based on lectures that Vujičić gave to both undergraduate and postgraduate students over a period of several decades. Its guiding principle is to develop the subject rigorously but economically, with minimal prerequisites and with plenty of geometric intuition. The book offers a practical system of studies with an abundance of worked examples coordinated in such a way as to permit the diligent student to progress continuously from the first easy lessons to a real mastery of the subject. Throughout this book, the author has succeeded in maintaining rigour while giving the reader an intuitive understanding of the subject. He has imbued the book with the same good sense and helpfulness that characterized his teaching during his lifetime. Sadly, having just completed the book, Milan Vujičić suddenly died in December 2005.

Having known Milan well, as my thesis advisor, a colleague and a dear friend, I am certain that he would wish this book to be dedicated to his wife Radmila and his sons Boris and Andrej for their patience, support and love.

Belgrade, July 2007   *Djordje Šijački,*

Acknowledgements

Thanks are due to several people who have helped in various ways to bring Professor Vujičić's manuscript to publication. Vladislav Pavlovič produced the initial Latex copy, and subsequently, Dr. Patricia Heggie provided timely and invaluable technical help in this area. Professors John Cornwell and Nikola Ruskuc of the University of St. Andrews read and made helpful comments upon the manuscript in the light of which Professor Milan Damnjanovič of the University of Belgrade made some amendments. Finally, it is a pleasure to thank Professor Djordje Šijački of the University of Belgrade and the Serbian Academy of Sciences for writing the Foreword.

Contents

Chapter 1
Vector Spaces

1.1 Introduction

The idea of a *vector* is one of the greatest contributions to mathematics, which came directly from physics. Namely, vectors are basic mathematical objects of classical physics since they describe physical quantities that have both magnitude and direction (displacement, velocity, acceleration, forces, e.g. mechanical, electrical, magnetic, gravitational, etc.).

Geometrical vectors (arrows) in two-dimensional planes and in the three-dimensional space (in which we live) form real vector spaces defined by the addition of vectors and the multiplication of numbers with vectors. To be able to describe lengths and angles (which are essential for physical applications), real vector spaces are provided with the *dot product* of two vectors. Such vector spaces are then called *Euclidean spaces*.

The theory of real vector spaces can be generalized to include other sets of objects: the set of all real matrices with m rows and n columns, the set of all real polynomials, the set of all real polynomials whose order is smaller than $n \in \mathbb{N}$, the set of all real functions which have the same domain of definition, the sets of all continuous, differentiable or integrable functions, the set of all solutions of a given homogeneous system of linear equations, etc. Most of these generalized vector spaces are many-dimensional.

The most typical and very useful are the vector spaces of matrix-columns $\bar{x} = [x_1\ x_2 \ldots x_n]^T$ of n rows and one column, where $n = 2, 3, 4, \ldots$, and the components x_i, $i = 1, 2, \ldots, n$, are real numbers. We denote these spaces by $\mathbb{R}^n$, which is the usual notation for the sets of ordered n-tuples of real numbers. (The ordered n-tuples can, of course, be represented by matrix-rows $[x_1,\ x_2 \ldots x_n]$ as well, but the matrix columns are more appropriate when we deal with matrix transformations in $\mathbb{R}^n$, which are applied to the left $A\bar{x}$, where A is an $m \times n$ real matrix.) We shall call the elements of $\mathbb{R}^n$ n-vectors.

The vector spaces $\mathbb{R}^n$ for $n = 2, 3$, play an important role in geometry, describing lines and planes, as well as the area of triangles and parallelograms and the volume of a parallelepiped.

The vector spaces $\mathbb{R}^n$ for $n > 3$ have no geometrical interpretation. Nevertheless, they are essential for many problems in mathematics (e.g. for systems of linear equations), in physics ($n = 4$, space–time events in the special theory of relativity), as well as in economics (linear economic models).

Modern physics, in particular Quantum Mechanics, as well as the theory of elementary particles, uses complex vector spaces. As far as Quantum Mechanics is concerned, there were at first two approaches: the wave mechanics of Schrödinger and the matrix mechanics of Heisenberg. Von Neumann proved that both are isomorphic to the infinite dimensional unitary (complex) vector space (called *Hilbert space*). The geometry of Hilbert space is now universally accepted as the mathematical model for Quantum Mechanics.

The *Standard Model* of elementary particles treats several sets of particles. One set comprises quarks, which initially formed a set of only three particles, described by means of the $SU(3)$ group (unitary-complex-3×3 matrices with unit determinant) but became a set of six particles, described by the $SU(6)$ group.

1.2 Geometrical Vectors in a Plane

A geometrical vector in a Euclidean plane E_2 is defined as a directed line segment (an *arrow*).

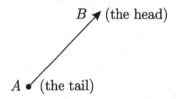

It has its initial point A (the tail) and its terminal point B (the arrow-head). The usual notation for the vector is $\overrightarrow{AB}$. Vectors have two characteristic properties—the length $\| \overrightarrow{AB} \|$ (a positive number, also called the *norm*) and the direction (this means that we know the line on which the segment lies and the direction in which the arrow points).

Two vectors are considered *"equal"* if they have the same length and are lying on parallel lines having the same direction. In other words, they are equal if they can be placed one on the top of the other by a translation in the plane.

$$\overrightarrow{AB} = \overrightarrow{CD}$$

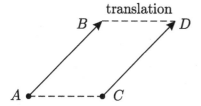

This relation in the set of all vectors in the plane is obviously reflexive, symmetric and transitive (an *equivalence relation*)(verify), so it produces a partition in this set into equivalence classes of equal vectors. We shall denote the set of all equivalence classes by V_2 and choose a representative of each class as we find it convenient.

The most convenient way to choose a representative is to select a point in E_2 and declare it as the origin O. Then, we define as the representative of each class that vector $\bar{a}$ from the class whose initial point is O.

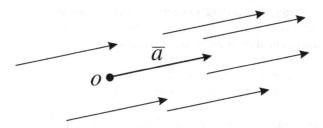

Several vectors from the class $[\bar{a}]$ represented by the vector $\bar{a}$ which starts at O.

We can now define a binary operation in V_2 ($V_2 \times V_2 \rightarrow V_2$) called the *addition of classes* by defining the addition of representatives by the parallelogram rule: the representatives $\bar{a}$ and $\bar{b}$ of two classes $[\bar{a}]$ and $[\bar{b}]$ form the two sides of a parallelogram.

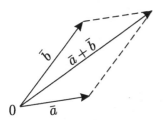

The diagonal from O represents the class $[\bar{a}+\bar{b}]$. That this is the correct definition of the addition of classes becomes obvious when we verify that the sum of any other vector from the class $[\bar{a}]$ with any vector from the class $[\bar{b}]$ will be in the class $[\bar{a}+\bar{b}]$. Take any vector from the class $[\bar{a}]$, and any vector from the class $[\bar{b}]$, and bring by translation the vector from $[\bar{b}]$ to the terminal point of the vector from $[\bar{a}]$:

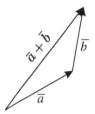

Now, connect the initial point of the first vector to the terminal point of the second vector (we "add" the second vector to the first one). This is the triangle rule for the addition of individual vectors, and it clearly shows that the sum belongs to the class $[\bar{a}+\bar{b}]$.

The addition of all vectors from the class $[\bar{a}]$ with all vectors from the class $[\bar{b}]$ will give precisely all vectors from the class $[\bar{a}+\bar{b}]$. (We have already proved that the sum of any vector from $[\bar{a}]$ with any vector from $[\bar{b}]$ will be a vector from $[\bar{a}+\bar{b}]$). Now we can prove that all vectors from $[\bar{a}+\bar{b}]$ are indeed sums of the above kind, since every vector from $[\bar{a}+\bar{b}]$ can be immediately decomposed as such a sum: from its initial point draw a vector from $[\bar{a}]$ and at the terminal point of this new vector start a vector from $[\bar{b}]$, whose terminal point will coincide with the terminal point of the original vector. Δ

(One denotes both the addition of numbers and the addition of vectors by the same sign "+," since there is no danger of confusion—one cannot add numbers to vectors).

It is obvious that the above binary operation is defined for every two representatives of equivalence classes (it is a closed operation—the *first* property).

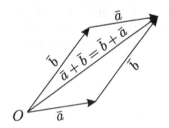

The addition of vectors is commutative

$$\bar{a}+\bar{b}=\bar{b}+\bar{a},$$

as can be seen from the diagram. This is the *second* property.

This operation is also associative (see the diagram), so it can be defined for three or more vectors:

$$(\bar{a}+\bar{b})+\bar{c}=\bar{a}+(\bar{b}+\bar{c})=\bar{a}+\bar{b}+\bar{c}.$$

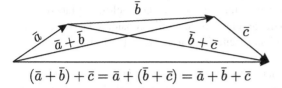

$$(\bar{a} + \bar{b}) + \bar{c} = \bar{a} + (\bar{b} + \bar{c}) = \bar{a} + \bar{b} + \bar{c}$$

We simply "add" one vector after another. This is the *third* property.

Each vector $\bar{a}$ has its unique negative $-\bar{a}$ *(the additive inverse)*, which has the same length, lies on any of parallel lines, and has the opposite direction of the arrow:

This is the *fourth property*.

When we add $\bar{a}$ and $-\bar{a}$, we get a unique vector whose initial and terminal points coincide—the zero vector $\bar{0}$: $\bar{a} + (-\bar{a}) = \bar{0}$. The vector $\bar{0}$ has length equal to zero and has no defined direction. It is the *additive identity (neutral)* since $\bar{a} + \bar{0} = \bar{a}$. This is the *fifth* property of vector addition.

It follows that the addition of vectors makes V_2 an *Abelian group* since all five properties of this algebraic structure are satisfied in vector addition: it is a closed operation which is commutative and associative, each vector has a unique inverse, and there exists a unique identity.

1.3 Vectors in a Cartesian (Analytic) Plane $\mathbb{R}^2$

Any pair of perpendicular axes (directed lines) passing through the origin O, with marked unit lengths on them, is called a *rectangular coordinate system*.

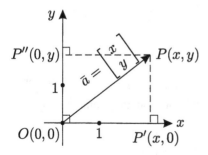

Each point P in the plane has now two coordinates (x, y), which are determined by the positions of two orthogonal projections P' and P'' of P on the coordinate axes x and y, respectively.

Thus, each rectangular coordinate system transforms the Euclidean plane into a Cartesian (analytic) plane. Analytic geometry is generally considered to have been founded by the French mathematician Rene Descartes (Renatus Cartesius) in the first half of the 17th century.

This means that every coordinate system introduces a *bijection* (1-1 and onto mapping) between the set of all points in the plane E_2 and the set $\mathbb{R}^2 = \mathbb{R} \times \mathbb{R}$ of all ordered pairs of real numbers. We usually identify E_2 with $\mathbb{R}^2$, having in mind that this identification is different for each coordinate system.

We have defined as the most natural representative of each equivalence class of equal geometrical vectors that vector $\bar{a}$ from the class which has its initial point at the origin $O(0,0)$. Such a vector is determined only by the coordinates (x,y) of its terminal point $P(x,y)$. We say that $\bar{a}$ is the position vector of $P(x,y)$ and denote $\bar{a} = \overrightarrow{OP}$ by the coordinates of P arranged as a matrix-column $\bar{a} = \begin{bmatrix} x \\ y \end{bmatrix}$ (this arrangement is more convenient than a matrix-row $[x \ y]$ when we apply different matrix transformations from the left). We call x and y components of $\bar{a}$, and we say that the matrix-column $\begin{bmatrix} x \\ y \end{bmatrix}$ represents $\bar{a}$ in the given coordinate system.

From now on, we shall concentrate our attention on the set of all matrix-columns $\begin{bmatrix} x \\ y \end{bmatrix}$, $x, y \in \mathbb{R}$, which can also be denoted by $\mathbb{R}^2$ (ordered pairs of real numbers arranged as number columns). We call each $\begin{bmatrix} x \\ y \end{bmatrix}$ a 2-vector, since it represents one equivalence class of geometrical vectors in E_2 (one element from V_2).

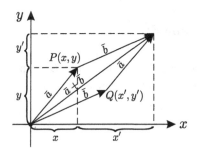

The addition in $\mathbb{R}^2$ of two position vectors $\bar{a} = \begin{bmatrix} x \\ y \end{bmatrix}$ and $\bar{b} = \begin{bmatrix} x \\ y \end{bmatrix}$ can be performed component-wise as the addition between the corresponding components (see the diagram)

$$\bar{a} + \bar{b} = \begin{bmatrix} x \\ y \end{bmatrix} + \begin{bmatrix} x' \\ y' \end{bmatrix} = \begin{bmatrix} x + x' \\ y + y' \end{bmatrix}.$$

(Note that this is the general rule for addition of matrices of the same size).

Since the components are real numbers, and the addition of real numbers makes $\mathbb{R}$ an Abelian group, we immediately see that $\mathbb{R}^2$ is also an Abelian group with respect to this addition of matrix-columns:

$\mathbb{R}^2$ is obviously closed under $+$, since every two matrix-columns can be added to give the third one;

This addition is commutative $\bar{a} + \bar{b} = \begin{bmatrix} x + x' \\ y + y' \end{bmatrix} = \begin{bmatrix} x' + x \\ y' + y \end{bmatrix} = \bar{b} + \bar{a}$;

It is also associative, let $\bar{c} = \begin{bmatrix} x'' \\ y'' \end{bmatrix}$, then

$$(\bar{a} + \bar{b}) + \bar{c} = \begin{bmatrix} (x + x') + x'' \\ (y + y') + y'' \end{bmatrix} = \begin{bmatrix} x + (x' + x'') \\ y + (y' + y'') \end{bmatrix} = \bar{a} + (\bar{b} + \bar{c}) = \bar{a} + \bar{b} + \bar{c};$$

There is a unique additive identity (neutral) called the *zero vector* $\bar{0} = \begin{bmatrix} 0 \\ 0 \end{bmatrix}$, such that

$$\bar{a} + \bar{0} = \begin{bmatrix} x + 0 \\ y + 0 \end{bmatrix} = \begin{bmatrix} x \\ y \end{bmatrix} = \bar{a};$$

Every vector $\bar{a} = \begin{bmatrix} x \\ y \end{bmatrix}$ has a unique additive inverse (the *negative vector*) $-\bar{a} = \begin{bmatrix} -x \\ -y \end{bmatrix}$, so that $\bar{a} + (-\bar{a}) = \begin{bmatrix} x + (-x) \\ y + (-y) \end{bmatrix} = \begin{bmatrix} 0 \\ 0 \end{bmatrix} = \bar{0}.$

1.4 Scalar Multiplication (The Product of a Number with a Vector)

It is natural to denote $\bar{a} + \bar{a}$ as $2\bar{a}$ and so on, motivating introduction of the number-vector product, which gives another vector: for every $c \in \mathbb{R}$ and for every $\bar{a} \in \mathbb{R}^2$, we define the product $c\bar{a} = c \begin{bmatrix} x \\ y \end{bmatrix} = \begin{bmatrix} cx \\ cy \end{bmatrix}$, which is a vector parallel to $\bar{a}$, and having the same $(c > 0)$ or the opposite $(c < 0)$ direction as $\bar{a}$. Since the length of vector $\bar{a}$ is obviously $||\bar{a}|| = \sqrt{x^2 + y^2}$, the length of $c\bar{a}$ is $||c\bar{a}|| = \sqrt{c^2 x^2 + c^2 y^2} = |c|\, ||\bar{a}||$.

This is an $\mathbb{R} \times \mathbb{R}^2 \to \mathbb{R}^2$ mapping usually called scalar multiplication, since real numbers are called scalars in tensor algebra.

Scalar multiplication is a closed operation (defined for every $c \in \mathbb{R}$ and for every $\bar{a} \in \mathbb{R}^2$). Since it is an $\mathbb{R} \times \mathbb{R}^2 \to \mathbb{R}^2$ mapping, it must be related to defining operations in $\mathbb{R}$ (which is a field) and in $\mathbb{R}^2$ (which is an Abelian group). These operations are the addition and multiplication of numbers in $\mathbb{R}$, and the addition of vectors in $\mathbb{R}^2$.

(i) The distributive property of the addition of numbers with respect to scalar multiplication:

$$(c+d)\bar{a} = (c+d)\begin{bmatrix} x \\ y \end{bmatrix} = \begin{bmatrix} (c+d)x \\ (c+d)y \end{bmatrix} = \begin{bmatrix} cx+dx \\ cy+dy \end{bmatrix} = \begin{bmatrix} cx \\ cy \end{bmatrix} + \begin{bmatrix} dx \\ dy \end{bmatrix} = c\bar{a} + d\bar{a};$$

(ii) The associative property of the multiplication of numbers with respect to scalar multiplication:

$$(cd)\bar{a} = (cd)\begin{bmatrix} x \\ y \end{bmatrix} = \begin{bmatrix} cdx \\ cdy \end{bmatrix} = c\begin{bmatrix} dx \\ dy \end{bmatrix} = c(d\bar{a});$$

(iii) The distributive property of the addition of vectors with respect to scalar multiplication:

$$c(\bar{a}+\bar{b}) = c\left(\begin{bmatrix} x \\ y \end{bmatrix} + \begin{bmatrix} x' \\ y' \end{bmatrix}\right) = c\begin{bmatrix} x+x' \\ y+y' \end{bmatrix} = \begin{bmatrix} cx+cx' \\ cy+cy' \end{bmatrix} = c\begin{bmatrix} x \\ y \end{bmatrix} + c\begin{bmatrix} x' \\ y' \end{bmatrix} = c\bar{a} + c\bar{b};$$

for all $c, d \in \mathbb{R}$ and all $\bar{a}, \bar{b} \in \mathbb{R}^2$.

(iv) $1\bar{a} = \bar{a}$ (the number 1 is neutral for both the multiplication of numbers, $1c = c$, and the multiplication of numbers with vectors, $1\bar{a} = \bar{a}$).

Definition Vector addition (with the five properties of an Abelian group) and scalar multiplication (with the four properties above) make $\mathbb{R}$ an algebraic structure called a *real vector space*.

Since $\mathbb{R}^2$ represents V_2, the set of the equivalence classes of equal vectors, it means that V_2 is also a real vector space.

When the two operations (vector addition and scalar multiplication) that define real vector spaces are combined, for instance in $c_1\bar{a}_1 + c_2\bar{a}_2 + \ldots + c_n\bar{a}_n = \sum_{i=1}^{n} c_i\bar{a}_i$, $c_1, c_2, \ldots, c_n \in \mathbb{R}$, $\bar{a}_1, \bar{a}_2, \ldots, \bar{a}_n \in \mathbb{R}^2$, then one is talking about a *linear combination*. This is the most general operation that can be performed in a real vector space, and it characterizes that algebraic structure.

1.5 The Dot Product of Two Vectors (or the Euclidean Inner Product of Two Vectors in $\mathbb{R}^2$)

Note: This subject is treated in detail in Sect. 3.1.

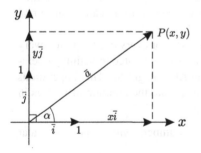

If we choose two unit vectors (orts) $\bar{i}$ and $\bar{j}$ in the directions of the x and y axes, respectively, then we immediately see that $\bar{a}$ is a linear combination of two vector components $x\bar{i}$ and $y\bar{j}$: $\bar{a} = x\bar{i} + y\bar{j}$ or $\begin{bmatrix} x \\ y \end{bmatrix} = x \begin{bmatrix} 1 \\ 0 \end{bmatrix} + y \begin{bmatrix} 0 \\ 1 \end{bmatrix}$. This is a unique expansion of $\bar{a}$ in terms of $\bar{i}$ and $\bar{j}$, so we call $\{\bar{i}, \bar{j}\}$ a basis in $\mathbb{R}^2$, and since $\bar{i}$ and $\bar{j}$ are orthogonal unit vectors, we say that it is an orthonormal (ON) basis.

The scalar projection x of $\bar{a}$ on the direction of ort $\bar{i}$ is the result of the obvious formula $x/||\bar{a}|| = \cos \alpha \Rightarrow x = ||\bar{a}|| \cos \alpha$.

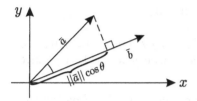

Similarly, the scalar projection of $\bar{a}$ on any other vector $\bar{b}$ is obtained as $||\bar{a}|| \cos \theta$, where θ is the smaller angle ($0° \leq \theta \leq 180°$) between $\bar{a}$ and $\bar{b}$.

But, in physics, if $\bar{a}$ is a force and we want to calculate the work W done by this force in producing a displacement $\bar{b}$, then this work is the product of the scalar projection of the force on the direction of displacement ($||\bar{a}|| \cos \theta$) by the length of this displacement ($||\bar{b}||$): $W = ||\bar{a}|| \cos \theta \cdot ||\bar{b}|| == ||\bar{a}|| \, ||\bar{b}|| \cos \theta$. This expression for W is denoted as $\bar{a} \cdot \bar{b}$ and called the *dot product* of the force and the displacement:

$$W = \bar{a} \cdot \bar{b} = ||\bar{a}|| \, ||\bar{b}|| \cos \theta.$$

The dot product is a $\mathbb{R}^2 \times \mathbb{R}^2 \to \mathbb{R}$ map, since the result is a number.

The principal properties of the dot product are

1. The dot product is commutative: $\bar{a} \cdot \bar{b} = \bar{b} \cdot \bar{a}$ (obvious);
2. It is distributive with regard to vector addition

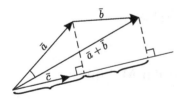

$$(\bar{a} + \bar{b}) \cdot \bar{c} = \bar{a} \cdot \bar{c} + \bar{b} \cdot \bar{c},$$

since the scalar projection of $\bar{a} + \bar{b}$ along the line of vector $\bar{c}$ is the sum of the projections of $\bar{a}$ and $\bar{b}$;

3. It is associative with respect to scalar multiplication $k(\bar{a} \times \bar{b}) = (k\bar{a}) \cdot \bar{b} = \bar{a} \cdot (k\bar{b})$. For $k > 0$ it is obvious, since $k\bar{a}$ and $k\bar{b}$ have the same direction as $\bar{a}$ and $\bar{b}$, respectively.

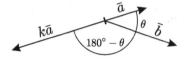

For $k < 0$, we have $(k\bar{a}) \cdot \bar{b} = |k| \, ||\bar{a}|| \, ||\bar{b}|| \cos(180° - \theta) = k(\bar{a} \cdot \bar{b})$, since $|k| = -k$ and $\cos(180° - \theta) = -\cos\theta$;

4. The dot product is strictly positive $\bar{a} \cdot \bar{a} = ||\bar{a}||^2 > 0$ if $\bar{a} \neq \bar{0}$ and $\bar{a} \cdot \bar{a} = 0$ iff $\bar{a} = \bar{0}$ (obvious), so only the zero vector $\bar{0}$ has zero length, other vectors have positive lengths.

Note that the two nonzero vectors $\bar{a}$ and $\bar{b}$ are perpendicular (orthogonal) if and only if their dot product is zero:

$$||\bar{a}|| \, ||\bar{b}|| \cos\theta = 0 \Leftrightarrow \cos\theta = 0 \Leftrightarrow \theta = 90°.$$

Making use of the above properties 2 and 3, as well as of the dot-multiplication table for $\bar{i}$ and $\bar{j}$

$$\begin{array}{c|cc} \cdot & \bar{i} & \bar{j} \\ \hline \bar{i} & 1 & 0 \\ \bar{j} & 0 & 1 \end{array},$$

one can express the dot product of $\bar{a} = x\bar{i} + y\bar{j}$ and $\bar{b} = x'\bar{i} + y'\bar{j}$ in terms of their components:

$$\bar{a} \cdot \bar{b} = (x\bar{i} + y\bar{j}) \cdot (x'\bar{i} + y'\bar{j}) = xx'\bar{i} \cdot \bar{i} + xy'\bar{i} \cdot \bar{j} + yx'\bar{j} \cdot \bar{i} + yy'\bar{j} \cdot \bar{j} = xx' + yy'.$$

This expression is how the dot product is usually defined in $\mathbb{R}^2$. It should be emphasized that in another coordinate system this formula will give the same value, since it is always equal to $||\bar{a}|| \, ||\bar{b}|| \cos\theta$.

Note that the dot product for three vectors is meaningless.

1.6 Applications of the Dot Product and Scalar Multiplication

A. The length (norm) $||\bar{a}||$ of a vector $\bar{a} = \begin{bmatrix} x \\ y \end{bmatrix} = x\bar{i} + y\bar{j}$ can be expressed by the *dot product*:

since $\bar{a} \cdot \bar{a} = ||\bar{a}||^2 \cos 0° = ||\bar{a}||^2 = x^2 + y^2$, it follows that

$$||\bar{a}|| = (\bar{a} \cdot \bar{a})^{1/2} = \sqrt{x^2 + y^2}.$$

The cosine of the angle θ between $\bar{a} = \begin{bmatrix} x \\ y \end{bmatrix}$ and $\bar{b} = \begin{bmatrix} x' \\ y' \end{bmatrix}$ is obviously

$$\cos\theta = \frac{\bar{a}\cdot\bar{b}}{||\bar{a}||\,||\bar{b}||} = \frac{xx'+yy'}{\sqrt{x^2+y^2}\sqrt{x'^2+y'^2}}, \; 0° \le \theta \le 180°.$$

The unit vector (*ort*) $\bar{a}_0$ in the direction of $\bar{a}$ is obtained by dividing $\bar{a}$ by its length $\bar{a}_0 = \frac{\bar{a}}{||\bar{a}||}$, so that $\bar{a} = ||\bar{a}||\bar{a}_0$ and $||\bar{a}_0|| = 1$.

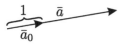

The components (scalar projections) x, y of $\bar{a} = \begin{bmatrix} x \\ y \end{bmatrix}$ are the result of dot-multiplication of $\bar{a}$ with $\bar{i}$ and $\bar{j}$, respectively:

$$x = \bar{a}\cdot\bar{i} = x\cdot 1 + y\cdot 0, \; y = \bar{a}\cdot\bar{j} = x\cdot 0 + y\cdot 1 \Rightarrow \bar{a} = (\bar{a}\cdot\bar{i})\bar{i} + (\bar{a}\cdot\bar{j})\bar{j}.$$

The distance $d(A,B)$ between two points $A(x,y)$ and $B(x',y')$ is the length of the difference $\bar{a} - \bar{b} = \begin{bmatrix} x-x' \\ y-y' \end{bmatrix}$ of their position vectors $\bar{a} = \begin{bmatrix} x \\ y \end{bmatrix}$ and $\bar{b} = \begin{bmatrix} x' \\ y' \end{bmatrix}$:

$$d(A,B) = ||\bar{a}-\bar{b}|| = \sqrt{(x-x')^2 + (y-y')^2}.$$

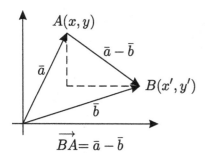

B. The Cauchy–Schwarz inequality is an immediate consequence of the definition of the dot product: since $|\cos\theta| \le 1$ for any angle θ, we have that $\bar{a}\cdot\bar{b} = ||\bar{a}||\,||\bar{b}||\cos\theta$ implies $|\bar{a}\cdot\bar{b}| \le ||\bar{a}||\,||\bar{b}||$.

The *triangle inequality* is a direct consequence of the Cauchy–Schwarz inequality $-||\bar{a}||\,||\bar{b}|| \le \bar{a}\cdot\bar{b} \le ||\bar{a}||\,||\bar{b}|| : (*)$

$$\|\bar{a}+\bar{b}\|^2 = (\bar{a}+\bar{b})\cdot(\bar{a}+\bar{b}) =$$

$$= \|\bar{a}\|^2 + 2(\bar{a}\cdot\bar{b}) + \|\bar{b}\|^2 \overset{(*)}{\leq} \|\bar{a}\|^2 +$$
$$+ 2\|\bar{a}\|\,\|\bar{b}\| + \|\bar{b}\|^2 = (\|\bar{a}\|+\|\bar{b}\|)^2,$$

that implies $\|\bar{a}+\bar{b}\|\leq\|\bar{a}\|+\|\bar{b}\|$, which means that the length of a side of a triangle does not exceed the sum of the lengths of the other two sides.

C. One of the most important theorems in trigonometry—*the cosine rule*—can be easily obtained by means of the dot product. If the sides of a triangle are represented by vectors $\bar{a},\bar{b},\bar{c}$, such that $\bar{c} = \bar{a}+\bar{b}$, then

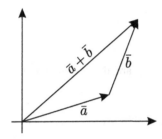
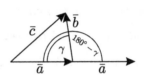

$\bar{c}\cdot\bar{c} = (\bar{a}+\bar{b})\cdot(\bar{a}+\bar{b})$ which implies $\|\bar{c}\|^2 = \|\bar{a}\|^2 + \|\bar{b}\|^2 + 2(\bar{a}\cdot\bar{b}) = \|\bar{a}\|^2 + \|\bar{b}\|^2 + 2\|\bar{a}\|\,\|\bar{b}\|\cos(180°-\gamma)$, and finally $c^2 = a^2 + b^2 - 2ab\cos\gamma$, where $\|\bar{a}\| = a, \|\bar{b}\| = b, \|\bar{c}\| = c$, and $\cos(180°-\gamma) = -\cos\gamma$.

D. One can easily prove (using the dot product) that the three altitudes in a triangle *ABC* are concurrent.

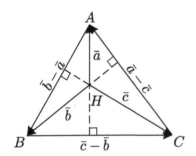

Let altitudes through A and B intersect at H, and let us take H as the origin. Then let $\bar{a},\bar{b},\bar{c}$ be the position vectors of A,B,C. Since $\overrightarrow{HA}= \bar{a}$ and $\overrightarrow{BC}= \bar{c}-\bar{b}$ are perpendicular to each other, we have $\bar{a}\cdot(\bar{c}-\bar{b}) = 0$ or $\bar{a}\cdot\bar{c} = \bar{a}\cdot\bar{b}$. Similarly, $\overrightarrow{HB}\cdot\overrightarrow{CA}= 0 \Rightarrow \bar{b}\cdot(\bar{a}-\bar{c}) = 0$ or $\bar{b}\cdot\bar{c} = \bar{a}\cdot\bar{b}$. Subtracting these equations, one gets $(\bar{b}-\bar{a})\cdot\bar{c} = 0$ or $\overrightarrow{AB}\cdot\overrightarrow{HC}= 0$ or $\overrightarrow{AB}\perp\overrightarrow{HC}$. Therefore, H lies on the third altitude (through C), and the three altitudes in ABC are concurrent (at H which is called the *orthocenter*).

E. (1) As two simple and useful applications of scalar multiplication, let us consider the section formula (the ratio theorem) and the position vector of the centroid of a triangle.

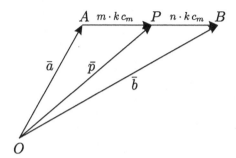

The section formula gives us the position vector $\bar{p}$ of point P specified by its position ratio with respect to two fixed points A and B:

$$\frac{AP}{PB} = \frac{m \cdot kc_m}{n \cdot kc_m} = \frac{m}{n} \Rightarrow AP = \frac{m}{n} PB.$$

Since vectors $\overrightarrow{AP} = \bar{p} - \bar{a}$ and $\overrightarrow{PB} = \bar{b} - \bar{p}$ lie on the same line, one is the scalar multiple of the other

$$\bar{p} - \bar{a} = \frac{m}{n}(\bar{b} - \bar{p}) \Rightarrow n\bar{p} - n\bar{a} = m\bar{b} - m\bar{p} \Rightarrow$$

$$\Rightarrow (m+n)\bar{p} = m\bar{b} + n\bar{a}, \text{ and finally}$$

$$\bar{p} = \frac{m\bar{b} + n\bar{a}}{m+n} \quad \text{(the section formula)}.$$

The mid point of AB $(m = n)$ has the position vector $\bar{p} = \frac{\bar{a} + \bar{b}}{2}$.

(2) Consider an arbitrary triangle ABC. Let D, E, F be the mid-points of the sides BC, CA, AB, respectively. The medians of the triangle are the lines AD, BE, CF. We shall show, by the methods of vector algebra [see (1) above], that these three lines are concurrent.

Let G be defined as the point on the median AD such that $\frac{AG}{GD} = \frac{2}{1}$, and hence, by the section formula

$$\bar{g} = \frac{2\bar{d} + \bar{a}}{3}.$$

As D is the mid-point of BC, its position vector is

$$\bar{d} = \frac{\bar{b} + \bar{c}}{2}.$$

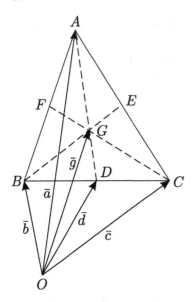

Substituting this vector in the expression for $\bar{g}$, we have

$$g = \frac{\bar{a} + \bar{b} + \bar{c}}{3}.$$

Because this expression for $\bar{g}$ is completely symmetrical in $\bar{a}, \bar{b}, \bar{c}$, we would obtain the same answer if we calculated the position vectors of the points on the other two medians BE and CF corresponding to the ratio $2 : 1$ (verify). Therefore, the point G lies on all three medians. It is called the *centroid* of ABC.

F. Let us prove that the line segment that joins the mid-points of two sides of a triangle is parallel to and precisely one-half the length of the third side. Thus,

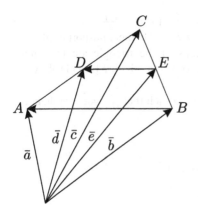

$\bar{d} = \frac{\bar{a}+\bar{c}}{2}$ and $\bar{e} = \frac{\bar{b}+\bar{c}}{2} \Rightarrow \bar{d} - \bar{e} = \frac{1}{2}(\bar{a} - \bar{b})$, so $\overrightarrow{ED} = \frac{1}{2}\,\overrightarrow{BA}$. Therefore, $\overrightarrow{ED}$ is

parallel to $\overrightarrow{BA}$, since it is a scalar multiple of $\overrightarrow{BA}$, and its length $\| \overrightarrow{ED} \|$ is $\frac{1}{2}$ of
$\| \overrightarrow{BA} \|$.

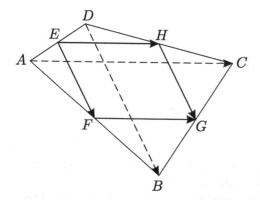

Using this result, one can easily prove that the mid-points of the sides of any quadrilateral $ABCD$ are the vertices of a parallelogram:

$$\overrightarrow{EH} = \frac{1}{2}\,\overrightarrow{AC} = \overrightarrow{FG},$$

$$\overrightarrow{EF} = \frac{1}{2}\,\overrightarrow{DB} = \overrightarrow{HG}.$$

1.7 Vectors in Three-Dimensional Space (Spatial Vectors)

The notion of a (geometric) vector in three-dimensional Euclidean space E_3 is the same as in two-dimensional space (plane)—it is a directed line segment (an arrow).

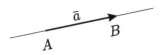

It has a length and a direction defined by the line in space on which the segment lies and the arrow which points to one or the other side. We denote it by its end points as $\overrightarrow{AB}$ or simply by $\bar{a}$. The vectors that have the same length and lie on parallel lines (with the same direction of the arrow) are considered to be equal. This means that the set of all vectors in E_3 is partitioned into equivalence classes of equal vectors. We shall denote the set of all these classes as V_3. Any choice of three mutually perpendicular axes with unit lengths on them and concurrent at the origin O

introduces a rectangular coordinate system in E_3 and a bijection between the set of all points in E_3 and the set $\mathbb{R}^3$ of all ordered triples (x, y, z) of real numbers.

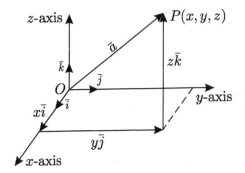

(We choose a right-handed coordinate system - x, y, z axes point as the thumb, the index finger and the middle finger of the right hand.)

The natural representative of a class of equal vectors is the vector from the class whose initial point is at the origin $O(0,0,0)$. This vector is denoted by the coordinates of its terminal point

$$P(x,y,z) : \bar{a} = \overrightarrow{OP} = \begin{bmatrix} x \\ y \\ z \end{bmatrix}$$

arranged as a matrix-column. Here, x, y, z are called the scalar-components of $\bar{a}$.

We can use $\mathbb{R}^3$ to denote both the set of all points in E_3 and the set of their position vectors.

Note that the representative column $\begin{bmatrix} x \\ y \\ z \end{bmatrix}$ of $\bar{a}$ depends on the choice of the rectangular coordinate system (with the same origin O). Later on (in Sect. 4.4), we shall prove the basic transformation formula: In another rectangular coordinate system obtained by the orthogonal replacement matrix $\mathcal{R}$ $(\mathcal{R}^{-1} = \mathcal{R}^T)$

$$\begin{bmatrix} \bar{i}_1 \\ \bar{j}_1 \\ \bar{k}_1 \end{bmatrix} = \mathcal{R} \begin{bmatrix} \bar{i} \\ \bar{j} \\ \bar{k} \end{bmatrix}.$$

The representative matrix-column of $\bar{a}$ changes analogously

$$\begin{bmatrix} x_1 \\ y_1 \\ z_1 \end{bmatrix} = \mathcal{R} \begin{bmatrix} x \\ y \\ z \end{bmatrix}.$$

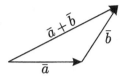

We define the addition in V_3 analogously as in V_2 by "adding" $\bar{b}$ to the terminal point of $\bar{a}$ (the *triangle rule*) or in a component-wise manner by adding the corresponding components of the natural representatives of the classes in the chosen rectangular coordinate system

$$\bar{a} + \bar{b} = \begin{bmatrix} x \\ y \\ z \end{bmatrix} + \begin{bmatrix} x' \\ y' \\ z' \end{bmatrix} = \begin{bmatrix} x+x' \\ y+y' \\ z+z' \end{bmatrix}.$$

This addition of vectors makes V_3, as well as $\mathbb{R}^3$, an Abelian group (the properties and proofs are the same as in the two-dimensional cases).

The scalar multiplication and the dot product are also defined by analogy with $\mathbb{R}^2$:

$$c\bar{a} = c \begin{bmatrix} x \\ y \\ z \end{bmatrix} = \begin{bmatrix} cx \\ cy \\ cz \end{bmatrix} \text{ and } \bar{a} \cdot \bar{b} = xx' + yy' + zz'.$$

Example

$$\bar{v}_1 = \begin{bmatrix} 5 \\ -1 \\ 3 \end{bmatrix} \text{ and } \bar{w} = \begin{bmatrix} 4 \\ 2 \\ 1 \end{bmatrix}, \text{ then } \bar{v} + \bar{w} = \begin{bmatrix} 9 \\ 1 \\ 4 \end{bmatrix},$$

$$\bar{v} \cdot \bar{w} = 20 - 2 + 3 = 21, \, 2\bar{v} = \begin{bmatrix} 10 \\ -2 \\ 6 \end{bmatrix}, \, -\bar{w} = \begin{bmatrix} -4 \\ -2 \\ -1 \end{bmatrix}.$$

The principal properties of scalar multiplication and of the dot product are the same as in the two-dimensional case. This means that both $\mathbb{R}^3$ and V_3 are real vector spaces. Due to the dot product having the mentioned properties, $\mathbb{R}^3$ and V_3 are called *Euclidean vector spaces* (see Sect. 3.1).

The rectangular coordinate system is determined by three perpendicular (orthogonal) unit vectors (orts) $\bar{i} = \begin{bmatrix} 1 \\ 0 \\ 0 \end{bmatrix}, \bar{j} = \begin{bmatrix} 0 \\ 1 \\ 0 \end{bmatrix}, \bar{k} = \begin{bmatrix} 0 \\ 0 \\ 1 \end{bmatrix}$, so that each vector $a = \begin{bmatrix} x \\ y \\ z \end{bmatrix}$ can be written as the unique sum of its vector-components (or unique linear combination of $\bar{i}, \bar{j}, \bar{k}$)—see the diagram on p. 15:

$$\bar{a} = x\bar{i} + y\bar{j} + z\bar{k}, \text{ where}$$

$x = \bar{a} \cdot \bar{i}, \, y = \bar{a} \cdot \bar{j}, \, z = \bar{a} \cdot \bar{k}$ and

$$\begin{array}{c|ccc}
\cdot & \bar{i} & \bar{j} & \bar{k} \\
\hline
\bar{i} & 1 & 0 & 0 \\
\bar{j} & 0 & 1 & 0 \\
\bar{k} & 0 & 0 & 1
\end{array}.$$

We say that the vectors $\bar{i}, \bar{j}, \bar{k}$ form an orthonormal (ON) basis in $\mathbb{R}^3$.

1.8 The Cross Product in $\mathbb{R}^3$

The *cross product* is a binary vector operation (a mapping $\mathbb{R}^3 \times \mathbb{R}^3 \to \mathbb{R}^3$), which is only meaningful in $\mathbb{R}^3$. To every ordered pair of vectors in $\mathbb{R}^3$

$$\bar{a} = \begin{bmatrix} x \\ y \\ z \end{bmatrix} = x\bar{i} + y\bar{j} + z\bar{k} \text{ and } \bar{b} = \begin{bmatrix} x' \\ y' \\ z' \end{bmatrix} = x'\bar{i} + y'\bar{j} + z'\bar{k},$$

we associate a vector $\bar{c} = \bar{a} \times \bar{b} = x''\bar{i} + y''\bar{j} + z''\bar{k}$ that is perpendicular to each of them:

$$\bar{a} \cdot \bar{c} = 0 \text{ and } \bar{b} \cdot \bar{c} = 0.$$

This is a system of two homogeneous linear equations in three unknowns x'', y'', z'':

$$xx'' + yy'' + zz'' = 0 \text{ and } x'x'' + y'y'' + z'z'' = 0.$$

Here, we have fewer equations than unknowns, so we shall have to introduce a free parameter s for one of the unknowns (say z'') and x'' and y'' will be expressed in terms of s. Geometrically, solving such a homogeneous system boils down to finding the *kernel* of the coefficient matrix $A = \begin{bmatrix} x & y & z \\ x' & y' & z' \end{bmatrix}$. This is normally done by reducing this matrix to the unique row-echelon Gauss–Jordan modified (GJM) form (see the end of Sect. 2.18).

In this system, this takes the form

$$\begin{bmatrix} 1 & 0 & \dfrac{zy' - yz'}{xy' - yx'} \\ 0 & 1 & \dfrac{xz' - zx'}{xy' - yx'} \\ 0 & 0 & -1 \end{bmatrix},$$

where the last column is a unique basis vector of ker A.

The general solution of the system, i.e. the *kernel* of A is the line in $\mathbb{R}^3$

$$\begin{bmatrix} x'' \\ y'' \\ z'' \end{bmatrix} = \left\{ s \begin{bmatrix} \dfrac{zy' - yz'}{xy' - yx'} \\ \dfrac{xz' - zx'}{xy' - yx'} \\ -1 \end{bmatrix} : s \in \mathbb{R} \right\}$$

To simplify this expression, we can replace the free parameter s by another $s = -k(xy' - yx')$, $k \in \mathbb{R}$, and finally get

$$\left.\begin{array}{l} x'' = k(yz' - zy') \\ y'' = k(zx' - xz') \\ z'' = k(xy' - yx') \end{array}\right\} k \in \mathbb{R}.$$

Obviously $k = 1$ is the simplest solution, so

$$\bar{c} = (yz' - zy')\bar{i} + (zx' - yz')\bar{j} + (xy' - yx')\bar{k}.$$

A much less sophisticated and more transparent method can be performed as follows: Multiply the first equation by $(-x')$ and the second by x. Now, add the expressions so obtained giving

$$\frac{y''}{zx' - xz'} = \frac{z''}{xy' - yx'}.$$

Then, multiply the first equation by y' and the second by $(-y)$. Adding the expressions so obtained we have

$$\frac{x''}{yz' - zy'} = \frac{z''}{xy' - yx'}.$$

Since these three quotients are equal, but arbitrary, we introduce a free parameter $k \in \mathbb{R}$:

$$\frac{x''}{yz' - zy'} = \frac{y''}{zx' - xz'} = \frac{z''}{xy' - yx'} = k \in \mathbb{R}.$$

Naturally, we have the same situation as with the GJM method, and the simplest solution is again $k = 1$.

The components of the vector $\bar{c}$ can be written as determinants of 2×2 matrices:

$$\bar{c} = \bar{a} \times \bar{b} = \begin{vmatrix} y & z \\ y' & z' \end{vmatrix}\bar{i} - \begin{vmatrix} x & z \\ x' & z' \end{vmatrix}\bar{j} + \begin{vmatrix} x & y \\ x' & y' \end{vmatrix}\bar{k}.$$

Note: From now on we shall need several statements from the theory of determinants (see Appendix A).

For instance, the above expression for $\bar{c}$ can be interpreted as a symbolic determinant (in the first row are vectors and not numbers, but we can apply the rule for expanding a determinant about its first row since in vector algebra we have the two essential operations: the addition of vectors and the multiplication of vectors with scalars):

$$\bar{c} = \bar{a} \times \bar{b} = \begin{vmatrix} \bar{i} & \bar{j} & \bar{k} \\ x & y & z \\ x' & y' & z' \end{vmatrix}.$$

This new operation is not commutative, but instead it is anticommutative: $\bar{a} \times \bar{b} = -(\bar{b} \times \bar{a})$, since the interchange of two rows in a determinant changes its sign.

It is also not associative: $\bar{a} \times (\bar{b} \times \bar{c}) \neq (\bar{a} \times \bar{b}) \times \bar{c}$. The last property follows immediately if we calculate $\bar{a} \times (\bar{b} \times \bar{c})$ for three arbitrary vectors $\bar{a} = \begin{bmatrix} x \\ y \\ z \end{bmatrix}$,

$\bar{b} = \begin{bmatrix} x' \\ y' \\ z' \end{bmatrix}, \bar{c} = \begin{bmatrix} x'' \\ y'' \\ z'' \end{bmatrix}$:

$$\bar{a} \times (\bar{b} \times \bar{c}) = \begin{vmatrix} \bar{i} & \bar{j} & \bar{k} \\ x & y & z \\ \begin{vmatrix} y'z' \\ y''z'' \end{vmatrix} & \begin{vmatrix} z'x' \\ z''x'' \end{vmatrix} & \begin{vmatrix} x'y' \\ x''y'' \end{vmatrix} \end{vmatrix} =$$
$$= \bar{i}[y(x'y'' - y'x'') - z(z'x'' - x'z'')] + \bar{j} + \bar{k} \text{ components} =$$
$$= \bar{i}[x'(xx'' + yy'' + zz'') - x''(xx' + yy' + zz')] + \bar{j} + \bar{k} \text{ components} =$$
$$= (\bar{a} \cdot \bar{c})\bar{b} - (\bar{a} \cdot \bar{b})\bar{c}.$$

Now consider

$$(\bar{a} \times \bar{b}) \times \bar{c} = -\bar{c} \times (\bar{a} \times \bar{b}) = -[(\bar{c} \cdot \bar{b})\bar{a} - (\bar{c} \cdot \bar{a})\bar{b}] =$$
$$= (\bar{a} \cdot \bar{c})\bar{b} - (\bar{c} \cdot \bar{b})\bar{a}.$$

Obviously, the first terms $(\bar{a} \cdot \bar{c})\bar{b}$ agree, while the second ones $(\bar{a} \cdot \bar{b})\bar{c}$ and $(\bar{c} \cdot \bar{b})\bar{a}$ differ. So, $\bar{a} \times (\bar{b} \times \bar{c}) \neq (\bar{a} \times \bar{b}) \times \bar{c}$. Δ It follows that one cannot define the cross product $\bar{a} \times \bar{b} \times \bar{c}$ for three vectors, because it is not an associative binary operation.

The relation of the cross product with vector addition is $\bar{a} \times (\bar{b} + \bar{c}) = \bar{a} \times \bar{b} + \bar{a} \times \bar{c}$ (the distributive law), which can be obtained from

$$\begin{vmatrix} \bar{i} & \bar{j} & \bar{k} \\ x & y & z \\ x'+x'' & y'+y'' & z'+z'' \end{vmatrix} = \begin{vmatrix} \bar{i} & \bar{j} & \bar{k} \\ x & y & z \\ x' & y' & z' \end{vmatrix} + \begin{vmatrix} \bar{i} & \bar{j} & \bar{k} \\ x & y & z \\ x'' & y'' & z'' \end{vmatrix}.$$

It follows from an analogous argument that

$$(\bar{a} + \bar{b}) \times \bar{c} = \bar{a} \times \bar{c} + \bar{b} \times \bar{c}.$$

The relation of the cross product with scalar multiplication is

$$k(\bar{a} \times \bar{b}) = (k\bar{a}) \times \bar{b} = \bar{a} \times (k\bar{b}), \; k \in \mathbb{R}, \text{ (the associative law)},$$

since $k\begin{vmatrix} \bar{i} & \bar{j} & \bar{k} \\ x & y & z \\ x' & y' & z' \end{vmatrix}$ is equal to the same determinant in which the second or the third row is multiplied by the scalar k.

Finally, $\bar{a} \times \bar{a} = \bar{0}$ follows from $\begin{vmatrix} \bar{i} & \bar{j} & \bar{k} \\ x & y & z \\ x & y & z \end{vmatrix} = 0\bar{i} + 0\bar{j} + 0\bar{k}$. An important consequence

is that two nonzero vectors $\bar{a}$ and $\bar{b}$ are parallel iff $\bar{a} \times \bar{b} = \bar{0}$. This can be proved by observing that if $\bar{a}$ and $\bar{b}$ are parallel, then $\bar{a} = k\bar{b}$, so that $\bar{a} \times \bar{b} = k(\bar{b} \times \bar{b}) = \bar{0}$.

On the other hand, $\bar{a} \times \bar{b} = \bar{0} \Rightarrow \|\bar{a} \times \bar{b}\| = \|\bar{a}\| \, \|\bar{b}\| \sin \theta = 0 \Rightarrow \sin \theta = 0 \Rightarrow \theta = 0°$ or $\theta = 180°$ (since $0° \le \theta \le 180°$) (for $\|\bar{a} \times \bar{b}\|$ see Sect. 1.9).

As far as the three orthogonal unit vectors $\bar{i} = \begin{bmatrix} 1 \\ 0 \\ 0 \end{bmatrix}, \bar{j} = \begin{bmatrix} 0 \\ 1 \\ 0 \end{bmatrix}, \bar{k} = \begin{bmatrix} 0 \\ 0 \\ 1 \end{bmatrix}$ are

concerned, their cross-product table is as follows

$\times$	$\bar{i}$	$\bar{j}$	$\bar{k}$
$\bar{i}$	0	$\bar{k}$	$-\bar{j}$
$\bar{j}$	$-\bar{k}$	0	$\bar{i}$
$\bar{k}$	$\bar{j}$	$-\bar{i}$	0

, since

$$\bar{i} \times \bar{j} = \begin{vmatrix} \bar{i} & \bar{j} & \bar{k} \\ 1 & 0 & 0 \\ 0 & 1 & 0 \end{vmatrix} = \bar{k}, \bar{i} \times \bar{k} = \begin{vmatrix} \bar{i} & \bar{j} & \bar{k} \\ 1 & 0 & 0 \\ 0 & 0 & 1 \end{vmatrix} = -\bar{j} \text{ and } \bar{j} \times \bar{k} = \begin{vmatrix} \bar{i} & \bar{j} & \bar{k} \\ 0 & 1 & 0 \\ 0 & 0 & 1 \end{vmatrix} = \bar{i}.$$

This table has zeros on the main diagonal and is *skew symmetric* with respect to this diagonal since the cross product is anticommutative.

From $\bar{i} \times \bar{j} = \bar{k}$, it follows that the direction of $\bar{a} \times \bar{b}$ is such that $\bar{a}, \bar{b}, \bar{a} \times \bar{b}$ form a right-handed system.

The cross product of vectors in $\mathbb{R}^3$ plays an essential role in the theoretical formulation of Mechanics and Electromagnetism.

1.9 The Mixed Triple Product in $\mathbb{R}^3$. Applications of the Cross and Mixed Products

Since both the dot and the cross products do not exist for three vectors, we can consider only the mixed triple product (the dot product in relation to the cross product):

$$\bar{a} \cdot (\bar{b} \times \bar{c}) = x \begin{vmatrix} y' & z' \\ y'' & z'' \end{vmatrix} + y \begin{vmatrix} z' & x' \\ z'' & x'' \end{vmatrix} + z \begin{vmatrix} x' & y' \\ x'' & y'' \end{vmatrix} = \begin{vmatrix} x & y & z \\ x' & y' & z' \\ x'' & y'' & z'' \end{vmatrix},$$

$$\text{where } \bar{a} = \begin{bmatrix} x \\ y \\ z \end{bmatrix}, \bar{b} = \begin{bmatrix} x' \\ y' \\ z' \end{bmatrix}, \bar{c} = \begin{bmatrix} x'' \\ y'' \\ z'' \end{bmatrix}.$$

The mixed triple product is a number, since this is a proper determinant. The above equality represents the first-row expansion of the determinant. The third-row expansion of the same determinant gives

$$\begin{vmatrix} y & z \\ y' & z' \end{vmatrix} x'' + \begin{vmatrix} z & x \\ z' & x' \end{vmatrix} y'' + \begin{vmatrix} x & y \\ x' & y' \end{vmatrix} z'' = (\bar{a} \times \bar{b}) \cdot \bar{c}.$$

From this, we conclude that the signs $\cdot$ and $\times$ can be interchanged: $\bar{a} \cdot (\bar{b} \times \bar{c}) = (\bar{a} \times \bar{b}) \cdot \bar{c}$ (note that the cross product must be kept in brackets). For this reason, the mixed triple product is often denoted as

$$[\bar{a}\bar{b}\bar{c}] = \begin{vmatrix} x & y & z \\ x' & y' & z' \\ x'' & y'' & z'' \end{vmatrix}.$$

Obvious additional properties are

$$[\bar{a}\bar{b}\bar{c}] = [\bar{c}\bar{a}\bar{b}] = [\bar{b}\bar{c}\bar{a}] = -[\bar{c}\bar{b}\bar{a}] = -[\bar{a}\bar{c}\bar{b}] = -[\bar{b}\bar{a}\bar{c}],$$

since every interchange of two rows in a determinant changes its sign.

Applications of the Cross and Mixed Products

A. *The area of a parallelogram and of a triangle.*
    The length of $\bar{a} \times \bar{b}$ can be determined as follows:

$$\begin{aligned}
||\bar{a} \times \bar{b}||^2 &= (\bar{a} \times \bar{b}) \cdot (\bar{a} \times \bar{b}) = \bar{a} \cdot [\bar{b} \times (\bar{a} \times \bar{b})] = \\
&= \bar{a} \cdot [(\bar{b} \cdot \bar{b})\bar{a} - (\bar{a} \cdot \bar{b})\bar{b}] = ||\bar{a}||^2 ||\bar{b}||^2 - (\bar{a} \cdot \bar{b})^2 = \\
&= ||\bar{a}||^2 ||\bar{b}||^2 - ||\bar{a}||^2 ||\bar{b}||^2 \cos^2 \theta = ||\bar{a}||^2 ||\bar{b}||^2 \sin^2 \theta, \text{ and finally} \\
&||\bar{a} \times \bar{b}|| = ||\bar{a}|| \, ||\bar{b}|| \sin \theta, \text{ since } \sin \theta \geq 0 \text{ for } 0° \leq \theta \leq 180°.
\end{aligned}$$

Now, we can calculate the area A of the parallelograms determined by two vectors $\bar{a}$ and $\bar{b}$:

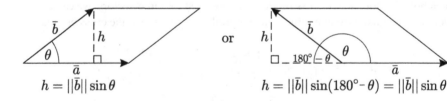

$$h = ||\bar{b}|| \sin \theta \qquad\qquad h = ||\bar{b}|| \sin(180° - \theta) = ||\bar{b}|| \sin \theta$$

$$A = ||\bar{a}|| \, h = ||\bar{a}|| \, ||\bar{b}|| \sin \theta = ||\bar{a} \times \bar{b}||.$$

The area A_Δ of triangles determined by vectors $\bar{a}$ and $\bar{b}$ is calculated analogously:

$$A_\Delta = \frac{1}{2} ||\bar{a}|| \, h = \frac{1}{2} ||\bar{a}|| \, ||\bar{b}|| \sin \theta = \frac{1}{2} ||\bar{a} \times \bar{b}||.$$

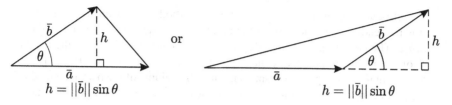

$$h = \|\bar{b}\| \sin\theta \qquad\qquad h = \|\bar{b}\| \sin\theta$$

B. *The sine rule.*

The sine rule can be easily obtained by means of a cross product. The three cross products $\bar{a} \times \bar{b}, \bar{b} \times \bar{c}, \bar{c} \times \bar{a}$ in the triangle ABC have the same lengths equal to twice the area of $\triangle ABC$:

$$2A_{\triangle ABC} = \|\bar{a} \times \bar{b}\| = \|\bar{b} \times \bar{c}\| = \|\bar{c} \times \bar{a}\| \text{ or}$$
$$ab\sin\gamma = bc\sin\alpha = ca\sin\beta,$$
$$\text{where } a = \|\bar{a}\|, b = \|\bar{b}\|, c = \|\bar{c}\|.$$

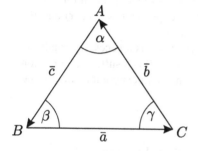

Finally, $ab\sin\gamma = ac\sin\beta \Rightarrow \dfrac{b}{\sin\beta} = \dfrac{c}{\sin\gamma}$, and

$$ca\sin\beta = cb\sin\alpha \Rightarrow \dfrac{a}{\sin\alpha} = \dfrac{b}{\sin\beta},$$

giving the *sine rule* $\dfrac{a}{\sin\alpha} = \dfrac{b}{\sin\beta} = \dfrac{c}{\sin\gamma}.$

C. *The volume of a parallelepiped.*

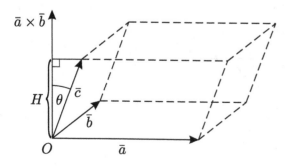

We introduced $\bar{a} \times \bar{b}$ as a vector perpendicular to both $\bar{a}$ and $\bar{b}$, and consequently, it is perpendicular to the base of the parallelepiped determined by the vectors $\bar{a}, \bar{b}, \bar{c}$. The positive projection $||\bar{c}||\,|\cos\theta|$ of $\bar{c}$ on the line of $\bar{a} \times \bar{b}$ is the height H of the parallelepiped. (Note that $0° \leq \theta \leq 180°$, so that $1 \geq \cos\theta \geq -1$, but the height must be a positive number). Since the volume of the parallelepiped is the product of the area of the base $||\bar{a} \times \bar{b}||$ with the height $||\bar{c}||\,|\cos\theta|$, we have

$$V = ||\bar{a} \times \bar{b}||\,||\bar{c}||\,|\cos\theta| = |(\bar{a} \times \bar{b}) \cdot \bar{c}| = |[\bar{a}\bar{b}\bar{c}]|, \text{ where}$$

$$[\bar{a}\bar{b}\bar{c}] = \begin{vmatrix} x & y & z \\ x' & y' & z' \\ x'' & y'' & z'' \end{vmatrix}, \ \bar{a} = \begin{vmatrix} x \\ y \\ z \end{vmatrix}, \ \bar{b} = \begin{vmatrix} x' \\ y' \\ z' \end{vmatrix}, \ \bar{c} = \begin{vmatrix} x'' \\ y'' \\ z'' \end{vmatrix}$$

(the absolute value of the mixed triple product).

If the three vectors $\bar{a}, \bar{b}, \bar{c}$ lie in the same plane (i.e. if they are coplanar), then they cannot form a parallelepiped and $[\bar{a}\bar{b}\bar{c}] = 0$. This is not only a necessary condition for coplanarity of three vectors but also a sufficient condition: if three vectors lie in one plane, then they are linearly dependent (see later) and the determinant of their column vectors is always zero.

More concisely, coplanarity $\Rightarrow [\bar{a}\bar{b}\bar{c}] = 0$ ($[\bar{a}\bar{b}\bar{c}] = 0$ is a necessary condition for coplanarity), $[\bar{a}\bar{b}\bar{c}] = 0 \Rightarrow$ coplanarity ($[\bar{a}\bar{b}\bar{c}] = 0$ is a sufficient condition for linear dependence of $\bar{a}, \bar{b}, \bar{c}$, i.e., any of them is a linear combination of the others, so they lie in the same plane).

1.10 Equations of Lines in Three-Dimensional Space

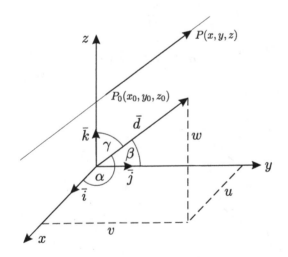

Given a fixed point $P_0(x_0, y_0, z_0)$ on a line and the direction vector

$$\bar{d} = \begin{bmatrix} u \\ v \\ w \end{bmatrix}$$

parallel to the line, we see that

$$\overrightarrow{P_0P} = t\bar{d},$$

where $P(x, y, z)$ is any point on the line, so that the parameter t takes any real value $(-\infty < t < \infty)$. This is the vector equation of the line.

Since the components of the vector $\overrightarrow{P_0P}$ are

$$x - x_0, \ y - y_0, \ z - z_0,$$

one has the parametric equations of the line $(x - x_0 = tu, \ldots)$:

$$x = x_0 + tu, \ y = y_0 + tv, \ z = z_0 + tw,$$

or the symmetric form of the equations of the line

$$\frac{x - x_0}{u} = \frac{y - y_0}{v} = \frac{z - z_0}{w} = t.$$

Taking into account that $u = \bar{d} \cdot \bar{i} = ||\bar{d}|| \cos \alpha$, $v = \bar{d}\bar{j} = ||\bar{d}|| \cos \beta$, $w = \bar{d} \cdot \bar{k} = ||\bar{d}|| \cos \gamma$, we can multiply these equations by $||\bar{d}||$ and obtain

$$\frac{x - x_0}{\cos \alpha} = \frac{y - y_0}{\cos \beta} = \frac{z - z_0}{\cos \gamma}.$$

These direction cosines are the components of the unit direction vector $\frac{\bar{d}}{||\bar{d}||} = \bar{d}_0 = \begin{bmatrix} \cos \alpha \\ \cos \beta \\ \cos \gamma \end{bmatrix}$. The above equations with cosines are the simplest equations of a line.

[Note that $||\bar{d}_0|| = 1$ implies $\cos^2 \alpha + \cos^2 \beta + \cos^2 \gamma = 1$, which is the three-dimensional analogue of the famous trigonometric identity

$$\sin^2 \alpha + \cos^2 \alpha = 1 \text{ or } \cos^2 \alpha + \cos^2 \beta = 1.]$$

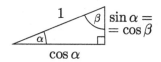

$$\sin \alpha = \\ = \cos \beta$$

Instead of giving a point on the line together with the direction vector, one can use two points $P_1(x_1, y_1, z_1)$ and $P_2(x_2, y_2, z_2)$ on the line to derive its equations.

Obviously, $\overrightarrow{P_1P} = t \overrightarrow{P_1P_2}$, which gives the symmetric form of the equations:

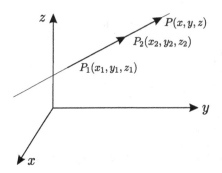

$$\frac{x - x_1}{x_2 - x_1} = \frac{y - y_1}{y_2 - y_1} = \frac{z - z_1}{z_2 - z_1}.$$

Remember that two points determine a line.

Also, two nonparallel planes (see Sect. 1.11) $a_1x + b_1y + +c_1z + d_1 = 0$ and $a_2x + b_2y + c_2z + d_2 = 0$, with the normals $\bar{n}_1 = \begin{bmatrix} a_1 \\ b_1 \\ c_1 \end{bmatrix}$ and $\begin{bmatrix} a_2 \\ b_2 \\ c_2 \end{bmatrix}$ under condition $\bar{n}_2 \neq k\bar{n}_1$, determine the parametric equations of the intersecting line by solving the above system for x and y in terms of the parameter $t = z$.

Example Find the parametric equations of the intersection line of the two planes

$$2x - y + 3z - 1 = 0$$
$$5z + 4y - z - 7 = 0.$$

The result is $x = \frac{11}{13} - \frac{11}{13}t$, $y = \frac{9}{13} + \frac{17}{13}t$, $z = t$, so that

$$P_0(\frac{11}{13}, \frac{9}{13}, 0) \text{ and } \bar{d} = \begin{bmatrix} -\frac{11}{13} \\ \frac{17}{13} \\ 1 \end{bmatrix}.$$

1.11 Equations of Planes in Three-Dimensional Space

The vector equation of the plane passing through the given point $P_0(x_0, y_0, z_0)$ and having the nonzero vector $\bar{n} = \begin{bmatrix} a \\ b \\ c \end{bmatrix}$ as its normal is obviously

$$\bar{n} \cdot \overrightarrow{P_0P} = 0,$$

where $P(x, y, z)$ is an arbitrary point on the plane, so that the vector $\overrightarrow{P_0P}$ lies in the plane and it is perpendicular to the normal $\bar{n}$.

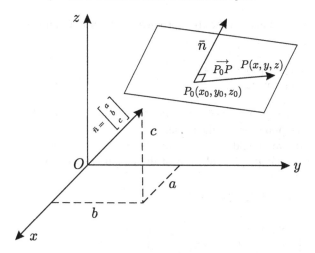

In terms of the components of the vectors $\bar{n}$ and $\overrightarrow{P_0P}=\begin{bmatrix} x-x_0 \\ y-y_0 \\ z-z_0 \end{bmatrix}$, we have

$$a(x-x_0)+b(y-y_0)+c(z-z_0)=0,$$

which is called the point-normal equation of the plane.

If a,b,c are real constants, and a,b,c are not all zero (a requirement equivalent to $a^2+b^2+c^2>0$), then the graph of the linear equation

$$ax+by+cz+d=0$$

is a plane having the vector $\bar{n}=\begin{bmatrix} a \\ b \\ c \end{bmatrix}$ as its normal, and (if $a\neq 0$) passing through the point $(-\frac{d}{a},0,0)$, since it can be written as $a(x+\frac{d}{a})+b(y-0)+c(z-0)=0$. It also passes through the point $(0,-\frac{d}{b},0)$ if $b\neq 0$ or through the point $(0,0,-\frac{d}{c})$ if $c\neq 0$.

This linear equation in x,y,z is called the general form of the equation of a plane.

The three points A,B,C determine a plane uniquely if they are not collinear (i.e., if they do not lie on the same line). The obvious condition for these points to be collinear is that the vectors $\overrightarrow{AB}$ and $\overrightarrow{AC}$ are parallel or $\overrightarrow{AC}=k\overrightarrow{AB}$. In this case, their cross product is zero: $\overrightarrow{AB}\times\overrightarrow{AC}=k(\overrightarrow{AB}\times\overrightarrow{AB})=0$, so that the procedure which follows is not applicable.

The point-normal equations of the plane passing through three noncollinear points

$$A(x_1,y_1,z_1),B(x_2,y_2,z_2),C(x_3,y_3,z_3)$$

can be obtained easily. We make use of the fact that the vectors $\overrightarrow{AB}$ and $\overrightarrow{AC}$ lie in the plane, so that the vector $\overrightarrow{AB}\times\overrightarrow{AC}$ is normal to the plane: we calculate the components a, b, c of $\overrightarrow{AB} \times \overrightarrow{AC}$ by means of the cross product symbolic determinant

$$\begin{vmatrix} \bar{i} & \bar{j} & \bar{k} \\ x_2 - x_1 & y_2 - y_1 & z_2 - z_1 \\ x_3 - x_1 & y_3 - y_1 & z_3 - z_1 \end{vmatrix} = a\bar{i} + b\bar{j} + c\bar{k},$$

and write the required equations $a(x - x_i) + b(y - y_i) + c(z - z_i) = 0$, where i can be either 1 or 2 or 3, since all three points are on the plane.

To save unnecessary effort, we should know that all these three point-normal forms are going to produce one single general form of the equation of the plane through the three given points.

Example $A(1,2,3)$, $B(0,-1,1)$, $C(4,3,-2)$

$$\overrightarrow{AB} \times \overrightarrow{AC} = \begin{vmatrix} \bar{i} & \bar{j} & \bar{k} \\ -1 & -3 & -2 \\ 3 & 1 & -5 \end{vmatrix} = \bar{i}\begin{vmatrix} -3 & -2 \\ 1 & -5 \end{vmatrix} - \bar{j}\begin{vmatrix} -1 & -2 \\ 3 & -5 \end{vmatrix} + \bar{k}\begin{vmatrix} -1 & -3 \\ 3 & 1 \end{vmatrix} =$$

$$= 17\bar{i} - 11\bar{j} + 8\bar{k}.$$

$$\left.\begin{array}{l} \text{Point } A : 17(x-1) - 11(y-2) + 8(z-3) = 0 \\ \text{Point } B : 17(x-0) - 11(y+1) + 8(z-1) = 0 \\ \text{Point } C : 17(x-4) - 11(y-3) + 8(z+2) = 0 \end{array}\right\} \begin{array}{l} \text{the unique general form} \\ 17x - 11y + 8z - 19 = 0. \end{array}$$

the point-normal equations of the plane passing through A, B, C.

1.12 Real Vector Spaces and Subspaces

The definition of a *real vector space* involves the field $\mathbb{R}$ of real numbers (sometimes called scalars) and a nonempty set V whose elements (called vectors) can be of very different nature (e.g. geometrical vector-arrows, matrix-columns, other kinds of matrices, polynomials, all sorts of functions). In V, there are operations of addition and scalar multiplication which assign to every ordered pair $\bar{u}, \bar{v} \in V$ the sum $\bar{u} + \bar{v} \in V$, and to any $\bar{u} \in V, c \in \mathbb{R}$ the product $c\bar{u} \in V$, respectively. (In short, addition is a $V \times V \to V$ map, and scalar multiplication is a $\mathbb{R} \times V \to V$ map.)

Then V is called a *real vector space*, and often denoted by $V(\mathbb{R})$, if the following nine axioms hold:

1. V is a commutative (Abelian) group under addition (five axioms);
2. Scalar multiplication is distributive over additions in both V and $\mathbb{R}$—$a, b, c \in \mathbb{R}$, $\bar{u}, \bar{v} \in V$, $c(\bar{u} + \bar{v}) = c\bar{u} + c\bar{v}$; $(a + b)\bar{u} = a\bar{u} + b\bar{u}$ (two axioms);
3. Scalar multiplication is associative with respect to multiplication in $\mathbb{R}$—$a, b \in \mathbb{R}, \bar{u} \in V$,

$$(ab)\bar{u} = a(b\bar{u});$$

4. The multiplicative identity 1 in $\mathbb{R}$ remains the identity of scalar multiplication: $1\bar{u} = \bar{u}$.

One should remember that the algebraic structure of the field $\mathbb{R}$ is determined by two binary operations—the addition "$+$" and the multiplication "$\times$" of real numbers, so that $(\mathbb{R}, +)$ is an Abelian group (five axioms), $(\mathbb{R} \setminus \{0\}, \times)$ is also an Abelian group, and multiplication is distributive over addition—$(a + b)c = ac + bc$. Thus, $V(\mathbb{R})$ is determined by $(5 + 4) + (5 + 5 + 1) = 20$ axioms altogether.

Any subset W of the vector space $V(\mathbb{R})$ which is a real vector space by itself is called a *subspace* of $V(\mathbb{R})$. ($V(\mathbb{R})$ and $\{\bar{0}\}$ are the trivial subspaces.) To check whether a nonempty subset W is a subspace, it is sufficient to verify that it is closed under addition and scalar multiplication (other properties are inherited from $V(\mathbb{R})$): (*i*) for every two vectors $\bar{u}, \bar{v} \in W$, their sum should also belong to W, i.e., $\bar{u} + \bar{v} \in W$; (*ii*) for every $k \in \mathbb{R}$ and every $\bar{u} \in W$, their product must be a vector in W, i.e., $k\bar{u} \in W$. Evidently, each subspace contains $\bar{0}$—the zero vector.

If we are given a set of k vectors $\{\bar{v}_1, \bar{v}_2, \dots, \bar{v}_k\}$ from $V(\mathbb{R})$ and a set of k scalars (real numbers) $\{a_1, a_2, \dots, a_k\}$, then we can form a *linear combination* of these two sets by using the above-defined operations—the addition of vectors and the multiplication of scalars with vectors:

$$a_1\bar{v}_1 + a_2\bar{v}_2 + \dots + a_k\bar{v}_k = \sum_{i=1}^{k} a_i\bar{v}_i \in V(\mathbb{R}).$$

This is the most general operation with vectors and scalars that can be performed in $V(\mathbb{R})$.

The set of all linear combinations of vectors from the subset $S = \{\bar{v}_1, \bar{v}_2, \dots, \bar{v}_n\} \subseteq V(\mathbb{R})$ is obviously closed under the addition of vectors:

$$\sum_{i=1}^{k} b_i\bar{v}_i + \sum_{i=1}^{n} c_i\bar{v}_i = \sum_{i=1}^{n} (b_i + c_i)\bar{v}_i,$$

as well as under scalar multiplication:

$$k\sum_{i=1}^{n} d_i\bar{v}_i = \sum_{i=1}^{n} (k\,d_i)\bar{v}_i.$$

Thus, it is a subspace of $V(\mathbb{R})$ called the *lineal* (or *linear span*) of S, and it is denoted by $\mathrm{LIN}(S)$. It is the most natural and frequently used simple way of making subspaces of $V(\mathbb{R})$. For example, $\mathbb{R}^3$ is the lineal of $\bar{i} = \begin{bmatrix} 1 \\ 0 \\ 0 \end{bmatrix}$, $\bar{j} = \begin{bmatrix} 0 \\ 1 \\ 0 \end{bmatrix}$, $\bar{k} = \begin{bmatrix} 0 \\ 0 \\ 1 \end{bmatrix}$,

since

$$\bar{a} = \begin{bmatrix} a_1 \\ a_2 \\ a_3 \end{bmatrix} = a_1\bar{i} + a_2\bar{j} + a_3\bar{k}, \text{ so } \mathbb{R}^3 = \mathrm{LIN}(\{\bar{i}, \bar{j}, \bar{k}\}).$$

1.13 Linear Dependence and Independence. Spanning Subsets and Bases

Let $X = \{\bar{x}_1, \bar{x}_2, \ldots, \bar{x}_k\}$ be a set of k nonzero vectors in $V(\mathbb{R})$. Then, X is said to be a linearly dependent set if one of its vectors, say $\bar{x}_j$, is a linear combination of the others

$$\bar{x}_j = \sum_{i=1(i \neq j)}^{k} c_i \bar{x}_i.$$

Not all c_i can be 0, since $\bar{x}_j \neq \bar{0}$. This expression can be written as $\sum_{i=1}^{k} c_i \bar{x}_i = \bar{0}$, with $c_j = -1$. We see that this new linear combination gives as a result the zero vector $\bar{0}$, but at least two coefficients (including c_j) are different from zero. This is the usual definition of a linearly dependent set of nonzero vectors:

A linear combination of the vectors from the set X of nonzero vectors can give $\bar{0}$ without all coefficients being zero. (Note that at least two coefficients must be different from 0, since all vectors are nonzero, so that $c_l \bar{x}_l = \bar{0}$, $c_l \neq 0$, would imply a contradiction $\bar{x}_l = \bar{0}$).

The visualization of linear dependence in two- and three-dimensional spaces $\mathbb{R}^2$ and $\mathbb{R}^3$ is straightforward: Two vectors $\{\bar{x}_1, \bar{x}_2\}$ in $\mathbb{R}^2$ are linearly dependent if and only if one of them is a multiple of the other, which means that they both lie on the same line passing through the origin (i.e., iff they are *collinear*). In $\mathbb{R}^3$, three vectors $\{\bar{x}_1, \bar{x}_2, \bar{x}_3\}$ are linearly dependent if and only if one is in the subspace spanned by the other two, which is the plane through the origin determined by these two. Hence, the three vectors in $\mathbb{R}^3$ are linearly dependent iff they all lie in the same plane through the origin (i.e., iff they are *coplanar*).

Remark Any set of vectors from $V(\mathbb{R})$ that contains the zero vector $\bar{0}$ can form a linear combination that is equal to $\bar{0}$ without all coefficients being zero, since the zero vector can have any coefficient. This means that such sets satisfy only a formal definition of linear dependence, since nonzero elements in them need not be linearly dependent at all.

A more useful test for linear dependence of the ordered set $x = \{\bar{x}_1, \bar{x}_2, \ldots, \bar{x}_k\}$ of nonzero vectors from $V(\mathbb{R})$ is the following: The set X is linearly dependent if and only if one of the vectors $\bar{x}_i$, $2 \leq i \leq k$, is a linear combination of the preceding ones (its predecessors).

Proof (sufficiency) Assume that $\bar{x}_i = a_1 \bar{x}_1 + a_2 \bar{x}_2 + \ldots + a_{i-1} \bar{x}_{i-1}$, $2 \leq i \leq k$. This implies $a_1 \bar{x}_1 + a_2 \bar{x}_2 + \ldots + a_{i-1} \bar{x}_{i-1} - \bar{x}_i + 0\bar{x}_{i+1} + \ldots + 0\bar{x}_k = \bar{0}$, so that one can make a linear combination of the vectors from X equal $\bar{0}$, but such that some coefficients (e.g., $a_i = -1$) in this linear combination are different from $\bar{0}$. The set X is linearly dependent. (necessity) If the vectors from X are linearly dependent, this means $\sum_{j=1}^{k} a_j \bar{x}_j = \bar{0}$ without all a_j being zero. Let l be the largest index such that $a_l \neq 0$. (Note that $l \neq 1$, since $l = 1$ would imply $a_1 \bar{x}_1 = \bar{0} \Rightarrow \bar{x}_1 = \bar{0}$, which is impossible, since all vectors in X are nonzero. For the same reason, a_l cannot be the only nonzero coefficient.) Since $a_l \neq 0$, we can write

$$\bar{x}_l = -\frac{1}{a_l}(a_1\bar{x}_1 + a_2\bar{x}_2 + \cdots + a_{l-1}\bar{x}_{l-1}),$$

i.e., $\bar{x}_l$ is a linear combination of the preceding vectors. Δ

If none of the vectors from the set $Y = \{\bar{y}_1, \bar{y}_2, \ldots, \bar{y}_m\}$ of nonzero vectors from $V(\mathbb{R})$ is a linear combination of the others from Y, then we say that the set Y is *linearly independent*. In this case, we can make their linear combination to be equal to the zero vector $\bar{0}$ only with all coefficients being zero:

$$\sum_{i=1}^{m} b_i \bar{y}_i = \bar{0} \Rightarrow \forall b_i = 0.$$

Note that the set consisting only of one nonzero vector is a linearly independent set.

If we have a subset $Z = \{\bar{z}_1, \bar{z}_2, \ldots, \bar{z}_r\}$ of vectors from $V(\mathbb{R})$ such that every vector $\bar{v} \in V(\mathbb{R})$ is a linear combination of the vectors from Z, then we say that Z is a *spanning subset* of $V(\mathbb{R})$ or

$$V(\mathbb{R}) = \text{LIN}(Z).$$

There is a very important relation between the number n of elements in a linearly independent (LI) set $L = \{\bar{b}_1, \bar{b}_2, \ldots, \bar{b}_n\}$ in $V(\mathbb{R})$ and the number m of vectors in a set $S = \{\bar{a}_1, \bar{a}_2, \ldots, \bar{a}_m\}$, which spans $V(\mathbb{R})$, i.e., $V(\mathbb{R}) = \text{LIN}(S)$.

The *replacement theorem* states that $n \leq m$, i.e., the number of LI vectors in $V(\mathbb{R})$ can only be smaller or equal to the number of vectors in any spanning (SP) set.

Proof We transfer $\bar{b}_1$ at the first place of S, thus obtaining the new set $S_1 = \{\bar{b}_1, \bar{a}_1, \bar{a}_2, \ldots, \bar{a}_m\}$. This new set is also an SP set [*] (the coefficient in front of $\bar{b}_1$ in any expansion can be 0), but it is certainly linearly dependent, since $\bar{b}_1$ is a linear combination of $\bar{a}$s. This means that one of the $\bar{a}$s, say $\bar{a}_j$, is a linear combination of the preceding ones. We can remove this vector and the remaining set $S'_1 = \{\bar{b}_1, \bar{a}_1, \ldots, \bar{a}_{j-1}, \bar{a}_{j+1}, \ldots, \bar{a}_m\}$ will still be an SP one.

Indeed, $\text{LIN}(S'_1) = \text{LIN}(S_1)$, i.e., $\bar{a}_j$ is redundant in the spanning function of S_1. To prove this, let us remember that in $\text{LIN}(S_1)$ every vector $\bar{x} \in V(\mathbb{R})$ is of the form $\bar{x} = 0\bar{b}_1 + \sum_{i=1}^{m} c_i \bar{a}_i$. But since $\bar{a}_j = d_1\bar{a}_1 + d_2\bar{a}_2 + \ldots + d_{j-1}\bar{a}_{j-1} + f_1\bar{b}_1$, we can replace $c_j\bar{a}_j$ in the expansion of $\bar{x}$ and obtain

$$\bar{x} = (c_1 + c_j d_1)\bar{a}_1 + \ldots + (c_{j-1} + c_j d_{j-1})\bar{a}_{j-1} + c_{j+1}\bar{a}_{j+1} + \ldots + c_m\bar{a}_m + c_j f_1\bar{b}_1.$$

Therefore, every vector $\bar{x}$ from $\text{LIN}(S_1)$ is a linear combination of vectors from S'_1.

Thus, in the SP set S'_1 one of the $\bar{a}$s is replaced by one of the $\bar{b}$s.

Now transferring $\bar{b}_2$ at the first place of S'_1, we get the set $S_2 = \{\bar{b}_2, \bar{b}_1, \bar{a}_1, \ldots, \bar{a}_{j-1}, \bar{a}_{j+1}, \ldots, \bar{a}_m\}$, which also spans $V(\mathbb{R})$, but it is linearly dependent for sure. One of the $\bar{a}$s is a linear combination of the preceding vectors, so that we can remove it. The new set S'_2 has two $\bar{a}$s replaced by two $\bar{b}$s, and it is still an SP set. (Note that $\bar{b}_1$ could not be a linear combination of $\bar{b}_2$, since they are LI).Continuing this replacement procedure, we are faced with two mutually exclusive alternatives: either

$m < n$ or $n \leq m$. Assume the first one $m < n$. In this case, we eventually get two sets: $S'_m = \{\bar{b}_m, \bar{b}_{m-1}, \ldots, \bar{b}_2, \bar{b}_1\}$ and $\{\bar{b}_{m+1}, \ldots, \bar{b}_n\}$, which is obviously an impossible situation – S'_m spans $V(\mathbb{R})$, so that all vectors in $V(\mathbb{R})$ are linear combinations of vectors in S'_m, including $\bar{b}_{m+1}, \ldots, \bar{b}_n$, which are by assumption linearly independent of the vectors in S'_m. We conclude that $n \leq m$ remains the only possibility. Δ

Any ordered set of nonzero vectors $X = \{\bar{x}_1, \bar{x}_2, \ldots, \bar{x}_n\}$ in $V(\mathbb{R})$, which is linearly independent (LI) and spanning (SP) at the same time is called a *basis* of $V(\mathbb{R})$. Bases play the most important role in $V(\mathbb{R})$.

The main property of every basis (actually an equivalent definition) is that every vector $\bar{a} \in V(\mathbb{R})$ is a *unique* linear combination of its vectors:

$$\bar{a} = \sum_{i=1}^{n} a_i \bar{x}_i$$

(all a_i are uniquely determined by $\bar{a}$). The uniqueness of this expansion is a direct consequence of the fact that the basis is an LI set – if there were two linear combinations $\bar{a} = \sum_{i=1}^{n} a_i \bar{x}_i$ and $\bar{a} = \sum_{i=1}^{n} a'_i \bar{x}_i$, then subtracting one from the other we get

$$\bar{0} = \sum_{i=1}^{n} (a_i - a'_i) \bar{x}_i,$$

which implies uniqueness $a_i = a'_i$, $i = 1, 2, \ldots, n$, due to LI property of the basis X.

On the other hand, the uniqueness of the expansion of every vector from $V(\mathbb{R})$ applies to the zero vector $\bar{0} \in V(\mathbb{R})$ as well: $\bar{0} = \sum_{i=1}^{n} a_i \bar{x}_i \Rightarrow \forall a_i = 0$, since $a_1 = a_2 = \ldots = a_n = 0$ is obviously this unique set of coefficients that gives $\bar{0}$. This means that the basis X is an LI set.

Therefore, we have two equivalent definitions of a basis: A basis is any ordered subset of nonzero vectors in $V(\mathbb{R})$ that is at the same time SP and LI or, by the other definition, a basis is an ordered subset in $V(\mathbb{R})$ such that every vector $\bar{a} \in V(\mathbb{R})$ is a unique linear combination of its vectors [uniquely spanning (USP) subset in $V(\mathbb{R})$].

All bases in $V(\mathbb{R})$ have the same number of vectors, and the number n is called the *dimension* of $V(\mathbb{R})$. It is written as $\dim V(\mathbb{R}) = n$ or $V_n(\mathbb{R})$.

The *proof* is very simple: if $B_1 = \{\bar{x}_1, \bar{x}_2, \ldots, \bar{x}_n\}$ and $B_2 = \{\bar{y}_1, \bar{y}_2, \ldots, \bar{y}_m\}$ are two bases, then $n \leq m$ since B_1 is an LI and B_2 an SP set. Also, $m \leq n$ because B_1 is at the same time an SP set and B_2 is an LI set. Thus $m = n$. Δ

In conclusion, we can say that

the number of LI vectors in $V_n(\mathbb{R})$	$\leq$	the dimension n of $V(\mathbb{R})$	$\leq$	the number of SP vectors in $V_n(\mathbb{R})$.

(the number of vectors
that are LI and SP at
the same time)

So, a basis is a set with the maximal number ($= n$) of LI vectors in $V_n(\mathbb{R})$, equal to the dimension of $V_n(\mathbb{R})$. It is also a set with the minimal number ($= n$) of SP vectors in $V_n(\mathbb{R})$.

Also, any set of n LI vectors in $V_n(\mathbb{R})$ is an SP set as well, i.e., they are a basis. Likewise, any set of n SP vectors in $V_n(\mathbb{R})$ is simultaneously an LI set, i.e., a basis. The *proof* of this statement is as follows: Let $L = \{\bar{x}_1, \bar{x}_2, \ldots, \bar{x}_n\}$ be an LI set in $V_n(\mathbb{R})$. Let us form its linear span LIN(L). It is a subspace of $V_n(\mathbb{R})$, with L being its basis. Since L has n elements, the dimension of the subspace LIN(L) is n, so that LIN(L) $= V_n(\mathbb{R}) \Rightarrow L$ spans $V_n(\mathbb{R})$. If $S = \{\bar{y}_1, \bar{y}_2, \ldots, \bar{y}_n\}$ is a set of n SP vectors in $V_n(\mathbb{R})$, it must be LI as well, because it cannot just be a linearly dependent set. If it were linearly dependent, that would mean that one of its vectors is a linear combination of its predecessors. We could take out that vector from S, and the remaining $(n-1)$ vectors would continue to be SP, which is impossible. $\varDelta$

1.14 The Three Most Important Examples of Finite-Dimensional Real Vector Spaces

1.14.1 The Vector Space $\mathbb{R}^n$ (Number Columns)

The most typical and very useful are the vector spaces of matrix columns

$$\bar{x} = \begin{bmatrix} x_1 \\ x_2 \\ \vdots \\ x_n \end{bmatrix}$$ of n rows and one column, where $n = 2, 3, 4, \ldots$, and the components

x_i, $i = 1, 2, \ldots, n$, are real numbers. We shall denote these spaces by $\mathbb{R}^n$, which is the usual notation for the set of ordered n-tuples of real numbers. We shall call the elements of $\mathbb{R}^n$ n-vectors.

Remark The ordered n-tuples can, of course, be represented by the matrix rows $[x_1 \, x_2 \ldots x_n]$ as well, but the matrix columns are more appropriate when we deal with matrix transformations applied from the left, e.g. $A\bar{x}$, where A is an $m \times n$ matrix [see Sect. 1.15.2(iv)].

The addition of n-vectors is defined as the addition of the corresponding components:

$$\begin{bmatrix} a_1 \\ a_2 \\ \vdots \\ a_n \end{bmatrix} + \begin{bmatrix} b_1 \\ b_2 \\ \vdots \\ b_n \end{bmatrix} = \begin{bmatrix} a_1 + b_1 \\ a_2 + b_2 \\ \vdots \\ a_n + b_n \end{bmatrix}.$$

(We use the same sign $+$ for these two different operations.) With this binary operation $\mathbb{R}^n \times \mathbb{R}^n \to \mathbb{R}^n$, the set $\mathbb{R}^n$ becomes an Abelian group. This is obvious since the addition in $\mathbb{R}$ makes $\mathbb{R}$ an Abelian group.

The n-vectors can be multiplied with real numbers (scalars), so that each component of an n-vector is multiplied by the same real number:

$$c \begin{bmatrix} a_1 \\ a_2 \\ \vdots \\ a_n \end{bmatrix} = \begin{bmatrix} ca_1 \\ ca_2 \\ \vdots \\ ca_n \end{bmatrix}.$$

This scalar multiplication obviously satisfies four basic properties:

1. $(c+d)\bar{a} = c\bar{a} + d\bar{a}$,
2. $c(\bar{a} + \bar{b}) = c\bar{a} + c\bar{b}$,
3. $(cd)\bar{a} = c(d\bar{a})$,
4. $1\bar{a} = \bar{a}$.

With these two operations the set $\mathbb{R}^2$ becomes a *real vector space*.

In $\mathbb{R}^n$, we can define the dot (or inner) product as an $\mathbb{R}^n \times \mathbb{R}^n \to \mathbb{R}$ mapping:

$$\bar{a} \cdot \bar{b} = a_1 b_1 + a_2 b_2 + \ldots + a_n b_n = \sum_{i=1}^{n} a_i b_i.$$

This product satisfies four obvious properties:

1. $\bar{a} \cdot \bar{b} = \bar{b} \cdot \bar{a}$ (commutativity),
2. $(\bar{a} + \bar{b}) \cdot \bar{c} = \bar{a} \cdot \bar{c} + \bar{b} \cdot \bar{c}$ (distributivity with respect to addition in $\mathbb{R}^n$),
3. $c(\bar{a} \cdot \bar{b}) = (c\bar{a}) \cdot \bar{b} = \bar{a} \cdot (c\bar{b})$ (associativity with respect to scalar multiplication in $\mathbb{R}^n$),
4. $\bar{a} \cdot \bar{a} > 0$ if $\bar{a} \neq \bar{0}$ and $\bar{a} \cdot \bar{a} = 0$ iff $\bar{a} = \bar{0}$ (positive definiteness).

With the dot product the space $\mathbb{R}^n$ becomes an (inner product) Euclidean space.

Making use of the dot product, we can define the length (or norm) of vectors in $\mathbb{R}^n$ as

$$||\bar{a}|| = (\bar{a} \cdot \bar{a})^{1/2} = \sqrt{a_1^2 + a_2^2 + \cdots + a_n^2},$$

as well as the angle between vectors $\bar{a}$ and $\bar{b}$

$$\cos \theta = \frac{\bar{a} \cdot \bar{b}}{||\bar{a}|| \, ||\bar{b}||},$$

even though such concepts can be visualized only for $n = 2$ and $n = 3$, since the vector spaces $\mathbb{R}^n$ for $n > 3$ have no geometrical interpretation. Nevertheless, they are essential for many problems in mathematics, in physics, as well as in economics.

We can see that $\mathbb{R}^n$ is an n-dimensional space since there is a natural (standard) basis of n vectors in $\mathbb{R}^n$:

$$B_S = \{\bar{e}_1, \bar{e}_2, \ldots, \bar{e}_n\} = \left\{ \begin{bmatrix} 1 \\ 0 \\ \vdots \\ 0 \end{bmatrix}, \begin{bmatrix} 0 \\ 1 \\ \vdots \\ 0 \end{bmatrix}, \ldots, \begin{bmatrix} 0 \\ 0 \\ \vdots \\ 1 \end{bmatrix} \right\}$$

(all columns have n components, one 1 and the rest are zeros). This is a trivial illustration that every vector $\bar{a} \in \mathbb{R}^n$ is a unique linear combination of vectors from B_S:

$$\bar{a} = \begin{bmatrix} a_1 \\ a_2 \\ \vdots \\ a_n \end{bmatrix} = a_1 \begin{bmatrix} 1 \\ 0 \\ \vdots \\ 0 \end{bmatrix} + a_2 \begin{bmatrix} 0 \\ 1 \\ \vdots \\ 0 \end{bmatrix} + \cdots + a_n \begin{bmatrix} 0 \\ 0 \\ \vdots \\ 1 \end{bmatrix} = \sum_{i=1}^{n} a_i \bar{e}_i.$$

A more general problem when a set of n vectors in $\mathbb{R}^n$ is its basis has a particularly nice solution. Take $n = 3$ (for other n the solution is analogous) and consider three vectors $\bar{x} = \begin{bmatrix} x_1 \\ x_2 \\ x_3 \end{bmatrix}, \bar{y} = \begin{bmatrix} y_1 \\ y_2 \\ y_3 \end{bmatrix}, \bar{z} = \begin{bmatrix} z_1 \\ z_2 \\ z_3 \end{bmatrix}$. They form a basis in $\mathbb{R}^3$ if every given vector $\bar{a} = \begin{bmatrix} a_1 \\ a_2 \\ a_3 \end{bmatrix}$ has a unique expansion $\bar{a} = c_1\bar{x} + c_2\bar{y} + c_3\bar{z}$.

This is a 3×3 system of linear equations with three unknowns c_1, c_2, c_3:

$$c_1 x_1 + c_2 y_1 + c_3 z_1 = a_1$$
$$c_1 x_2 + c_2 y_2 + c_3 z_2 = a_2$$
$$c_1 x_3 + c_2 y_3 + c_3 z_3 = a_3,$$

and it has a unique solution $\Rightarrow \{\bar{x}, \bar{y}, \bar{z}\}$ is a basis, if the determinant of the coefficient matrix is nonzero:

$$\begin{vmatrix} x_1 & y_1 & z_1 \\ x_2 & y_2 & z_2 \\ x_3 & y_3 & z_3 \end{vmatrix} \neq 0.$$

1.14.2 The Vector Space $\mathbb{R}_{n \times n}$ (Matrices)

A real matrix $A = \begin{bmatrix} a_{11} & a_{12} & \cdots & a_{1n} \\ a_{21} & a_{22} & \cdots & a_{2n} \\ \vdots & \vdots & \ddots & \vdots \\ a_{m1} & a_{m2} & \cdots & a_{mn} \end{bmatrix} = [a_{ij}]_{m \times n} \left(\begin{array}{c} \text{a very useful} \\ \text{short notation} \end{array} \right)$ is a rectangular

array of real numbers a_{ij}, $i = 1, 2, \ldots, m$; $j = 1, 2, \ldots, n$; the numbers a_{ij} are called the entries (or elements) of A. The size of the matrix A is defined by the number m of its rows and the number n of its columns, and written as $m \times n$.

Two matrices $A = [a_{ij}]_{m \times n}$ and $B = [b_{ij}]_{m \times n}$ of the same size are considered equal if the corresponding entries are equal: $a_{ij} = b_{ij}$, for all i and j.

The set of all real matrices of the size $m \times n$ is denoted by $\mathbb{R}_{m \times n}$.

Any two matrices $A = [a_{ij}]_{m \times n}$ and $B = [b_{ij}]_{m \times n}$ of the same size can be added elementwise:

$$A + B = [a_{ij}]_{m \times n} + [b_{ij}]_{m \times n} = [a_{ij} + b_{ij}]_{m \times n} = [c_{ij}]_{m \times n} = C.$$

This binary operation $\mathbb{R}_{m \times n} \times \mathbb{R}_{m \times n} \to \mathbb{R}_{m \times n}$ makes this set an Abelian group, because $\mathbb{R}$ is an Abelian group with respect to the addition of numbers. One can easily verify that the addition of matrices is (i) closed, (ii) commutative, (iii) associative, (iv) there exists the unique zero matrix $\hat{0} = [0]_{m \times n}$ (all elements are zero), which is the identity (neutral) for the addition $A + \hat{0} = \hat{0} + A = A$ for any A, (v) there exists a unique negative (the additive inverse) matrix $-A = [-a_{ij}]_{m \times n}$ for every A, so that $A + (-A) = \hat{0}$.

For any matrix $A = [a_{ij}]_{m \times n}$ and any real number (scalar) $c \in \mathbb{R}$, we can define scalar multiplication (the scalar-matrix product) as the matrix $cA = [ca_{ij}]_{m \times n}$ (every element of A is multiplied by c). This is a mapping of the type $\mathbb{R} \times \mathbb{R}_{m \times n} \to \mathbb{R}_{m \times n}$.

Scalar multiplication is related to addition and multiplication in the field $\mathbb{R}$ of scalars, and to the addition of matrices in $\mathbb{R}_{m \times n}$ as follows:

(i)

$$(c + d)A = [(c + d)a_{ij}]_{m \times n} = [ca_{ij} + da_{ij}]_{m \times n} = [ca_{ij}]_{m \times n} + [da_{ij}]_{m \times n}$$
$$= cA + dA;$$

(ii) $(cd)A = [(cd)a_{ij}]_{m \times n} = [c(da_{ij})]_{m \times n} = c(dA);$

(iii) $c(A + B) = [c(a_{ij} + b_{ij})]_{m \times n} = [ca_{ij} + cb_{ij}]_{m \times n} = cA + cB;$

(iv) $1A = [1 \cdot a_{ij}]_{m \times n} = [a_{ij}]_{m \times n} = A$ (the neutral of the multiplication of numbers remains the neutral for scalar multiplication).

The addition of matrices and scalar multiplication satisfying altogether $5 + 4$ properties listed above, make the set $\mathbb{R}_{m \times n}$ a real vector space of dimension $m \cdot n$. The statement about dimension follows from the fact that the standard (most natural) basis in $\mathbb{R}_{m \times n}$ consists of $m \cdot n$ matrices with only one element equal to 1, while all others are 0:

$$E_{pq} = [\delta_{pi} \delta_{qj}]_{m \times n}, \, p = 1, 2, \ldots, m; \, q = 1, 2, \ldots, n,$$

where

$$\delta_{ij} = \begin{cases} 1, \, i = j \\ 0, \, i \neq j \end{cases} \text{ (the Kronecker delta symbol).}$$

Obviously, every matrix $A = [a_{ij}]_{m \times n}$ is their unique linear combination:

$$A = \sum_{p=1}^{m} \sum_{q=1}^{n} a_{pq} E_{pq}.$$

Example of a basis.

Consider the four-dimensional vector space $\mathbb{R}_{2 \times 2}$. Prove that the following set of four 2×2 matrices

$$A = \begin{bmatrix} 3 & 6 \\ 3 & -6 \end{bmatrix}, B = \begin{bmatrix} 0 & -1 \\ -1 & 0 \end{bmatrix}, C = \begin{bmatrix} 0 & -8 \\ -12 & -4 \end{bmatrix}, D = \begin{bmatrix} 1 & 0 \\ -1 & 2 \end{bmatrix} \text{ is a basis.}$$

Since the number of matrices is equal to the dimension of the space, we have only to prove that the set is LI or SP. The test for LI is

$$aA + bB + cC + dD = \begin{bmatrix} 0 & 0 \\ 0 & 0 \end{bmatrix} \Rightarrow a = b = c = d = 0.$$

This is in fact a 4×4 homogeneous system of linear equations in a, b, c, d, which has only the trivial solution if the determinant of the coefficient matrix is different from 0. Writing the test explicitly

$$a\begin{bmatrix} 3 & 6 \\ 3 & -6 \end{bmatrix} + b\begin{bmatrix} 0 & -1 \\ -1 & 0 \end{bmatrix} + c\begin{bmatrix} 0 & -8 \\ -12 & -4 \end{bmatrix} + d\begin{bmatrix} 1 & 0 \\ -1 & 2 \end{bmatrix} = \begin{bmatrix} 0 & 0 \\ 0 & 0 \end{bmatrix},$$

we get the linear system of homogeneous equations

$$\begin{array}{rrrrl} 3a & & & +d & = 0 \\ 6a & -b & -8c & & = 0 \\ 3a & -b & -12c & -d & = 0 \\ -6a & & -4c & +2d & = 0, \end{array}$$

and the determinant of the coefficient matrix is

$$\begin{vmatrix} 3 & 0 & 0 & 1 \\ 6 & -1 & -8 & 0 \\ 3 & -1 & -12 & -1 \\ -6 & 0 & -4 & 2 \end{vmatrix} = 3\begin{vmatrix} -1 & -8 & 0 \\ -1 & -12 & -1 \\ 0 & -4 & 2 \end{vmatrix} - 1\begin{vmatrix} 6 & -1 & -8 \\ 3 & -1 & -12 \\ -6 & 0 & -4 \end{vmatrix} =$$

$$= 3(24 + 4 - 16) - (24 - 72 + 48 - 12) = 4 \times 12 = 48 \neq 0.$$

(by the Sarrus rule, see appendix A-1)

Therefore, $a = b = c = d = 0$ is the only solution of this system, so that A, B, C, D form a basis in $\mathbb{R}_{2 \times 2}$. Δ

1.14.3 The Vector Space P_3 (Polynomials)

Let us consider the set P_3 of all real polynomials whose degree is less than 3, i.e., the set

$$P_3 = \{ax^2 + bx + c \,|\, a, b, c, \in \mathbb{R}\}.$$

Let us define addition in P_3, as well as scalar multiplication with real numbers, and show that these operations satisfy all nine axioms (five for addition and four for scalar multiplication), so that P_3 becomes a real vector space of dimension 3.

If $p(x)$ and $p'(x)$ are two polynomials from P_3, then

$$p(x) + p'(x) = (ax^2 + bx + c) + (a'x^2 + b'x + c') =$$
$$= (a + a')x^2 + (b + b')x + (c + c') \in P_3,$$

as well as for $d \in \mathbb{R}$, $dp(x) = (ad)x^2 + (bd)x + (cd) \in P_3$. These are two mappings $P_3 \times P_3 \to P_3$ and $\mathbb{R} \times P_3 \to P_3$, respectively.

The addition is obviously a closed operation (axiom 1); it is commutative and associative since addition in $\mathbb{R}$ has the same properties (axioms 2 and 3); the number 0 ($a = b = c = 0$) is called the *zero polynomial* and plays the role of the additive identity $p(x) + 0 = p(x)$ (axiom 4); the additive inverse of every $p(x) \in P_3$ is obviously $-p(x) : p(x) + (-p(x)) = 0$ (axiom 5);

The four basic properties of scalar multiplication are

(i) $d(p(x) + p'(x)) = d(a + a')x^2 + d(b + b')x + d(c + c') = dp(x) + dp'(x)$;
(ii) $(d + e)p(x) = (d + e)ax^2 + (d + e)bx + (d + e)c = dp(x) + ep(x)$;
(iii) $(de)p(x) = (de)ax^2 + (de)bx + (de)c = d(ep(x))$;
(iv) $1p(x) = 1ax^2 + 1bx + 1c = p(x)$.

The most natural (standard) basis in P_3 is obviously $B = \{x^2, x, 1\}$ since every $p(x) \in P_3$ is a unique linear combination of these polynomials $p(x) = ax^2 + bx + c$. So, the dimension of P_3 is 3.

Example of a basis.

To verify that three other polynomials, e.g., $p_1(x) = 1$, $p_2(x) = x - 1$, $p_3 = x^2 - 2x + 1$ are also a basis in P_3, we shall prove only that they are linearly independent. Remember: In the three-dimensional space P_3 any three linearly independent (LI) vectors form a basis.

If they are a basis, we shall uniquely expand $p(x) = 2x^2 - 5x + 6$ in this basis.

To verify LI: $c_1 p_1(x) + c_2 p_2(x) + c_2 p_3(x) = 0$, this implies $c_1 + c_2(x - 1) + c_3(x^2 - 2x + 1) = 0 \Rightarrow (c_1 - c_2 + c_3) \cdot 1 + (c_2 - 2c_3)x + c_3 x^2 = 0$, and since $\{x^2, x, 1\}$ is the standard basis, all these coefficients must be zero:

$$c_1 - c_2 + c_3 = 0$$
$$c_2 - 2c_3 = 0$$
$$c_3 = 0,$$

so this homogeneous 3×3 linear system obviously has the unique solution $c_3 = c_2 = c_1 = 0$.

The coefficient matrix of the above linear system is $A = \begin{bmatrix} 1 & -1 & 1 \\ 0 & 1 & -2 \\ 0 & 0 & 1 \end{bmatrix}$, so that det $A = 1 \neq 0$, which verifies the uniqueness of the solution and consequently LI of the three polynomials $p_1(x), p_2(x), p_3(x)$.

To expand $p(x)$ in this new basis, we write the nonhomogeneous linear system [$p(x)$ instead of 0]:

$$c_1 - c_2 + c_3 = 6$$
$$c_2 - 2c_3 = -5$$
$$c_3 = 2,$$

which gives the unique solution by substitution

$$c_3 = 2, c_2 = -1, c_1 = 3.$$

Thus, $p(x) = 3p_1(x) - p_2(x) + 2p_3(x)$ (the unique expansion). [Verification: $2x^2 - 5x + 6 = 3(1) - (x - 1) + 2(x^2 - 2x + 1) = 3 - x + 1 + 2x^2 - 4x + 2 = 2x^2 - 5x + 6$].

1.15 Some Special Topics about Matrices

1.15.1 Matrix Multiplication

The two matrices $A = [a_{ij}]_{m \times r}$ and $B = [b_{ij}]_{r \times n}$ (the number of columns in A is equal to the number of rows in B) can be multiplied to produce an $m \times n$ matrix $C = [c_{ij}]_{m \times n}$ as follows:

$$c_{ij} = \sum_{k=1}^{r} a_{ik} b_{kj}.$$

(ij-entry in the product matrix $C = AB$ is the dot product of the i-th row of A with the j-column of B, both being considered as r-vectors).

This matrix multiplication is associative:

$$A(BC) = (AB)C = ABC,$$

as is easily seen if we denote $A = [a_{ir}]_{m \times p}, B = [b_{rs}]_{p \times q}, C = [c_{sj}]_{q \times n}$, then

$$A(BC) = [\sum_{r=1}^{p} a_{ir}(\sum_{s=1}^{q} b_{rs} c_{sj})]_{m \times n} = [\sum_{s=1}^{q}(\sum_{r=1}^{p} a_{ir} b_{rs}) c_{sj}]_{m \times n} =$$

$$= (AB)C = [\sum_{r=1}^{p} \sum_{s=1}^{q} a_{ir} b_{rs} c_{sj}]_{m \times n} = ABC.$$

The relation between scalar and matrix multiplications is

$$c(AB) = (cA)B = A(cB), c \in \mathbb{R}.$$

Matrix multiplication is distributive over matrix addition : $A(B + C) = AB + +AC; (B + C)A = BA + CA$ (obvious), but we need to prove both distributive laws since matrix multiplication is not commutative.

Equality of AB and BA can fail to hold for three reasons:

(i) AB is defined, but BA is undefined (e.g. A is a 2×3 matrix and B is a 3×4 matrix);

(ii) AB and BA are both defined, but have different sizes (e.g. A is a 2×3 matrix and B is a 3×2 matrix);

(iii) $AB \neq BA$ even when both AB and BA are defined and have the same size (e.g.
$$A = \begin{bmatrix} -1 & 0 \\ 2 & 3 \end{bmatrix}, B = \begin{bmatrix} 1 & 2 \\ 3 & 0 \end{bmatrix}, \text{ while } AB = \begin{bmatrix} -1 & -2 \\ 11 & 4 \end{bmatrix} \text{ and } BA = \begin{bmatrix} 3 & 6 \\ -3 & 0 \end{bmatrix}).$$

1.15.2 Some Special Matrices

(i) A matrix $A = [a_{ij}]_{n \times n}$, which has an equal number of rows and columns, is called a *square matrix* of order n. The elements $a_{11}, a_{22}, \ldots, a_{nn}$ form the main diagonal of A. The sum of the elements on the main diagonal of A is called the *trace* of A, and it is denoted by

$$\mathrm{tr}A = \sum_{i=1}^{n} a_{ii}.$$

A nonempty collection of matrices is called an *algebra of matrices* if it is closed under the operations of matrix addition, scalar, as well as matrix multiplication. Thus, the collection $\mathbb{R}_{n \times n}$ of all square $n \times n$ real matrices forms the *algebra of square real matrices of order n*.

A square matrix in which all the entries above (or below) the main diagonal are zero is called a *lower* (or an *upper*) triangular matrix.

A square matrix $A = [a_{ij}]_{n \times n}$ in which every entry off the main diagonal is zero, that is,

$$a_{ij} = 0 \text{ for } i \neq j,$$

is called a *diagonal* matrix. The sum, the scalar product, and the product of diagonal matrices are again diagonal matrices. (The collection of all diagonal real $n \times n$ matrices is thus closed under these operations.) So, all the real $n \times n$ diagonal matrices form an algebra of matrices (a subalgebra of $\mathbb{R}_{n \times n}$). In fact, the diagonal matrices form a *commutative* algebra, since the product of diagonal matrices is a commutative operation.

A diagonal matrix for which all the elements are equal, that is,

$$a_{ij} = c \text{ for } i = j \text{ and } a_{ij} = 0 \text{ for } i \neq j,$$

is called a *scalar matrix*. A scalar matrix $n \times n$ with $c = 1$ is called the *identity matrix* of order n and it is denoted by $I_n = [\delta_{ij}]_{n \times n}$.

(ii) The zero matrix has all its elements equal to zero:

$$\hat{0} = [0]_{m \times n}$$

(Note that it is not necessarily a square matrix).

It is interesting to observe that the product of two matrices A and B can be the zero matrix, that is, $AB = \hat{0}$, and that neither A nor B is the zero matrix itself. An example:

$$A = \begin{bmatrix} 1 & 2 \\ 2 & 4 \end{bmatrix}, \ B = \begin{bmatrix} 4 & -6 \\ -2 & 3 \end{bmatrix}, \text{ but } AB = \begin{bmatrix} 0 & 0 \\ 0 & 0 \end{bmatrix}.$$

One more interesting fact about matrix multiplication (as different from the properties of number multiplication) is that the cancelation law does not hold for matrices:

If $A \neq \hat{0}$ and $AB = AC$, then we cannot conclude that $B = C$. An example:

$$A = \begin{bmatrix} 1 & 2 \\ 2 & 4 \end{bmatrix}, \ B = \begin{bmatrix} 2 & 1 \\ 3 & 2 \end{bmatrix}, \ C = \begin{bmatrix} -2 & 7 \\ 5 & -1 \end{bmatrix}, \ AB = AC = \begin{bmatrix} 8 & 5 \\ 16 & 10 \end{bmatrix}, \text{ but } B \neq C.$$

(iii) The *transpose* of a matrix $A = [a_{ij}]_{m \times n}$ is the $n \times m$ matrix $A^T = [a_{ij}]^T{}_{n \times m}$, where

$$a_{ij}^T = a_{ji},$$

that is, the transpose of A is obtained by interchanging the rows and columns of A.

Transposition is an involutive operation since $(A^T)^T = A$. The relations of transposition to matrix addition and scalar multiplication are obvious:

$$(A + B)^T = [(a_{ij} + b_{ij})^T] = [a_{ij}^T + b_{ij}^T] = A^T + B^T,$$
$$\text{and } (cA)^T = [(ca_{ij})^T] = [ca_{ij}^T] = cA^T.$$

Less obvious, but frequently used, is the relationship between transposition and matrix multiplication: Let $A = [a_{ij}]_{m \times r}$ and $B = [b_{ij}]_{r \times n}$, then $AB = [\sum_{k=1}^r a_{ik}b_{kj}]_{m \times n}$. Its transpose is $(AB)^T = [\sum_{k=1}^r a_{jk}b_{ki}]_{n \times m} = [\sum_{k=1}^r b_{ik}^T a_{kj}^T]_{n \times m}$. On the other hand, $B^T = [b_{ij}^T]_{n \times r}$, $A^T = [a_{ij}^T]_{r \times m}$, so that $B^T A^T = [\sum_{k=1}^r b_{ik}^T a_{kj}^T]_{n \times m}$, and finally

$$(AB)^T = B^T A^T \text{ (the reversal rule for transposes)}.$$

A square matrix $A = [a_{ij}]_{n \times n}$ is called symmetric (skew-symmetric) if $A^T = A$ ($A^T = -A$). The elements of a symmetric matrix are in fact symmetric with respect to the main diagonal: $a_{ij} = a_{ji}$. Symmetric tensors, which are symmetric real 3×3 matrices which depend on the choice of coordinate system, play a very important role in the description of a number of physical phenomena in mechanics, as well in electrodynamics.

Symmetric matrices form a subspace in $\mathbb{R}_{n \times n}$ and the simplest (standard) basis in this subspace for $n = 3$ is

$$\left\{ \begin{bmatrix} 1 & 0 & 0 \\ 0 & 0 & 0 \\ 0 & 0 & 0 \end{bmatrix}, \begin{bmatrix} 0 & 0 & 0 \\ 0 & 1 & 0 \\ 0 & 0 & 0 \end{bmatrix}, \begin{bmatrix} 0 & 0 & 0 \\ 0 & 0 & 0 \\ 0 & 0 & 1 \end{bmatrix}, \begin{bmatrix} 0 & 1 & 0 \\ 1 & 0 & 0 \\ 0 & 0 & 0 \end{bmatrix}, \begin{bmatrix} 0 & 0 & 1 \\ 0 & 0 & 0 \\ 1 & 0 & 0 \end{bmatrix}, \begin{bmatrix} 0 & 0 & 0 \\ 0 & 0 & 1 \\ 0 & 1 & 0 \end{bmatrix} \right\},$$

so it is a six-dimensional subspace. Even though the set of symmetric matrices is closed under matrix addition and scalar multiplication, it is not closed under matrix multiplication (e.g. $A = \begin{bmatrix} 1 & 2 \\ 2 & 3 \end{bmatrix}$ and $B = \begin{bmatrix} 4 & 5 \\ 5 & 6 \end{bmatrix}$, but $AB = \begin{bmatrix} 14 & 17 \\ 23 & 28 \end{bmatrix}$), so symmetric matrices do not form a subalgebra of matrices.

Every square matrix A is the unique sum of a symmetric and a skew-symmetric matrix:

$$A = \frac{A + A^T}{2} + \frac{A - A^T}{2},$$

since $(A + A^T)^T = A + A^T$ and $(A - A^T)^T = -(A - A^T)$.

(iv) (see Sect. 1.14.1) The column matrices $n \times 1$ (or the matrix-columns or the number columns) are extremely important in all applications. We call them *n-vectors* and denote them as

$$\begin{bmatrix} x_1 \\ x_2 \\ \vdots \\ x_n \end{bmatrix} = [x_i]_{n \times 1}.$$

They form a real vector space $\mathbb{R}_{n \times 1}$, which is n-dimensional, with the standard basis $\{\bar{e}_1, \bar{e}_2, \ldots, \bar{e}_n\}$, in which all columns have n components, one 1 and the rest are zeros.

In Sect. 2.6.3 we shall show that any choice of a basis $v = \{\bar{v}_1, \bar{v}_2, \ldots, \bar{v}_n\}$ in a real n-dimensional vector space $V_n(\mathbb{R})$ induces an isomorphism between that space and the space $\mathbb{R}_{n \times 1}$:

$$\bar{x} \in V_n(\mathbb{R}), \ \bar{x} = \sum_{i=1}^{n} x_i \bar{v}_i \Leftrightarrow \begin{bmatrix} x_1 \\ x_2 \\ \vdots \\ x_n \end{bmatrix}.$$

This isomorphism is called the representation of vectors from $V_n(\mathbb{R})$ by matrix-columns from $\mathbb{R}_{n \times 1}$, induced by the basis v.

We have the analogous situation with complex vector spaces $V_n(\mathbb{C})$ and $\mathbb{C}_{n \times 1}$. On the other hand, in Sect. 4.5 (see also Sect. 4.1), we deal with *linear functionals* in $V_n(F)$ (where F can be $\mathbb{R}$ or $\mathbb{C}$), which are vectors in the dual space $V_n^*(F)$. To represent a linear functional $f \in V_n^*(F)$ in a basis $v = \{\bar{v}_1, \bar{v}_2, \ldots, \bar{v}_n\}$ in $V_n(F)$, we apply f to all these basis vectors and form the matrix-row of so-obtained n numbers from the field F

$$[f(\bar{v}_1) f(\bar{v}_2) \ldots f(\bar{v}_n)].$$

Therefore, for the representation of vectors from a dual space $V_n^*(F)$ we need $F_{1 \times n}$, the vector space of matrix-rows of numbers from the field F.

To simplify the notation, we shall denote $\mathbb{R}_{n \times 1}$ and $\mathbb{C}_{n \times 1}$ by $\mathbb{R}^n$ and $\mathbb{C}^n$, respectively, which is slightly incorrect because these two latter notations are mainly used for the sets of ordered n-tuples of real or complex numbers, regardless of their arrangement as columns or rows.

The arrangement of ordered n-tuples as matrix-columns is a common choice in physics and many other applications. In a number of mathematical text books the matrix-rows are preferred.

The main difference between these two choices is how we apply linear transformations (matrices):

In $\mathbb{R}_{n \times 1}$, we apply an $m \times n$ matrix A to vectors from the left

$$A\bar{x} = \bar{y}, \; \bar{x} \in \mathbb{R}_{n \times 1}, \; \bar{y} \in \mathbb{R}_{m \times 1},$$

or
$$\begin{bmatrix} a_{11} & a_{12} & \cdots & a_{1n} \\ a_{21} & a_{22} & \cdots & a_{2n} \\ \vdots & \vdots & \ddots & \vdots \\ a_{m1} & a_{m2} & \cdots & a_{mn} \end{bmatrix} \begin{bmatrix} x_1 \\ x_2 \\ \vdots \\ x_n \end{bmatrix} = \begin{bmatrix} a_{11}x_1 + a_{12}x_2 + \cdots + a_{1n}x_n \\ a_{21}x_1 + a_{22}x_2 + \cdots + a_{2n}x_n \\ \vdots \;\; \vdots \;\; \vdots \\ a_{m1}x_1 + a_{m2}x_2 + \cdots + a_{mn}x_n \end{bmatrix} = \begin{bmatrix} y_1 \\ y_2 \\ \vdots \\ y_n \end{bmatrix},$$

and obtain the vector $\bar{y}$ in $\mathbb{R}_{m \times 1}$ (see Sect. 2.6.2).

In the space of matrix-rows $\mathbb{R}_{1 \times n}$, we have a similar situation, but the $n \times m$ matrix A is applied to vector $\bar{x}^T$ from the right to produce the matrix-row $\bar{y}^T$ in $\mathbb{R}_{1 \times m}$

$$\bar{x}^T A = \bar{y}^T, \; \bar{x}^T \in \mathbb{R}_{1 \times n}, \; \bar{y}^T \in \mathbb{R}_{1 \times m},$$

or
$$[x_1 \, x_2 \ldots x_n] \begin{bmatrix} a_{11} & a_{12} & \cdots & a_{1m} \\ a_{21} & a_{22} & \cdots & a_{2m} \\ \vdots & \vdots & \ddots & \vdots \\ a_{n1} & a_{n2} & \cdots & a_{nm} \end{bmatrix} = [y_1 \, y_2 \ldots y_n],$$

where $y_i, \, i = 1, 2, \ldots, m$ are given as

$$y_i = \sum_{j=1}^{n} a_{ji} x_j.$$

(v) An $n \times n$ real square matrix is called *invertible (or nonsingular or regular)* if there exists another $n \times n$ matrix B such that

$$AB = BA = I_n.$$

It can be proved (by using the elementary matrices—see Sect. 2.12) that the equations $AB = I_n$ and $BA = I_n$ always imply each other.

This *proof* is based on the fact that in an invertible square matrix A all n columns (as well as the n rows) are linearly independent and that its GJ reduced form is just I_n (see the theorem which states when a linear system $A\bar{x} = \bar{b}$, $A \in \mathbb{R}_{n \times n}$ has a unique solution for every $\bar{b} \in \mathbb{R}^n$, Sect. 2.16.1). This form is achieved by multiplying A from the left by an appropriate set of elementary matrices $\{F_q, \ldots, F_2, F_1\}$ (this set is not necessarily unique):

$$F_q \ldots F_2 F_1 A = I_n \text{ (see the end of Sect. 2.15).}$$

Therefore,

$$AB = I_n \Rightarrow F_q \ldots F_2 F_1 AB = F_q \ldots F_2 F_1 \Rightarrow$$
$$\Rightarrow B = F_q \ldots F_2 F_1 \Rightarrow BA = F_q \ldots F_2 F_1 \Rightarrow BA = I_n \ \triangle$$

The matrix B is called an *inverse* of A.

But, if a matrix A has an inverse then this inverse is unique. This is obvious, since if we suppose that there are two inverses B and C, i.e. $BA = AC = I_n$, then $B = BI_n = B(AC) = (BA)C = I_nC = C$. We denote this unique inverse of A as A^{-1}.

All invertible real $n \times n$ matrices form the group $GL(n, \mathbb{R})$, where GL stands for *general linear group*. It is a subset of $\mathbb{R}_{n \times n}$, which is, as we have proved, a real vector space. But the group $GL(n, \mathbb{R})$ is not a vector space, since the sum of two invertible matrices is not necessarily an invertible matrix.

Some useful properties of invertible matrices are

1. If A is an invertible matrix, then A^{-1} is also invertible, and

$$(A^{-1})^{-1} = A.$$

This follows from $A^{-1}A = AA^{-1} = I_n$.
2. If A and B are invertible, then AB is also invertible, and

$$(AB)^{-1} = B^{-1}A^{-1}.$$

This follows from $(AB)(B^{-1}A^{-1}) = A(BB^{-1})A^{-1} = AI_nA^{-1} = AA^{-1} = I_n$ (the reversal rule for matrix inverses).
3. If A is invertible, then A^T is also invertible and

$$(A^T)^{-1} = (A^{-1})T.$$

This follows from $AA^{-1} = I_n \Rightarrow (AA^{-1})^T = I_n^T = I_n \Rightarrow (A^{-1})^T A^T = I_n$.
4. $(I_n)^{-1} = I_n$. This is from $I_n I_n = I_n$.

The above properties 1, 2, and 4 show immediately that invertible matrices in $\mathbb{R}_{n \times n}$ form a group $GL(n, \mathbb{R})$ (closure under matrix inversion, and matrix multiplication, as well as the existence of the unity I_n).

The inverse matrix A^{-1} of A is calculated in the appendix A.2(4), which follows.

Appendix A
Determinants

A.1 Definitions of Determinants

The determinant is a very important scalar function defined on square $n \times n$ real matrices

$$\det : \mathbb{R}_{n \times n} \rightarrow \mathbb{R}.$$

As we have shown, determinants are relevant for solving consistent linear systems with a small number of unknowns (Cramer's rule-remark 2 in Sect. 2.16.1), for finding areas and volumes in $\mathbb{R}^2$ and $\mathbb{R}^3$ (see Sect. 1.9), and they are also of interest in the study of advanced calculus for functions of several variables.

We define the determinant of any square matrix using mathematical induction. Another method for defining determinants is by using permutations. We shall present both methods, but we prefer the first one, since it is easier for proving properties of determinants.

We shall start with determinants of 2×2 matrices. The simplest motivation for defining such determinants is to be found in solving linear systems with two unknowns

$$a_1 x + b_1 y = c_1$$
$$a_2 x + b_2 y = c_2$$

or, in matrix notation

$$\begin{bmatrix} a_1 & b_1 \\ a_2 & b_2 \end{bmatrix} \begin{bmatrix} x \\ y \end{bmatrix} = \begin{bmatrix} c_1 \\ c_2 \end{bmatrix}.$$

It is quite an elementary task to show that such a system has a unique solution if and only if

$$a_1 b_2 - a_2 b_1 \neq 0,$$

and this solution is

$$x = \frac{c_1 b_2 - c_2 b_1}{a_1 b_2 - a_2 b_1} \text{ and } y = \frac{a_1 c_2 - a_2 c_1}{a_1 b_2 - a_2 b_1}.$$

Here, we note the significance of the scalar

$$a_1b_2 - a_2b_1$$

for the coefficient matrix $A = \begin{bmatrix} a_1 & b_1 \\ a_2 & b_2 \end{bmatrix}$. To find a more compact form for this scalar,

we define the *determinant* of this matrix as $\begin{vmatrix} a_1 & b_1 \\ a_2 & b_2 \end{vmatrix} = a_1b_2 - a_2b_1 = \det A$.

(Note the difference in notation: $[\ ]$ for a matrix, and $|\ |$ for a determinant).

Similarly, $c_1b_2 - c_2b_1 = \begin{vmatrix} c_1 & b_1 \\ c_2 & b_2 \end{vmatrix}$ and $a_1c_2 - a_2c_1 = \begin{vmatrix} a_1 & c_1 \\ a_2 & c_2 \end{vmatrix}$.

Finally,

$$x = \frac{\begin{vmatrix} c_1 & b_1 \\ c_2 & b_2 \end{vmatrix}}{\begin{vmatrix} a_1 & b_1 \\ a_2 & b_2 \end{vmatrix}} \text{ and } y = \frac{\begin{vmatrix} a_1 & c_1 \\ a_2 & c_2 \end{vmatrix}}{\begin{vmatrix} a_1 & b_1 \\ a_2 & b_2 \end{vmatrix}}.$$

The denominator in both quotients is the determinant $\det A \neq 0$ of the coefficient

matrix $A = \begin{bmatrix} a_1 & b_1 \\ a_2 & b_2 \end{bmatrix}$. The numerators are obtained as determinants of the coefficient

matrix where the first and second columns, respectively, are replaced by the column

of the constant terms $\begin{bmatrix} c_1 \\ c_2 \end{bmatrix}$. This is the well-known *Cramer's rule* for solving linear

systems with unique solutions defined in Sect. 2.16.1, Remark 2.

It would be quite a challenge to perform the similar task for a 3×3 linear system

$$a_{11}x + a_{12}y + a_{13}z = b_1$$

$$a_{21}x + a_{22}y + a_{23}z = b_2$$

$$a_{31}x + a_{32}y + a_{33}z = b_3$$

or, in matrix notation,

$$\begin{bmatrix} a_{11} & a_{12} & a_{13} \\ a_{21} & a_{22} & a_{23} \\ a_{31} & a_{32} & a_{33} \end{bmatrix} \begin{bmatrix} x \\ y \\ z \end{bmatrix} = \begin{bmatrix} b_1 \\ b_2 \\ b_3 \end{bmatrix}.$$

But, to make this long task short by Cramer's rule, let us say that the result is

$$x = \frac{\begin{vmatrix} b_1 & a_{12} & a_{13} \\ b_2 & a_{22} & a_{23} \\ b_3 & a_{32} & a_{33} \end{vmatrix}}{\begin{vmatrix} a_{11} & a_{12} & a_{13} \\ a_{21} & a_{22} & a_{23} \\ a_{31} & a_{32} & a_{33} \end{vmatrix}}, y = \frac{\begin{vmatrix} a_{11} & b_1 & a_{13} \\ a_{21} & b_2 & a_{23} \\ a_{31} & b_3 & a_{33} \end{vmatrix}}{\begin{vmatrix} a_{11} & a_{12} & a_{13} \\ a_{21} & a_{22} & a_{23} \\ a_{31} & a_{32} & a_{33} \end{vmatrix}}, z = \frac{\begin{vmatrix} a_{11} & a_{12} & b_1 \\ a_{21} & a_{22} & b_2 \\ a_{31} & a_{32} & b_3 \end{vmatrix}}{\begin{vmatrix} a_{11} & a_{12} & a_{13} \\ a_{21} & a_{22} & a_{23} \\ a_{31} & a_{32} & a_{33} \end{vmatrix}},$$

provided the determinant $\begin{vmatrix} a_{11} & a_{12} & a_{13} \\ a_{21} & a_{22} & a_{23} \\ a_{31} & a_{32} & a_{33} \end{vmatrix}$ of the coefficient matrix is nonzero.

When we calculate these solutions explicitly, this determinant turns out to be

$$\begin{vmatrix} a_{11} & a_{12} & a_{13} \\ a_{21} & a_{22} & a_{23} \\ a_{31} & a_{32} & a_{33} \end{vmatrix} = a_{11} a_{22} a_{33} - a_{11} a_{23} a_{32} - a_{12} a_{21} a_{33} +$$

$$+ a_{12} a_{23} a_{31} + a_{13} a_{21} a_{32} - a_{13} a_{22} a_{31}. \quad (*)$$

The row indices are $1\,2\,3$ in all $6 (= 3!)$ terms, while the column indices are the three even permutations of $1, 2, 3$: 123, 231, 312 and the three odd permutations 132, 213, 321. Terms with even permutations are positive and those with odd permutations are negative. (A permutation is even (odd) if there is an even (odd) number of inversions in it. An inversion occurs when a larger integer among 1, 2, 3 precedes a smaller one.) So, we have the sum of six terms (the number of permutations of three elements 3!=6), each term is a product of three matrix elements one from each row and one from each column.

We can now generalize this kind of definition of determinants by use of permutations and say:

Definition The determinant of an $n \times n$ real matrix $A = \begin{bmatrix} a_{11} & a_{12} & \cdots & a_{1n} \\ a_{21} & a_{22} & \cdots & a_{2n} \\ \vdots & \vdots & \ddots & \vdots \\ a_{31} & a_{32} & \cdots & a_{3n} \end{bmatrix}$ has $n!$

terms, i.e., products of n matrix elements one from each row and one from each column, with row indices in the natural order $1, 2, \ldots, n$, and the column indices are $n!/2$ even permutations of $1, 2, \ldots, n$, with $+$ sign, and $n!/2$ odd permutations with $-$ sign. The formal definition by permutations is

$$\det A = |A| = \sum_{\sigma \in S_n} (sign\,\sigma) a_{1\sigma(1)} a_{2\sigma(2)} \cdots a_{n\sigma(n)}$$

The sum goes over all $n!$ permutations σ of n numbers $1, 2, \ldots, n$, and these permutations form the symmetric group S_n.

$$sign\,\sigma = \begin{cases} 1 \text{ if } \sigma \text{ is an even permutation} \\ -1 \text{ if } \sigma \text{ is an odd permutation,} \end{cases}$$

while $\sigma(i)$, $i = 1, 2, \ldots, n$, is the image of i under the permutation σ.

Going back to the expansion $(*)$ we notice that it can be written as

$$\begin{vmatrix} a_{11} & a_{12} & a_{13} \\ a_{21} & a_{22} & a_{23} \\ a_{31} & a_{32} & a_{33} \end{vmatrix} = a_{11} \begin{vmatrix} a_{22} & a_{23} \\ a_{32} & a_{33} \end{vmatrix} - a_{12} \begin{vmatrix} a_{21} & a_{23} \\ a_{31} & a_{33} \end{vmatrix} + a_{13} \begin{vmatrix} a_{21} & a_{22} \\ a_{31} & a_{32} \end{vmatrix}. \quad (**)$$

It is called the *Laplace expansion* by the first row of the original determinant. The three smaller determinants are called cofactors, and they are the signed determinants of 2×2 submatrices obtained when we delete the row and column in the original matrix that correspond to the row and column of the elements (from the first row) a_{11}, a_{12}, a_{13}. The signs of cofactors are $(-1)^{1+1}, (-1)^{1+2}, (-1)^{1+3}$, i.e., (-1) to the power which is the sum of indices of the elements from the first row.

We can now generalize this kind of definition of the determinant by *mathematical induction*.

Definition Let us consider the determinant of a 2×2 matrix $\begin{bmatrix} a_{11} & a_{12} \\ a_{21} & a_{22} \end{bmatrix}$, i.e., $\det A = |A| = a_{11}a_{22} - a_{12}a_{21}$, as the induction base. The expansion $(**)$ of the determinant of a 3×3 matrix (the Laplace expansion) in terms of signed determinants of 2×2 matrices is the first induction step.

Let $A = [a_{ij}]$ be a real square $n \times n$ matrix with $n > 3$. The cofactor of a_{ij} in A is $c_{ij} = (-1)^{i+j} \det A_{ij}$, where A_{ij} is the *minor matrix* in A. It is the $(n-1) \times (n-1)$ submatrix obtained by crossing out the i-th row and j-th column of A:

$$A_{ij} = \begin{bmatrix} a_{11} & \cdots & a_{1\,j-1} & a_{a\,j+1} & \cdots & a_{1n} \\ \cdots & \cdots & \cdots & & \cdots & \cdots \\ a_{i-11} & \cdots & a_{i-1\,j-1} & a_{i-1\,j+1} & \cdots & a_{i-1n} \\ a_{i+11} & \cdots & a_{i+1\,j-1} & a_{i+1\,j+1} & \cdots & a_{i+1n} \\ \cdots & \cdots & \cdots & & \cdots & \cdots \\ a_{n1} & \cdots & a_{n\,j-1} & a_{n\,j+1} & \cdots & a_{nn} \end{bmatrix} \;\rightarrow\; i\text{-th row out}$$

$$\downarrow$$
$$j\text{-th column out}$$

[by induction hypothesis, we are supposed to know how to calculate determinants of $(n-1) \times (n-1)$ matrices.]

The determinant of A is the real number

$$\det A = |A| = a_{11}c_{11} + a_{12}c_{12} + \cdots + a_{1n}c_{1n} =$$
$$= \sum_{k=1}^{n} a_{1k}c_{1k} = \sum_{k=1}^{n} (-1)^{1+k} a_{1k} \det A_{1k}.$$

This is the Laplace expansion by cofactors of A along the first row of A.

Remark The Sarrus rule for calculating only the determinants of 3×3 matrices:

$$\begin{vmatrix} a_{11} & a_{12} & a_{13} \\ a_{21} & a_{22} & a_{23} \\ a_{31} & a_{32} & a_{33} \end{vmatrix} = a_{11}a_{22}a_{33} + a_{12}a_{23}a_{31} + a_{13}a_{21}a_{32} -$$

$$-a_{13}a_{22}a_{31} - a_{11}a_{23}a_{32} - a_{12}a_{21}a_{33} =$$

$$= + \begin{vmatrix} \boxtimes \end{vmatrix} - \begin{vmatrix} \boxtimes \end{vmatrix}$$

(the three positive products are calculated by following the multiplications in the first scheme, and the negative ones according to the second scheme).

Example

$$\begin{vmatrix} 6 & -1 & -8 \\ 3 & -1 & -12 \\ -6 & 0 & -4 \end{vmatrix} = 24 - 72 + 48 - 12 = 72 - 72 - 12 = -12.$$

A.2 Properties of Determinants

1. The column-interchange property (also the row-interchange property).

Statement *If two different columns of a square $n \times n$ matrix A are interchanged, the determinant of the resulting matrix B is*

$$\det B = -\det A.$$

Also, since $\det A^T = \det A$ *(see the next transpose property), we can conclude that the interchange of two different rows in A gives the matrix C, such that*

$$\det C = -\det A$$

Proof (for the interchange of columns).
For the case $n = 2$, this property is obvious (this is the induction base):

$$\det A = \begin{vmatrix} a_{11} & a_{12} \\ a_{21} & a_{22} \end{vmatrix} = a_{11}a_{22} - a_{12}a_{21}, \text{ and in } B \text{ we interchange columns,}$$

$$\det B = \begin{vmatrix} a_{12} & a_{11} \\ a_{22} & a_{21} \end{vmatrix} = a_{12}a_{21} - a_{11}a_{22} = -(a_{11}a_{22} - a_{12}a_{21}) = -\det A.$$

Assume $n > 2$, and that this column-interchange property holds for determinants of matrices of size $(n-1) \times (n-1)$ (induction hypothesis). Then this will be true for the determinant of a matrix which is of $n \times n$ size.
Let A be an $n \times n$ matrix and let B be the matrix obtained from A by interchanging the two neighboring columns—the i-th and $(i+1)$-th ones, leaving the other columns unchanged: $A = [\ldots c_i c_{i+1} \ldots]$ and $B = [\ldots c_{i+1} c_i \ldots]$, where c_i and c_{i+1} are those neighboring columns. Now $b_{1i} = a_{1i+1}$, and also for the corresponding cofactors $B_{1i} = A_{1i+1}$. Similarly, $b_{1i+1} = a_{1i}$ and $B_{1i+1} = A_{1i}$. Choosing $k \neq i$, $i+1$ (because $n > 2$ we can choose a third column k), we have $b_{1k} = a_{1k}$, but B_{1k} is obtained from A_{1k} by the interchange of two neighboring columns i and $i+1$, which due to the induction hypothesis means $\det B_{1k} = -\det A_{1k}$, since B_{1k} and A_{1k} are $(n-1) \times (n-1)$ minor submatrices.
If by Σ' we denote the sum from 1 to n, in which the i-th and $(i+1)$-th terms are missing, we have

$$\det B = \sum_{k=1}^{n}(-1)^{1+k}b_{1k}\det B_{1k} =$$
$$= \sum'(-1)^{1+k}b_{1k}\det B_{1k} + (-1)^{1+i}b_{1i}\det B_{1i} + (-1)^{1+(i+1)}b_{1\,i+1}\det B_{1\,i+1} =$$
$$= -\sum'(-1)^{1+k}a_{1k}\det A_{1k} + (-1)^{1+i}a_{1\,i+1}\det A_{1\,i+1} + (-1)^{1+(i+1)}a_{1i}\det A_{1i} =$$
$$= -\sum'(-1)^{1+k}\det A_{1k} - (-1)^{1+(i+1)}a_{1\,i+1}\det A_{1\,i+1} - (-1)^{1+i}a_{1i}\det A_{1i} =$$
$$= -\sum_{k=1}^{n}(-1)^{1+k}a_{1k}\det A_{1k} = -\det A.$$

We can conclude that the interchange of two neighboring columns in an $n \times n$ matrix A gives a new matrix B, such that

$$\det B = -\det A.$$

Now, assume that we interchange the two columns i and $(i+p)$, with $p > 1$. We have

$$A = [\ldots c_{i-1}\, c_i\, c_{i+1} \ldots c_{i+p-1}\, c_{i+p}\, c_{i+p+1} \ldots] \text{ and}$$
$$B = [\ldots c_{i-1}\, c_{i+p}\, c_{i+1} \ldots c_{i+p-1}\, c_i\, c_{i+p+1} \ldots].$$

If we interchange c_i and c_{i+p-1} in B, and so on to the left until we interchange c_i with c_{i+p}, we in fact perform p such permutations in B and obtain the new matrix

$$B' = [\ldots c_{i-1}\, c_i\, c_{i+p}\, c_{i+1} \ldots c_{i+p-1}\, c_{i+p+1} \ldots],$$

so that $\det B' = (-1)^p \det B$. Furthermore, we interchange c_{i+p} with c_{i+1} in B', and so on to the right until c_{i+p} reaches its natural position. In so doing, we perform $(p-1)$ permutations and reach the original matrix A. Therefore,

$$\det A = (-1)^{p-1}\det B' = (-1)^{2p-1}\det B = -\det B. \;\Delta$$

2. The transpose property

   Statement *For any square matrix A, we have $\det A^T = \det A$, or, in other words, the matrix A and its transpose have equal determinants.*

   Proof Verification of this property is trivial for determinants of matrices of the size 2×2:

   $$A = \begin{bmatrix} a_{11} & a_{12} \\ a_{21} & a_{22} \end{bmatrix}, \; \det A = a_{11}a_{22} - a_{12}a_{21},$$
   $$A^T = \begin{bmatrix} a_{11} & a_{21} \\ a_{12} & a_{22} \end{bmatrix}, \; \det A^T = a_{11}a_{22} - a_{12}a_{21}, \qquad \Rightarrow \det A^T = \det A,$$

   (this is the induction base).

   Assume $n > 2$, and that this transpose property holds for determinants of matrices of size $(n-1) \times (n-1)$, (this is the induction hypothesis). Then this will be true for the determinant of a matrix which is of $n \times n$ size.

Indeed, we perform the Laplace expansion of det A along the first row

$$\det A = \sum_{k=1}^{n}(-1)^{1+k}a_{1k}\det A_{1k},$$

where A is an $n \times n$ matrix. For det $B = \det A^T$, we perform the cofactor expansion on the first column det $B = \sum_{k=1}^{n}(-1)^{k+1}b_{k1}\det B_{k1}$. We know that the first column in B is in fact the first row in A, i.e.,

$$b_{k1} = a_{1k}, k = 1,2,\ldots,n.$$

Furthermore, the minor matrix B_{k1} is obviously the transpose of A_{1k}, $B_{k1} = A_{1k}^T$, and since they are $(n-1) \times (n-1)$ submatrices, we have

$$\det B_{k1} = \det A_{1k}, \text{ by the induction hypothesis.}$$

Replacing these two last equalities in the expansion of det $B = \sum_{k=1}^{n}(-1)^{1+k}a_{1k}\det A_{1k} = \det A$, which is the transpose property det $A^T = \det A$.
Δ

3. **Theorem [on the general Laplace (cofactor) expansion]** The determinant det A of an $n \times n$ matrix A can be computed by multiplying the entries of any row or any column by their respective cofactors and adding the resulting product. More precisely:

$$\det A = a_{i1}c_{i1} + a_{i2}c_{i2} + \cdots + a_{in}c_{in} =$$

$$= \sum_{k=1}^{n}(-1)^{i+k}a_{ik}\det A_{ik}, \ 1 \leq i \leq n,$$

(the cofactor expansion along the i-th row);

$$\det A = a_{1j}c_{1j} + a_{2j}c_{2j} + \cdots + a_{nj}c_{nj} =$$

$$= \sum_{k=1}^{n}(-1)^{k+j}a_{kj}\det A_{kj}, \ 1 \leq j \leq n,$$

(the cofactor expansion along the j-th column).

Proof By the $(i-1)$ permutations of the i-th row, $i = 2,3,\ldots,n$, in

$$A = \begin{bmatrix} a_{11} & a_{12} & \cdots & a_{1n} \\ a_{21} & a_{22} & \cdots & a_{2n} \\ \cdots & \cdots & \cdots & \cdots \\ a_{i-11} & a_{i-12} & \cdots & a_{i-1n} \\ a_{i1} & a_{i2} & \cdots & a_{in} \\ a_{i+11} & a_{i+12} & \cdots & a_{i+1n} \\ \cdots & \cdots & \cdots & \cdots \\ a_{n1} & a_{n2} & \cdots & a_{nn} \end{bmatrix}, \text{ we get the matrix}$$

$$B = \begin{bmatrix} a_{i1} & a_{i2} & \cdots & a_{in} \\ a_{11} & a_{12} & \cdots & a_{1n} \\ \cdots & \cdots & \cdots & \cdots \\ a_{i-11} & a_{i-12} & \cdots & a_{i-1n} \\ a_{i+11} & a_{i+12} & \cdots & a_{i+1n} \\ \cdots & \cdots & \cdots & \cdots \\ a_{n1} & a_{n2} & \cdots & a_{nn} \end{bmatrix} = \begin{bmatrix} b_{11} & b_{12} & \cdots & b_{1n} \\ b_{21} & b_{22} & \cdots & b_{2n} \\ \vdots & \vdots & \ddots & \vdots \\ b_{n1} & b_{n2} & \cdots & b_{nn} \end{bmatrix},$$

so that det $B = (-1)^{i-1}$ det A (the row-interchange property).

But, det $B = \sum_{k=1}^{n}(-1)^{1+k}b_{1k}$ det B_{1k} (the basic definition of a determinant, i.e., the Laplace expansion along the first row), and we see that $b_{1k} = a_{ik}$ and $B_{1k} = A_{ik}$, $k = 1,2,\ldots,n$, so that det $B = \sum_{k=1}^{n}(-1)^{1+k}a_{ik}$ det A_{ik}. By multiplying both sides with $(-1)^{i-1}$, we get

$$(-1)^{i-1}\det B = \sum_{k=1}^{n}(-1)^{i+k}a_{ik}\det A_{ik} = \det A. \, (*)$$

which is the Laplace cofactor expansion of det A along the i-th row, $1 \le i \le n$. The Laplace cofactor expansion along columns of det A follows by analogous reasoning, starting with the transpose property det $A = $ det A^{T}. Δ

4. **Theorem (on the inverse matrix A^{-1} of A)** If det $A \ne 0$, then the inverse matrix of $A = [a_{ij}]_{n\times n}$ is given as

$$A^{-1} = \frac{\text{adj } A}{\det A},$$

where adj A (the *adjoint* of A) is the transpose of the matrix made of cofactors of A:

$$\text{adj } A = \begin{bmatrix} c_{11} & c_{21} & \cdots & c_{n1} \\ c_{12} & c_{22} & \cdots & c_{n2} \\ \vdots & \vdots & \ddots & \vdots \\ c_{1n} & c_{2n} & \cdots & c_{nn} \end{bmatrix}.$$

Proof If we multiply the elements of a row (or column) of A with cofactors of another row (or column) and sum these products, the result is always zero:

$$\sum_{k=1}^{n}(-1)^{i+k}a_{jk}\det A_{ik} = \sum_{k=1}^{n}a_{jk}c_{ik} = 0 \; (i \ne j)$$

(the elements of the j-th row are multiplied by cofactors of the i-row and these products are summed up) and, similarly, for columns

$$\sum_{k=1}^{n}(-1)^{k+i}a_{kj}\det A_{ki} = \sum_{k=1}^{n}a_{kj}c_{ki} = 0 \; (i \ne j).$$

We shall prove only the first of the above expressions:
Consider the matrix B, which is obtained from A by replacing in A the i-th row by the j-th one $(i \ne j)$, so that B has two identical j-th rows. As a consequence of

the row-exchange property, we have $\det B = -\det B$, or $\det B = 0$. The cofactor expansion of $\det B$, i.e.,

$$\det B = \sum_{k=1}^{n} (-1)^{i+k} b_{ik} \det B_{ik} = 0$$

implies, due to $B_{ik} = A_{ik}$ and $b_{ik} = a_{jk}$, or, in detail,

$$
A =
\begin{bmatrix}
a_{11} & a_{12} & \cdots & a_{1n} \\
\cdots & \cdots & \cdots & \cdots \\
a_{i-11} & a_{i-12} & \cdots & a_{i-1n} \\
a_{i1} & a_{i2} & \cdots & a_{in} \\
a_{i+11} & a_{i+12} & \cdots & a_{i+1n} \\
\cdots & \cdots & \cdots & \cdots \\
a_{j-11} & a_{j-12} & \cdots & a_{j-1n} \\
a_{j1} & a_{j2} & \cdots & a_{jn} \\
a_{j+11} & a_{j+12} & \cdots & a_{j+1n} \\
\cdots & \cdots & \cdots & \cdots \\
a_{n1} & a_{n2} & \cdots & a_{nn}
\end{bmatrix}
, B =
\begin{bmatrix}
a_{11} & a_{12} & \cdots & a_{1n} \\
\cdots & \cdots & \cdots & \cdots \\
a_{i-11} & a_{i-12} & \cdots & a_{i-1n} \\
a_{j1} & a_{j2} & \cdots & a_{jn} \\
a_{i+11} & a_{i+12} & \cdots & a_{i+1n} \\
\cdots & \cdots & \cdots & \cdots \\
a_{j-11} & a_{j-12} & \cdots & a_{j-1n} \\
a_{j1} & a_{j2} & \cdots & a_{jn} \\
a_{j+11} & a_{j+12} & \cdots & a_{j+1n} \\
\cdots & \cdots & \cdots & \cdots \\
a_{n1} & a_{n2} & \cdots & a_{nn}
\end{bmatrix}
=
\begin{bmatrix}
b_{11} & b_{12} & \cdots & b_{1n} \\
\cdots & \cdots & \cdots & \cdots \\
b_{i-11} & b_{i-12} & \cdots & b_{i-1n} \\
b_{i1} & b_{i2} & \cdots & b_{in} \\
b_{i+11} & b_{i+12} & \cdots & b_{i+1n} \\
\cdots & \cdots & \cdots & \cdots \\
b_{j-11} & b_{j-12} & \cdots & b_{j-1n} \\
b_{j1} & b_{j2} & \cdots & b_{jn} \\
b_{j+11} & b_{j+12} & \cdots & b_{j+1n} \\
\cdots & \cdots & \cdots & \cdots \\
b_{n1} & b_{n2} & \cdots & b_{nn}
\end{bmatrix}
,
$$

that $\sum_{k=1}^{n} (-1)^{i+k} a_{jk} \det A_{ik} = \sum_{k=1}^{n} a_{jk} C_{ik} = 0$. $(i \neq j)$ (**)

We can now make a statement that summarizes the above results (*) in 3 and (**) in 4

$$\sum_{k=1}^{n} (-1)^{i+k} a_{jk} \det A_{ik} = \delta_{ji} \det A, \text{ or}$$

$$\sum_{k=1}^{n} a_{jk} C_{ik} = \delta_{ji} \det A.$$

In terms of matrix multiplication, this can be written as $A \text{ adj } A = \det A I_n$, where adj A is the transpose of the matrix of cofactors.

If $\det A \neq 0$, then we can divide both sides by $\det A$ and get $A \frac{\text{adj } A}{\det A} = I_n$, which means that

$$A^{-1} = \frac{\text{adj } A}{\det A}. \quad \Delta$$

This is the most useful formula for the inverse matrix A^{-1}, and it clearly shows that the matrix A is invertible (nonsingular or regular) if and only if its determinant is different from 0:

$$\det A \neq 0.$$

5. The determinant of a triangular matrix.

Statement The determinant of any triangular matrix (lower, upper, or diagonal) is the product of all diagonal elements.

Proof Let us consider a lower triangular $n \times n$ matrix A and perform in succession the Laplace expansion along the first row in its determinant and in the resulting cofactors:

$$\det A = \begin{vmatrix} a_{11} & 0 & 0 & \cdots & 0 \\ a_{21} & a_{22} & 0 & \cdots & 0 \\ a_{31} & a_{32} & a_{33} & \cdots & 0 \\ \cdots & \cdots & \cdots & \cdots & \cdots \\ a_{n1} & a_{n2} & a_{n3} & \cdots & a_{nn} \end{vmatrix} = a_{11} \begin{vmatrix} a_{22} & 0 & \cdots & 0 \\ a_{32} & a_{33} & \cdots & 0 \\ \vdots & \vdots & \ddots & \vdots \\ a_{n2} & a_{n3} & \cdots & a_{nn} \end{vmatrix} =$$

$$= a_{11} a_{22} \begin{vmatrix} a_{33} & \cdots & 0 \\ \vdots & \ddots & \vdots \\ a_{n3} & \cdots & a_{nn} \end{vmatrix} = \ldots = a_{11} a_{22} a_{33} \ldots a_{nn}.$$

For an upper triangular $n \times n$ matrix the proof is analogous. A diagonal matrix is at the same time a lower and an upper triangular matrix. Δ

6. **Theorem (on the multiplicative property of determinants)** If A and B are $n \times n$ matrices, then

$$\boxed{\det (AB) = \det A \cdot \det B}.$$

Proof To prove this theorem is very easy if A is a diagonal matrix, because the product

$$AB = \begin{bmatrix} a_{11} & & & 0 \\ & a_{22} & & \\ & & \ddots & \\ 0 & & & a_{nn} \end{bmatrix} \begin{bmatrix} b_{11} & b_{12} & \cdots & b_{1n} \\ b_{21} & b_{22} & \cdots & b_{2n} \\ \vdots & \vdots & \ddots & \vdots \\ b_{n1} & b_{n2} & \cdots & b_{nn} \end{bmatrix} =$$

$$= \begin{bmatrix} a_{11}b_{11} & a_{11}b_{12} & \cdots & a_{11}b_{1n} \\ a_{22}b_{21} & a_{22}b_{22} & \cdots & a_{22}b_{2n} \\ \vdots & \vdots & \ddots & \vdots \\ a_{nn}b_{n1} & a_{nn}b_{n2} & \cdots & a_{nn}b_{nn} \end{bmatrix}$$

has its i-th row equal to a_{ii} times the i-th row of B. Using the scalar multiplication property (see the following property 7) in each of these rows, we obtain

$$\det (AB) = \begin{vmatrix} a_{11}b_{11} & a_{11}b_{12} & \cdots & a_{11}b_{1n} \\ a_{22}b_{21} & a_{22}b_{22} & \cdots & a_{22}b_{2n} \\ \vdots & \vdots & \ddots & \vdots \\ a_{nn}b_{n1} & a_{nn}b_{n2} & \cdots & a_{nn}b_{nn} \end{vmatrix} =$$

$$= (a_{11} a_{22} \ldots a_{nn}) \det B = \det A \cdot \det B,$$

since the determinant of the diagonal matrix A is equal to the product of its diagonal elements (property 5).

Now suppose that A is an invertible matrix. We can use the row-reduction method analogous to the GJ method (see Sect. 2.13), but without making the leading entries equal to 1, and finally, we reduce A to a diagonal matrix D with nonzero diagonal elements. We can write $D = EA$, where E is the *product of elementary matrices* F_{ij} (see Sect. 2.12), corresponding to the row exchange (property 1) [its determinant is (-1)] and $F_{ij}(\lambda)$, corresponding to the row-addition (see property 8) with determinant equal to 1. In other words, $|F_{ij}A| = -|A|$ and $|F_{ij}(\lambda)A| = |A|$, or $|F_{ij}A| = |F_{ij}||A|$ and $|F_{ij}(\lambda)A| = |F_{ij}(\lambda)||A|$. In particular, if F_1 and F_2 stand for either F_{ij} or $F_{ij}(\lambda)$, we have

$$|F_1 F_2| = |F_1||F_2|, \text{ or, more generally,}$$
$$|F_1 F_2 \dots F_p| = |F_1||F_2| \dots |F_p|.$$

Since $E = F_1 F_2 \dots F_p$, where each of $F_1, F_2, \dots, F_p$ is either F_{ij} or $F_{ij}(\lambda)$, we have $EA = (F_1 F_2 \dots F_p A) = D$ and $|EA| = |F_1 F_1 \dots F_p A| = |F_1||F_2 \dots F_p A| = |F_1||F_2||F_3 \dots F_p A| = \dots = |F_1||F_2| \dots |F_p||A| = (-1)^s |A| = |D|$, where s is the number of F_{ij} factors (the row interchanges) in E.

So, $|A| = (-1)^s |D|$. Also $E(AB) = (EA)B = DB$, so that $|EAB| = |F_1 F_2 \dots F_p AB| = (-1)^s |AB| = |DB|$, and, finally, $|AB| = (-1)^s |DB| = (-1)^s |D||B| = |A||B|$

or

$$\det(AB) = \det A \cdot \det B.$$

If A is a singular (noninvertible) matrix, which means that A maps some nonzero vectors from $\mathbb{R}^n$ onto $\bar{0}_n$, then this would be the case with AB as well, so AB must be singular. Consequently, $|A| = 0$ and $|AB| = 0$, and we again have

$$|AB| = |A||B| \quad \Delta$$

An obvious generalization is that

$$\det(A_1 A_2 \dots A_k) = \det A_1 \cdot \det A_2 \cdots \det A_k.$$

7. The scalar multiplication property.

Statement *If a single row (or a single column due to the transpose property $|A^T| = |A|$, see property 2) of a square matrix A is multiplied by a scalar λ, the determinant of the resulting matrix is*

$$\lambda \det A,$$

or by using the elementary matrix $F_i(\lambda)$ (see Sect. 2.12)

$$|F_i(\lambda)A| = |F_i(\lambda)||A| = \lambda |A|.$$

Proof Let $\lambda \neq 1$ be any scalar (a real number), and let $B = F_i(\lambda)A$ be the matrix obtained from A by replacing the i-th row $[a_{i1}\, a_{i2} \dots a_{in}]$ of A by $[\lambda a_{i1}\, \lambda a_{i2} \dots \lambda a_{1n}]$. Since the rows of B are equal to those of A except for the i-th row, it follows that the minor matrices A_{ij} and B_{ij} are equal for each $j = 1, 2, \ldots n$. Therefore,

$$(-1)^{i+j}|A_{ij}| = (-1)^{i+j}|B_{ij}|,$$

and computing det B by the Laplace expansion along the i-th row, we have

$$\det B = \sum_{j=1}^{n} (-1)^{i+j} b_{ij} |B_{ij}| = \sum_{j=1}^{n} (-1)^{i+j} \lambda a_{ij} |A_{ij}| =$$

$$= \lambda \sum_{j=1}^{n} (-1)^{i+j} a_{ij} |A_{ij}| = \lambda \det A. \quad \Delta$$

8. The row-addition property

Statement *If the product of one row of a square matrix A by a scalar λ is added to a different row of A, the determinant of the resulting matrix is the same as det A, or, in terms of elementary matrices $F_{ij}(\lambda)$ (see Sect. 2.12):*

$$|F_{ij}(\lambda)A| = |F_{ij}(\lambda)||A| = |A|.$$

Proof Let $\bar{r}_j = [a_{j1} \, a_{j2} \ldots a_{jn}]$ be the j-th row of A. Suppose that $\lambda \bar{r}_j$ is added to the i-th row $\bar{r}_i$ of A, where $\lambda \neq 0$ is any scalar and $i \neq j$.
We obtain the matrix $B = F_{ij}(\lambda)A$ whose rows are the same as the corresponding rows of A except for the i-th row, which is

$$\lambda \bar{r}_j + \bar{r}_i = [\lambda a_{j1} + a_{i1} \; \lambda a_{j2} + a_{i2} \ldots \lambda a_{jn} + a_{in}].$$

Clearly, the minor matrices A_{ik} and B_{ik} are equal for each k. Computing det B by the Laplace expansion along the i-th row, we have

$$\det B = \sum_{k=1}^{n} (-1)^{i+k} b_{ik} |B_{ik}| = \sum_{k=1}^{n} (-1)^{i+k} (\lambda a_{jk} + a_{ik}) |A_{ik}| =$$

$$= \sum_{k=1}^{n} (-1)^{i+k} \lambda a_{jk} |A_{ik}| + \sum_{k=1}^{n} (-1)^{i+k} a_{ik} |A_{ik}| =$$

$$= \lambda \det C + \det A,$$

where C is the matrix obtained from A by replacing the i-th row of A with the j-th row of A. Because C has two equal rows, its determinant is zero (the row-interchange property), so det $B = \det A$. $\quad \Delta$

Note. Due to the transpose property $|A^T| = |A|$, the analogous statement is valid for the column-addition property.

Remark The row and column additions can be used to simplify calculations of determinants, since they do not change the value of a determinant, but can produce rows and columns with more zeros to enable easy Laplace expansions.

Example Calculate the determinant of the 4×4 matrix:

$$
\begin{vmatrix}
2 & -1 & 4 & -3 \\
-1 & 1 & 0 & 2 \\
3 & 2 & 3 & -1 \\
1 & -2 & 2 & 3
\end{vmatrix}
\quad
\begin{array}{l}
\text{we add column 2 to column 1} \\
\text{and } (-2) \text{ column 3 to column 4,}
\end{array}
$$

$$
= \begin{vmatrix}
1 & -1 & 4 & -1 \\
0 & 1 & 0 & 0 \\
5 & 2 & 3 & -5 \\
-1 & -2 & 2 & 7
\end{vmatrix}
\quad
\begin{array}{l}
\text{and perform the Laplace expansion} \\
\quad\quad\text{along the second row,}
\end{array}
$$

$$
= 1 \begin{vmatrix}
1 & 4 & -1 \\
5 & 3 & -5 \\
-1 & 2 & 7
\end{vmatrix}
\quad
\begin{array}{l}
\text{we add } (-5) \text{ row 1 to row 2 and add} \\
\quad\quad\text{row 1 to row 3,}
\end{array}
$$

$$
= \begin{vmatrix}
1 & 4 & -1 \\
0 & -17 & 0 \\
0 & 6 & 6
\end{vmatrix}
\quad \text{the Laplace expansion along the first column,}
$$

$$
= 1 \begin{vmatrix}
-17 & 0 \\
6 & 6
\end{vmatrix} = -17 \cdot 6 = -102
$$

Chapter 2
Linear Mappings and Linear Systems

2.1 A Short Plan for the First 5 Sections of Chapter 2

Since the whole concept of modern mathematics is based on sets and their mappings, we shall now, after studying sets of vector spaces, continue with the investigation of their mappings.

We shall first review the general theory of mappings. We want to emphasize that every map

$$f : A \rightarrow B \text{ (where } A, B \text{ are two arbitrary sets)}$$

is actually a *bijection* between the subset $f(A)$ of all images of f in the codomain B and the set of their inverse images in the domain A (this set makes a partition of the domain A).

A linear mapping L(in short: *linmap*) $L : V_n \rightarrow W_m$ takes all vectors from one vector space V_n onto their unique images in another space W_m, so that any linear combination $\sum_{i=1}^{k} a_i \bar{v}_i$ of vectors from V_n goes onto the same linear combination $\sum_{i=1}^{k} a_i L(v_i)$ of their images.

Every linear map $L : V_n \rightarrow W_m$ defines the two most important subspaces, one in V_n and one in W_m. The first subspace consists of all vectors from V_n which are taken by L onto the zero vector $\bar{0}_w$ in W_m. It is called the *kernel* of L, and it is denoted as ker L. The second subspace is made of images in W_m of all vectors from V_n. It is called the *range* of L and it is denoted as ran L.

When L is a many-to-one linmap then every element $\bar{y} \in$ ran L has its *inverse image* Inv($\bar{y}$) in V_n which is actually a subset of V_n that consists of ker L plus a unique particular preimage $\bar{y}' \in V_n$. This set ker $L + \bar{y}'$ is called the *coset* of $\bar{y}$ [it is a neighbor set of ker $L =$ Inv($\bar{0}_w$) obtained by translating the kernel with $\bar{y}'$, which is uniquely determined by $\bar{y}$]. The set of all cosets forms a vector space (the quotient space $V_n/$ker L) which is isomorphic to ran L:

$$V_n/\text{ker } L \stackrel{\sim}{=} \text{ran } L.$$

To determine ker L and ran L as the essential objects of linmap L, we usually exploit representation theory in which vectors from V_n and W_m are replaced by matrix columns from $\mathbb{R}^n$ and $\mathbb{R}^m$, respectively, and L itself by an $m \times n$ matrix M'. Making use of matrix algebra, we can now easily solve these problems for L (finding ker L and ran L) and many other more complicated problems (e.g., the Eigen problem – Chap. 5). Thus, representation theory is extremely useful for all researchers using linear maps in their experiments (for example, physicists dealing with atomic and subatomic worlds).

Choosing a basis $\{\bar{v}_1, \bar{v}_2, \ldots, \bar{v}_n\}$ in V_n and a basis $\{\bar{w}_1, \bar{w}_2, \ldots, \bar{w}_m\}$ in W_m, we can achieve two isomorphisms, one v between V_n and $\mathbb{R}_n$, and the other w between w_m and $\mathbb{R}^m$. It is then easy to define the $m \times n$ matrix M (this matrix is also a linmap $M : \mathbb{R}^n \to \mathbb{R}^m$) that represents L in the sense that if L maps $\bar{x} \in V_n$ onto $\bar{y} = L(\bar{x}) \in W_m$, then M maps $[\bar{x}]_v$ (the matrix-column that represents $\bar{x}$ in $\mathbb{R}^n$ by isomorphism v) onto the matrix-column $[\bar{y}]_w = [L(\bar{x})]_w$:

$$\boxed{M[\bar{x}]_v = [L(\bar{x})]_w.}$$

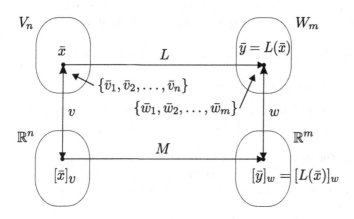

So, whichever problem we have to solve for $L : V_n \to W_m$, we first transfer it to $M : \mathbb{R}^n \to \mathbb{R}^m$ and perform the necessary matrix calculations. Taking these results back to V_n and W_m is then immediate by using the isomorphisms v and w.

2.2 Some General Statements about Mapping

Given two sets A and B, then a mapping (or a map) f from A to B is a rule which associates with each element $x \in A$ (called a *preimage*) a single element $f(x)$ in the set B. The set A is called the *domain* of f (Dom f) and the set B is called the *codomain* of f. One uses the notation $f : A \to B$.

The element $f(x) \in B$ is the *image* of $x \in A$ under f:

$$f : x \mapsto f(x).$$

(Note the difference in notation: $\mapsto$ for elements instead of $\rightarrow$ for sets)

The set of all images is a subset of B, and it is called the *range* of f (ran f):

$$\text{ran } f = f(A) = \{f(x) \mid x \in A\} \subseteq B.$$

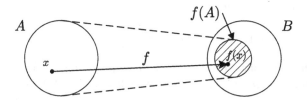

If $f(A) = B$, i.e., if every element from B is an image of elements from A, then f is an *onto* map (or *surjection*).

Generally, a map can be many-to-one, but if distinct elements from A have distinct images:

$$x_1 \neq x_2 \Rightarrow f(x_1) \neq f(x_2),$$

then f is a one-to-one (1-1) map (or an *injection*). The contrapositive statement (more often used for 1-1 maps) is that $f(x_1) = f(x_2) \Rightarrow x_1 = x_2$ (one image cannot have more than one preimage).

If a map f is at the same time 1-1 and onto, then it is called a *bijection*. In this case (and only in this case), the map f has the inverse map $f^{-1} : B \rightarrow A$ (also a bijection), so that $f \circ f^{-1} = I_B$ and $f^{-1} \circ f = I_A$, where I_A and I_B are the identity maps in A and B (e.g., $I_A(x) = x$ for all x from A), and $\circ$ is the composition of mappings.

If f is not a 1-1 map, then the set of all preimages of $y \in \text{ran } f$ is called its *inverse image*: $\text{Inv}(y) = \{x \mid x \in A \text{ and } f(x) = y\}$.

Since every element of A belongs to one and only one inverse image, the set of all inverse images makes a partition of A. In other words, the domain A is the union of all inverse images which are mutually disjoint subsets. So, the inverse images are the equivalence classes of the equivalence relation $\sim$ in A induced by this partition (two elements x_1 and x_2 are equivalent $x_1 \sim x_2$ if and only if they belong to the same inverse image, i.e., iff $f(x_1) = f(x_2)$. This relation is obviously reflexive, symmetric and transitive (SRT), so it is an equivalence relation). Obviously, if we consider the set $A/\sim$ of all inverse images, then there is a bijection f' between this set and ran $f = f(A)$. A convenient notation is

$$\boxed{A/_\sim \overset{f'}{\leftrightarrow} \text{ran } f}$$

Conclusion: Every map $f : A \rightarrow B$ gives rise to a bijection $\boxed{f : A/_\sim \leftrightarrow \text{ran } f}$, where ran $f = f(A)$ is the range of f (the subset of all images in the codomain B) and $A/_\sim$ is the set of all inverse images of f in the domain A.

2.3 The Definition of Linear Mappings (Linmaps)

Let V_n and W_m be two real vector spaces of dimensions n and m, respectively. A linear mapping (a *linmap*) from V_n to W_m is a map $L : V_n \to W_m$ which has two additional properties:

1. L preserves vector addition, i.e., it maps the sum of any two vectors $\bar{x}_1$ and $\bar{x}_2$ from V_n onto the sum of their images in W_m:

$$L(\bar{x}_1 + \bar{x}_2) = L(\bar{x}_1) + L(\bar{x}_2);$$

2. L preserves multiplication with scalars, i.e., it maps the multiple of any $\bar{x} \in V_n$ with the scalar c onto the multiple of the same c with the image of $\bar{x}$ in W_m:

$$L(c\bar{x}) = cL(\bar{x}), \quad \text{for all } c \in \mathbb{R}.$$

Graphically, it looks as if L enters into the above two brackets $(\bar{x}_1 + \bar{x}_2)$ and $(c\bar{x})$. It is a good mnemonic rule.

An equivalent and a more practical condition for the map $L : V_n \to W_m$ to be a linmap is that it should preserve any linear combination of vectors from V_n by mapping it onto the same linear combination of their images:

$$L(\sum_{i=1}^{k} c_i \bar{x}_i) = \sum_{i=1}^{k} c_i L(\bar{x}_i).$$

We can repeat the same mnemonic rule that it looks as if L enters the bracket. We shall often use this property in the other direction: L can be taken out in front of the bracket.

The linmap L is also called a vector space *homomorphism* because this term points out that the vector spaces V_n and W_m have the same algebraic structure based on the operations of linear combinations.

If L is a bijection (an injection and also a surjection), then we say that it is an *isomorphism* and that V and W are isomorphic (exactly corresponding in form). In this case obviously V and W must have the same dimension $m = n$.

A linmap $L : V_n \to V_m$ is given if we know the image of every $\bar{x} \in V_n$. A much more economical method is to know only the images $\{L(\bar{v}_1), L(\bar{v}_2), \ldots, L(\bar{v}_n)\}$ of any set $B = \{\bar{v}_1, \bar{v}_2, \ldots, \bar{v}_n\}$ of basis vectors in V_n. In this case, we expand $\bar{x}$ in this basis $\bar{x} = \sum_{i=1}^{n} x_i \bar{v}_i$ and apply the linmap L to this expansion:

$$L(\bar{x}) = L(\sum_{i=1}^{n} x_i \bar{v}_i) = \sum_{i=1}^{n} x_i L(\bar{v}_i),$$

so that the image $L(\bar{x})$ of $\bar{x}$ is determined by its components $x_1, x_2, \ldots, x_n$ in the basis B, and the images of this basis only.

Example Let $L: \mathbb{R}^3 \to \mathbb{R}^3$ be given by its action on the standard basis in $\mathbb{R}^3$

$$L\left(\begin{bmatrix} 1 \\ 0 \\ 0 \end{bmatrix}\right) = \begin{bmatrix} 1 \\ 5 \\ 7 \end{bmatrix}, \quad L\left(\begin{bmatrix} 0 \\ 1 \\ 0 \end{bmatrix}\right) = \begin{bmatrix} -1 \\ -4 \\ -6 \end{bmatrix}, \quad L\left(\begin{bmatrix} 0 \\ 0 \\ 1 \end{bmatrix}\right) = \begin{bmatrix} 3 \\ -4 \\ 2 \end{bmatrix}.$$

To find the image $L\left(\begin{bmatrix} x \\ y \\ z \end{bmatrix}\right)$ of any $\begin{bmatrix} x \\ y \\ z \end{bmatrix} \in \mathbb{R}^3$, we expand this vector in the standard basis

$$\begin{bmatrix} x \\ y \\ z \end{bmatrix} = x \begin{bmatrix} 1 \\ 0 \\ 0 \end{bmatrix} + y \begin{bmatrix} 0 \\ 1 \\ 0 \end{bmatrix} + z \begin{bmatrix} 0 \\ 0 \\ 1 \end{bmatrix} \quad \text{(trivial)}$$

and apply L:

$$L\left(\begin{bmatrix} x \\ y \\ z \end{bmatrix}\right) = x \begin{bmatrix} 1 \\ 5 \\ 7 \end{bmatrix} + y \begin{bmatrix} -1 \\ -4 \\ -6 \end{bmatrix} + z \begin{bmatrix} 3 \\ -4 \\ 2 \end{bmatrix} = \begin{bmatrix} x - y + 3z \\ 5x - 4y - 4z \\ 7x - 6y + 2z \end{bmatrix}.$$

If V and W have the same dimension n, then any choice of a basis $\{\bar{v}_1, \bar{v}_2, \ldots, \bar{v}_n\}$ in V_n and a basis $\{\bar{w}_1, \bar{w}_2, \ldots, \bar{w}_n\}$ in W_n establishes a particular isomorphism L:

$$\sum_{i=1}^{n} c_i \bar{v}_i \overset{L}{\leftrightarrow} \sum_{i=1}^{n} c_i \bar{w}_i.$$

When L is an isomorphism (a *bijective linmap*), it has the inverse map L^{-1}, which is also a bijection. In this case, the maximal number of linearly independent vectors in V and W must be the same $n = m$. Namely, the case $n > m$ is impossible because L would map any set of n linearly independent vectors from V_n onto n linearly independent ones in W_n but this space can have the maximum of m such vectors. Analogously, $n < m$ is impossible due to the L^{-1} map. In conclusion, an isomorphic map L can exist between v and w only if they have the same dimension.

2.4 The Kernel and the Range of L

Every linmap $L: V_n \to W_m$ determines two important subspaces—one in the domain V_n and one in the codomain W_m. The first subspace consists of all vectors from V_n which are mapped by L onto the zero vector $\bar{0}_w$ in W_m. It is called the *kernel* of L and denoted as ker L:

$$\ker L = \{\bar{x} \mid \bar{x} \in V_n \text{ and } L(\bar{x}) = \bar{0}_w\}.$$

The second subspace consists of images in W_m of all vectors from V_n. It is called the *range* of L and it is denoted as ran L:

$$\operatorname{ran} L = \{L(\bar{x}) \mid \bar{x} \in V_n\}.$$

We also write concisely ran $L = L(V_n)$.

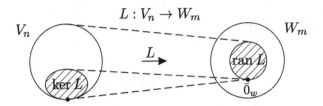

To prove that the above sets are indeed subspaces, we shall show that they are both *closed* when one takes any linear combination of their vectors. Let $\{\bar{x}_1, \bar{x}_2, \ldots, \bar{x}_p\}$ be any set of vectors from ker L, which means that $L(\bar{x}_1) = \bar{0}_w$, $L(\bar{x}_2) = \bar{0}_w, \ldots, L(\bar{x}_p) = \bar{0}_w$. Then, every linear combination of these vectors also belongs to ker L:

$$L\left(\sum_{i=1}^{k} a_i \bar{x}_i\right) = \sum_{i=1}^{k} a_i L(\bar{x}_i) = \sum_{i=1}^{k} a_i \bar{0}_w = \bar{0}_w.$$

The kernel of L is never empty since at least $\bar{0}_v$ belongs to it: $L(\bar{0}_v) = \bar{0}_w$.

Let us now take any set of vectors from ran L $\{\bar{y}_1, \bar{y}_2, \ldots, \bar{y}_q\}$. It follows that every one of them has at least one preimage in V_n : $L(\bar{x}'_1) = \bar{y}_1$, $L(\bar{x}'_2) = \bar{y}_2, \ldots, L(\bar{x}'_q) = \bar{y}_q$. Any linear combination of $\bar{y}$s $\sum_{i=1}^{q} b_i \bar{y}_i$ is also a vector in ran L since it has a preimage $\sum_{i=1}^{q} b_i \bar{x}'_i$ seen as follows

$$L\left(\sum_{i=1}^{q} b_i \bar{x}_i\right) = \sum_{i=1}^{q} b_i L(\bar{x}'_i) = \sum_{i=1}^{q} b_i \bar{y}_i. \;\; \Delta$$

Since we now know that ker L and ran L are subspaces, we shall investigate their relationship, in particular the connection between their dimensions.

We shall prove the very important relationship:

Theorem (dimension)

$$\boxed{\dim(\ker L) + \dim(\operatorname{ran} L) = \dim(\operatorname{Dom} L)}$$

Proof Assume that $\dim(\ker L) = k < n$; we call k the *defect* or *nullity* of L. (If $k = n$, then L maps all of V_n onto $\bar{0}_w$, so that ran $L = \bar{0}_w$ and its dimension is 0: $0 + n = n$). Choose any basis $\{\bar{v}_1, \bar{v}_2, \ldots, \bar{v}_k\}$ in ker L. Extend this basis by $(n - k)$ vectors to obtain the basis $\{\bar{v}_1, \bar{v}_2, \ldots, \bar{v}_k, \bar{v}_{k+1}, \ldots, \bar{v}_n\}$ in V_n. [This process of extension can be performed in the following way: since ker $L = \operatorname{LIN}(\{\bar{v}_1, \bar{v}_2, \ldots, \bar{v}_k\})$, take any vector $\bar{v}_{k+1}$ from V_n which is not in ker L. This vector is obviously linearly independent of $\{\bar{v}_1, \bar{v}_2, \ldots, \bar{v}_k\}$ since all vectors that are linearly dependent on them are in ker L. This means that the set of linearly independent vectors $\{\bar{v}_1, \bar{v}_2, \ldots, \bar{v}_k, \bar{v}_{k+1}\}$ is a basis in the subspace $\operatorname{LIN}(\{\bar{v}_1, \bar{v}_2, \ldots, \bar{v}_k, \bar{v}_{k+1}\})$ of V_n. Continue this process by choosing $\bar{v}_{k+2}$ outside this subspace, and so on. Finally, the subspace $\text{LIN}(\{\bar{v}_1, \bar{v}_2, \ldots, \bar{v}_k, \bar{v}_{k+1}, \bar{v}_{k+2}, \ldots, \bar{v}_n\})$ will be equal to V_n, since every subspace of V_n which has the dimension equal to n is identical to V_n.]

It is our task now to show that $(n-k)$ vectors $\{L(\bar{v}_{k+1}), L(\bar{v}_{k+2}), \ldots, L(\bar{v}_n)\}$ (the images in W_m of the above extension) form a basis in ran L, so that

$$\dim(\text{ran } L) = n - k,$$

which is in fact the dimension theorem. From now on we shall denote $\dim(\text{ran } L)$ as r and call it the *rank of L* so that the theorem can be expressed concisely as $r + k = n$.

In words, the rank and the defect of a linmap L add up to the dimension of its domain. The vectors $\{L(\bar{v}_{k+1}), L(\bar{v}_{k+2}), \ldots, L(\bar{v}_n)\}$ are linearly independent since

$$a_{k+1}L(\bar{v}_{k+1}) + a_{k+2}L(\bar{v}_{k+2}) + \ldots + a_n L(\bar{v}_n) = \bar{0}_w \Rightarrow$$

$$\Rightarrow L \sum_{i=k+1}^{n} a_i \bar{v}_i = \bar{0}_w \Rightarrow \sum_{i=k+1}^{n} a_i \bar{v}_i \in \ker L \Rightarrow$$

$$\Rightarrow \sum_{i=k+1}^{n} a_i \bar{v}_i = \sum_{i=1}^{k} c_i \bar{v}_i \Rightarrow \sum_{i=1}^{k} c_i \bar{v}_i + \sum_{i=k+1}^{n} (-a_i) \bar{v}_i = \bar{0}_v \Rightarrow$$

all c_i and all $a_i = 0$,

because $\{\bar{v}_1, \bar{v}_2, \ldots, \bar{v}_k, \bar{v}_{k+1}, \ldots, \bar{v}_n\}$ is a basis in V_n.

Furthermore, to show that the above images $\{L(\bar{v}_{k+1}), L(\bar{v}_{k+2}), \ldots, L(\bar{v}_n)\}$ form a spanning set in ran L, we notice that any vector $\bar{y}$ from ran L is of the form $\bar{y} = L(\bar{x})$ for some $\bar{x} \in V_n$. Expanding $\bar{x}$ in the above basis of V_n we get $\bar{x} = \sum_{i=1}^{n} x_i \bar{v}_i$, so that $\bar{y} = L(\bar{x}) = L(\sum_{i=1}^{n} x_i \bar{v}_i) = \sum_{i=1}^{n} x_i L(\bar{v}_i) = \sum_{i=k+1}^{n} x_i L(\bar{v}_i)$, since $L(\bar{v}_1) = L(\bar{v}_2) = \ldots = L(\bar{v}_n) = \bar{0}_w$. $\triangle$

2.5 The Quotient Space $V_n/\ker L$ and the Isomorphism $V_n/\ker L \overset{L'}{\cong} \text{ran } L$

When $L : V_n \to W_m$ is a many-to-one linmap, we investigate what is the nature of each inverse image, as well as what is the algebraic structure of the set $V_n/\sim_L$ of all of them, taking into account that both domain and codomain are vector spaces and the map L is a linear one.

Every $\bar{y} \in \text{ran } L$ has at least one preimage $\bar{x}$, $L(\bar{x}) = \bar{y}$. If we add to $\bar{x}$ any vector $\bar{k}$ from the ker L, this new vector $\bar{x} + \bar{k}$ will be mapped by L onto the same $\bar{y}$:

$$L(\bar{x} + \bar{k}) = L(\bar{x}) + L(\bar{k}) = L(\bar{x}) + \bar{0}_w = \bar{y}.$$

So, we see that the inverse image of $\bar{y}$ is at least $\bar{x} + \ker L$. (Adding the vector $\bar{x}$ to the subspace ker L means adding $\bar{x}$ to all vectors from ker L. This new subset in V_n is called the *neighboring set of* ker L represented by $\bar{x}$. A shorter name for $\bar{x} + \ker L$ is a *coset*). We have now to show that the coset $\bar{x} + \ker L$ is exactly the inverse image of $\bar{y}$ (there are no more elements from V_n in it).

Suppose that some $\bar{x}_1 \in V_n$ is also a preimage of $\bar{y}$: $L(\bar{x}_1) = \bar{y}$ and prove that $\bar{x}_1$ must be from $\bar{x} + \ker L$:

$$L(\bar{x}_1) = \bar{y} \text{ and } L(\bar{x}) = \bar{y} \Rightarrow$$
$$\Rightarrow L(\bar{x}_1) - L(\bar{x}) = \bar{0}_w \Rightarrow L(\bar{x}_1 - \bar{x}) = \bar{0}_w \Rightarrow \bar{x}_1 - \bar{x} \in \ker L \Rightarrow$$
$$\Rightarrow \bar{x}_1 \in \bar{x} + \ker L.$$

To see that the choice of $\bar{x}$ is arbitrary, i.e., that any other vector from $\bar{x} + \ker L$, say $\bar{x} + \bar{k}$, $k \in \ker L$, can serve as a coset representative instead of $\bar{x}$, we observe the equality:

$$(\bar{x} + \bar{k}) + \ker L = \bar{x} + (\bar{k} + \ker L) = \bar{x} + \ker L.$$

Later on, we shall prove (using any matrix representation of L) that every $\bar{y} \in$ ran L has a unique natural representative in its coset in V_n.

We already know that the set of all cosets (the inverse images of the linmap L considered as elements) is in bijective correspondence with ran L:

$$V_n/_{\sim L} \overset{L'}{\leftrightarrow} \text{ran } L,$$

where the new map L' is defined as

$$L'(\bar{x} + \ker L) = L(\bar{x} + \ker L) = L(\bar{x}) + L(\ker L) = L(\bar{x}) + \bar{0}_w = L(\bar{x}) = \bar{y}.$$

Notice the existence of the inverse map

$$(L')^{-1}(\bar{y}) = \bar{x} + \ker L$$

But, since V_n is a vector space and ran L is a subspace of W_n, one wonders if this bijection L' preserves some algebraic structures. To answer this question, we have to investigate more closely the possible algebraic structure of the set $V_n/_{\sim L}$.

It is easy to demonstrate that $\ker L$ [which is $\text{Inv}(\bar{0}_w)$] and all cosets (inverse images of all other elements from ran L) form a vector space (considering cosets as elements). Namely,

$$(\bar{x}_1 + \ker L) + (\bar{x}_2 + \ker L) = (\bar{x}_1 + \bar{x}_2) + (\ker L + \ker L) = (\bar{x}_1 + \bar{x}_2) + \ker L.$$

So, the operation of addition of cosets is determined by addition of their representatives. From this it follows that all five axioms characterizing an Abelian group are satisfied for this operation with cosets, since these axioms are satisfied for the vector addition of their representatives. The multiplication of a coset by a scalar c can obviously be defined as $c(\bar{x} + \ker L) = c\bar{x} + c\ker L = c\bar{x} + \ker L$, and consequently, all four properties of the multiplication of vectors with scalars are satisfied.

Thus, the set of all cosets considered as elements (including $\ker L = \bar{0}_v + \ker L$ as the zero vector) is a *vector space*. It is customary to denote this space by $V_n/\ker L$ and call it the *quotient space*.

(Notice that we replaced the symbol $V_n/{\sim_L}$ for the inverse images of L in V_n by $V_n/\ker L$ since inverse images are now the equivalence classes of the equivalence relation which is defined so that two vectors from V_n are equivalent if their difference is a vector from $\ker L$.)

The bijection L' between $V_n/\ker L$ and ran L obviously preserves the operation of linear combination of cosets

$$L'[\sum_{i=1}^{k} c_i(\bar{x}_i + \ker L)] = L[\sum_{i=1}^{k} c_i(\bar{x}_i + \ker L)] =$$

$$= \sum_{i=1}^{k} c_i L(\bar{x}_i + \ker L) = \sum_{i=1}^{k} c_i L'(\bar{x}_i + \ker L).$$

Therefore, L' is a linmap and being a bijection it is an isomorphism between vector spaces $V_n/\ker L$ and ran L:

$$\boxed{V_n/\ker L \overset{L'}{\cong} \operatorname{ran} L}.$$

They both have the dimension $r = n - k$.

Theorem (isomorphism) For every linmap $L : V_n \to W_m$, there is the quotient vector space $V_n/\ker L$ of inverse images (cosets of $\ker L$ in V_n) which has the dimension $\dim(\operatorname{ran} L) = n - \dim(\ker L)$ and which is isomorphic to ran L by the bijective linmap (isomorphism), $L'(\bar{x} + \ker L) = L(\bar{x})$, $\bar{x} \in V_n$,

$$\boxed{V_n/\ker L \overset{L'}{\cong} \operatorname{ran} L}.$$

When $L : V_n \to W_m$ is a $1-1$ map, then $\ker L = \{\bar{0}_v\}$, and L itself is an isomorphism between V_n and ran L:

$$V_n \overset{L}{\cong} \operatorname{ran} L, \ r = n, \ n \le m.$$

2.6 Representation Theory

To be able to determine the kernel $\ker L$, the range ran L and the quotient space $V_n/\ker L$, which comprises the inverse images of the linear map $L : V_n \to W_m$, and also to solve numerous applicative problems in which linmaps play decisive roles, we make use of *representation theory*. In this theory, vectors from V_n and W_m are replaced by matrix-columns from $\mathbb{R}^n$ and $\mathbb{R}^m$, respectively, using isomorphisms $V_n \overset{v}{\cong} \mathbb{R}^n$ and $W_m \overset{w}{\cong} \mathbb{R}^m$, which are induced by choices of the bases $v = \{\bar{v}_1, \bar{v}_2, \ldots, \bar{v}_n\}$ in V_n and $w = \{\bar{w}_1, \bar{w}_2, \ldots, \bar{w}_m\}$ in W_m. We use the same letters v and w for the bases and for isomorphisms induced by them.

The linmap $L : V_n \rightarrow W_m$ itself is then represented by an $m \times n$ real matrix M, which is also the linmap $M : \mathbb{R}^n \rightarrow \mathbb{R}^m$. Matrix algebra offers many methods for solving numerical problems that are now transferred from $L: V_n \rightarrow W_m$ to $M : \mathbb{R}^n \rightarrow \mathbb{R}^m$. The solutions obtained can be immediately taken back to V_n and W_n by the same isomorphisms v and w.

2.6.1 The Vector Space $\hat{L}(V_n, W_m)$

To prepare the ground for representation theory, we shall first show that all linmaps from V_n to W_n form a vector space, which may be denoted as

$$\hat{L}(V_n, W_m).$$

The *proof* of this statement is very simple: For two linmaps $L_1 : V_n \rightarrow W_m$ and $L_2 : V_n \rightarrow W_m$, we define their sum $L_1 + L_2$ as

$$[L_1 + L_2](\bar{x}) = L_1(\bar{x}) + L_2(\bar{x}), \text{ for any } \bar{x} \in V_n,$$

both $L_1(\bar{x})$ and $L_2(\bar{x})$ are vectors in W_m.
$L_1 + L_2$ is obviously a linmap itself, which can be easily demonstrated:

$$[L_1 + L_2](a\bar{x}_1 + b\bar{x}_2) = L_1(a\bar{x}_1 + b\bar{x}_2) + L_2(a\bar{x}_1 + b\bar{x}_2) =$$
$$= aL_1(\bar{x}_1) + bL_1(\bar{x}_2) + aL_2(\bar{x}_1) + bL_2(\bar{x}_2) =$$
$$= a[L_1 + L_2](\bar{x}_1) + b[L_1 + L_2](\bar{x}_2).$$

This addition of linmaps [elements in $\hat{L}(V_n, W_m)$] makes this set an Abelian group, since it is defined by making use of the addition of vectors in W_n, which operation satisfies all five axioms for an Abelian group.

We use the same sign $+$ for the addition of linmaps and for the addition in W_m since the elements that are added distinguish one sign from the other.

Similarly, we can define the multiplication of a linmap L with a scalar c using the analogous operation in W_m: $[cL](\bar{x}) = cL(\bar{x})$, $\forall \bar{x} \in V_n$, and prove that cL is a linmap. With this new operation, which satisfies all four axioms for the multiplication of scalars with vectors, the set $\hat{L}(V_n, W_m)$ becomes a vector space. $\triangle$

The dimension of the vector space $\hat{L}(V_n, W_m)$ is $m \cdot n$—the product of the dimensions of the common domain and codomain.

Proof We shall define $m \cdot n$ simplest linmaps in $\hat{L}(V_n, W_m)$ and prove that they form a basis in this vector space by showing that every $L : V_n \rightarrow W_m$ is their unique linear combination.

Let us define $m \cdot n$ elementary linmaps $\{E_{ij} \mid i = 1, 2, \ldots, m, \ j = 1, 2, \ldots, n\}$ from V_n to W_m by their action on chosen basis vectors $v = \{\bar{v}_1, \bar{v}_2, \ldots, \bar{v}_n\}$ in V_n so that the images are expressed in a basis $w = \{\bar{w}_1, \bar{w}_2, \ldots, \bar{w}_m\}$ in W_m

$$E_{ij}(\bar{v}_k) = \delta_{jk}\bar{w}_i, \ i = 1, 2, \ldots, m; \ j, k = 1, 2, \ldots, n. \ (*)$$
$$\text{(for example } E_{35}(\bar{v}_5) = \bar{w}_3, E_{35}(\bar{v}_k) = \bar{0}_w, \text{ if } k \neq 5\text{)}.$$

An arbitrary $L : V_n \to W_m$ is given if we know the images $\{L(\bar{v}_1), L(\bar{v}_2), \ldots, L(\bar{v}_n)\}$ of the basis vectors $\{\bar{v}_1, \bar{v}_2, \ldots, \bar{v}_n\}$ from V_n in W_m. We can uniquely expand these images in the basis $\{\bar{w}_1, \bar{w}_2, \ldots, \bar{w}_m\}$ in W_m,

$$L(\bar{v}_k) = \sum_{i=1}^{m} a_{ik} \bar{w}_i, \; k = 1, 2, \ldots, n$$

(the inverted order of indices in the expansion coefficients ik instead of ki will be explained later, see Sect. 2.6.4).

Furthermore,

$$L(\bar{v}_k) = \sum_{i=1}^{m} a_{ik} \bar{w}_i = \sum_{i=1}^{m} \sum_{j=1}^{n} a_{ij} \delta_{jk} \bar{w}_i$$

$$= [\sum_{i=1}^{m} \sum_{j=1}^{n} a_{ij} E_{ij}](\bar{v}_k), \; k = 1, 2, \ldots, n,$$

from $(*)$ above.

Two linmaps L and $[\sum_{i=1}^{m} \sum_{j=1}^{n} a_{ij} E_{ij}]$ are equal since they act identically on all vectors from a basis in V_n:

$$\boxed{L = \sum_{i=1}^{m} \sum_{j=1}^{n} a_{ij} E_{ij}}.$$

Thus, every $L : V_n \to W_n$ is a unique linear combination of $m \cdot n$ linmaps $\{E_{ij} \mid i = 1, 2, \ldots, m; \; j = 1, 2, \ldots, m\}$, so they are a basis in $\hat{L}(V_n, W_m)$. Δ

The unique coefficients a_{ij} in this expansion form the $m \times n$ matrix $M = [a_{ij}]_{m \times n}$, and we shall show that precisely this matrix represents the above linmap L with the help of the isomorphisms v and w.

2.6.2 The Linear Map $M : \mathbb{R}^n \to \mathbb{R}^m$

We have to do one more detailed study before formulating representation theory. Let us investigate the map $M : \mathbb{R}^n \to \mathbb{R}^m$, where M is an $m \times n$ real matrix. The matrix $M = [a_{ij}]_{m \times n}$ performs this map by matrix multiplication: an $m \times n$ matrix M can multiply every $n \times 1$ matrix-column $x \to \mathcal{X} = [x_i]_{n \times 1} = [x_1 \, x_2 \ldots x_n]^T$ from $\mathbb{R}^n$ to produce the unique $m \times 1$ matrix-column $\mathcal{Y} = [y_i]_{m \times 1} = [y_1 \, y_2 \ldots y_m]^T$ from $\mathbb{R}^m$

$$M\mathcal{X} = \begin{bmatrix} a_{11} & a_{12} & \cdots & a_{1n} \\ a_{21} & a_{22} & \cdots & a_{2n} \\ \vdots & \vdots & \ddots & \vdots \\ a_{m1} & a_{m2} & \cdots & a_{mn} \end{bmatrix} \begin{bmatrix} x_1 \\ x_2 \\ \vdots \\ x_n \end{bmatrix}$$

$$= \begin{bmatrix} a_{11}x_1 + a_{12}x_2 + \cdots + a_{1n}x_n \\ a_{21}x_1 + a_{22}x_2 + \cdots + a_{2n}x_n \\ \cdots \quad \cdots \quad \cdots \quad \cdots \\ a_{m1}x_1 + a_{m2}x_2 + \cdots + a_{mn}x_n \end{bmatrix} = \begin{bmatrix} y_1 \\ y_2 \\ \vdots \\ y_m \end{bmatrix} = \bar{\mathscr{Y}},$$

or, in short, $[a_{ij}]_{m \times n}[x_i]_{n \times 1} = [y_i]_{m \times 1}$, where

$$y_i = \sum_{i=1}^{n} a_{ik}x_k, \quad i = 1, 2, \ldots, m.$$

But matrix multiplication satisfies (among others) the following general rules (see Sect. 1.15.1)

$$M_1(M_2 + M_3) = M_1M_2 + M_1M_3 \text{ and } M_1(cM_2) = c(M_1M_2), \ c \in \mathbb{R}$$

(the size of M_1 must be compatible with those of M_2 and M_3, which, in their turn, must be of the same size).

These rules guarantee that the above map $M : \mathbb{R}^n \to \mathbb{R}^m$ is a *linear* one

$$M(\bar{\mathscr{X}}_1 + \bar{\mathscr{X}}_2) = M\bar{\mathscr{X}}_1 + M\bar{\mathscr{X}}_2 \text{ and } M(c\bar{\mathscr{X}}) = c(M\bar{\mathscr{X}}), \ c \in \mathbb{R}.$$

Therefore, the set $\hat{L}(\mathbb{R}^n, \mathbb{R}^m)$ of all matrix linmaps from $\mathbb{R}^n$ to $\mathbb{R}^m$ is a *vector space* of dimension $m \cdot n$. It can be easily shown that the matrices of size $m \times n$ are the only linmaps between $\mathbb{R}^n$ and $\mathbb{R}^m$ [We already know—see Sect. 1.14.2.— that the set $\mathbb{R}_{m \times n}$ of all real $m \times n$ matrices is a vector space of dimension $m \cdot n$. We shall use two different notations $\hat{L}(\mathbb{R}^n, \mathbb{R}^m)$ and $\mathbb{R}_{m \times n}$ because in the first set the elements are linmaps $M : \mathbb{R}^n \to \mathbb{R}^n$, and in the second $m \times n$ real matrices as such.].

2.6.3 The Three Isomorphisms v, w and v − w

First, we consider isomorphisms v and w. Since V_n and $\mathbb{R}^n$ have the same dimension n, we can define an isomorphism between them once the basis

$$v = \{\bar{v}_1, \bar{v}_2, \ldots, \bar{v}_n\}$$

is chosen in V_n. In $\mathbb{R}^n$, there is always the standard basis $\{\bar{e}_1, \bar{e}_2, \ldots, \bar{e}_n\}$ where

$$\bar{e}_1 = [\ \underbrace{10 \ldots 0}_{n \text{ numbers}}\ ]^T, \ \bar{e}_2 = [\ \underbrace{01 \ldots 0}_{n \text{ numbers}}\ ]^T, \ \ldots, \bar{e}_n = [\ \underbrace{00 \ldots 1}_{n \text{ numbers}}\ ]^T$$

(in short $\bar{e}_t = [\ \delta_{1t}\ \delta_{2t}\ \ldots\ \delta_{nt}]^T, \ t = 1, 2, \ldots, n$).

The isomorphism v induced by the choice of the basis v in V_n is the bijection which associates with every $\bar{x} \in V_n$, which must be first expanded in this basis— $\bar{x} = \sum_{i=1}^{n} x_i \bar{v}_i$, the matrix-column $\bar{\mathscr{X}} \in \mathbb{R}^n$, which is a linear combination of standard basis vectors in $\mathbb{R}^n$ with the same expansion coefficients as those of $\bar{x}$:

$$\bar{\mathscr{X}} = \sum_{i=1}^{n} x_i \bar{e}_i = [x_1 x_2 \ldots x_n]^T.$$

Thus, the bijection V_n is defined as

$$\bar{x} = \sum_{i=1}^{n} x_i \bar{v}_i \overset{v}{\leftrightarrow} \sum_{i=1}^{n} x_i \bar{e}_i = \bar{\mathscr{X}} = [x_1 x_2 \ldots x_n]^T.$$

In particular, $\bar{v}_1 \overset{v}{\leftrightarrow} \bar{e}_1$, $\bar{v}_2 \overset{v}{\leftrightarrow} \bar{e}_2$, $\ldots$, $\bar{v}_n \overset{v}{\leftrightarrow} \bar{e}_n$.

This bijection is obviously linear

$$a\bar{x}_1 + b\bar{x}_2 \overset{v}{\leftrightarrow} a\bar{\mathscr{X}}_1 + b\bar{\mathscr{X}}_2,$$

and thus it is an *isomorphism*. It can be called the v-*isomorphism*. It is in fact the *representation* of vectors from V_n by matrix-columns from $\mathbb{R}^n$ induced by the basis v:

$$\bar{\mathscr{X}} = [\bar{x}]_v.$$

Obviously, the matrix-column $\bar{\mathscr{X}}$ for the same $\bar{x} \in V_n$ will be different for another basis in V_n. This is an important separate problem, which will be discussed later on in the framework of tensor algebra (see Sect. 4.3.1).

Absolutely the same procedure can be repeated to get the w-*isomorphism* between vectors from W_m and matrix-columns of the length m (elements of $\mathbb{R}^m$), which is induced by the choice of the basis $w = \{\bar{w}_1, \bar{w}_2, \ldots, \bar{w}_m\}$ in W_m:

$$\bar{y} \in W_m, \ \bar{y} = \sum_{i=1}^{m} y_i \bar{w}_i \overset{w}{\leftrightarrow} \bar{\mathscr{Y}} = \sum_{i=1}^{m} y_i \bar{e}_i = [y_1 y_2 \ldots y_m]^T \in \mathbb{R}^m,$$

$$\text{or } \bar{\mathscr{Y}} = [\bar{y}]_w,$$

where now $\{\bar{e}_1, \bar{e}_2, \ldots, \bar{e}_m\}$ is the standard basis in $\mathbb{R}^m$

$$\bar{e}_t = [\delta_{1t} \delta_{2t} \ldots \delta_{mt}]^T, \ t = 1, 2, \ldots, m.$$

(It is usual to exploit the same symbols $\bar{e}_t$, $t = 1, 2, \ldots, k$ for vectors in standard bases regardless of the dimension k of the space.)

Second, to derive the v–w-*isomorphism* between the vector spaces $\hat{L}(V_n, W_m)$ and $\hat{L}(\mathbb{R}^n, \mathbb{R}^m)$, we make a natural requirement $M\bar{\mathscr{X}} = \bar{\mathscr{Y}}$ if $L\bar{x} = \bar{y}$. In other words

$$\boxed{M[\bar{x}]_v = [L(\bar{x})]_w}.$$

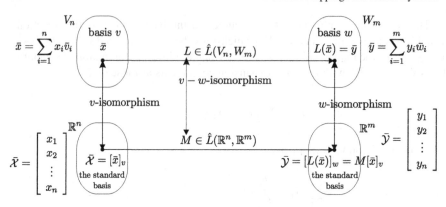

An illustration of representation theory.

From this requirement it immediately follows that M can be expressed as a composition of mappings

$$M = wLv^{-1}.$$

This $v - w$ mapping $L \mapsto M$ from $\hat{L}(V_n, V_m)$ to $\hat{L}(\mathbb{R}^n, \mathbb{R}^m)$ is obviously a bijection since there is the inverse map $L = w^{-1}Mv$.

This bijection is a linear map because

$$\begin{aligned}
[w(L_1 + L_2)v^{-1}]\vec{\mathcal{X}} &= [w(L_1 + L_2)](\bar{x}) = w(L_1(\bar{x}) + L_2(\bar{x})) = \\
&= wL_1(\bar{x}) + wL_2(\bar{x}) = [wL_1v^{-1}]\vec{\mathcal{X}} + [wLv^{-1}]\vec{\mathcal{X}} = \\
&= [wL_1v^{-1} + wL_2v^{-1}]\vec{\mathcal{X}} \Rightarrow w(L_1 + L_2)v^{-1} = wL_1v^{-1} \\
&\quad + wL_2v^{-1},
\end{aligned}$$

since $\vec{\mathcal{X}}$ is an arbitrary matrix-column from $\mathbb{R}^n$ and $v^{-1}\vec{\mathcal{X}} = \bar{x} \in V_n$.

Similarly, $[w(aL)v^{-1}]\vec{\mathcal{X}} = [w(aL)](\bar{x}) = w(aL(x)) = a(wL(\bar{x})) = a[wLv^{-1}]\vec{\mathcal{X}} \Rightarrow w(aL)v^{-1} = a(wLv^{-1})$.

Consequently, this isomorphism $M = wLv^{-1}$ between $\hat{L}(V_n, W_m)$ and $\hat{L}(\mathbb{R}^n, \mathbb{R}^m)$, which is generated directly by the v- and w-isomorphisms, can be called the v–w-isomorphism. It is in fact the *representation* of linmaps $L : V_n \to W_m$ by $m \times n$ matrices $M : \mathbb{R}^n \to \mathbb{R}^m$ induced by the v- and w-isomorphisms.

2.6.4 How to Calculate the Representing Matrix M

To calculate the representing matrix M, we start with the above requirement which defines M

$$M[\bar{x}]_v = [L(\bar{x})]_w.$$

The most economical derivation of M is to replace $\bar{x}$ in this requirement by n basis vectors $\{\bar{v}_1, \bar{v}_2, \ldots, \bar{v}_n\}$:

$$[L(\bar{v}_1)]_w = M[\bar{v}_1]_v = M\bar{e}_1 = \begin{bmatrix} a_{11} & a_{12} & \cdots & a_{1n} \\ a_{21} & a_{22} & \cdots & a_{2n} \\ \vdots & \vdots & \ddots & \vdots \\ a_{m1} & a_{m2} & \cdots & a_{mn} \end{bmatrix} \begin{bmatrix} 1 \\ 0 \\ \vdots \\ 0 \end{bmatrix} =$$

$$= \begin{bmatrix} a_{11} \\ a_{21} \\ \vdots \\ a_{m1} \end{bmatrix}, \quad \text{the first column of matrix } M;$$

$$[L(\bar{v}_2)]_v = [a_{12}\, a_{22} \ldots a_{m2}]^T, \quad \text{the second column of } M$$

$$\cdots\cdots\cdots\cdots\cdots\cdots\cdots\cdots\cdots\cdots\cdots\cdots\cdots\cdots\cdots$$

$$[L(\bar{v}_n)]_w = [a_{1n}\, a_{2n} \ldots a_{mn}]^T, \quad \text{the } n\text{-th column of } M.$$

Conclusion The n columns of the $m \times n$ matrix M that represents the linmap $L : V_n \to W_m$ in the $v-w$-isomorphism $M = wLv^{-1}$ are the columns of n images by L of the basis vectors $\{\bar{v}_1, \bar{v}_2, \ldots, \bar{v}_n\}$ in V_n represented in the basis $\{\bar{w}_1, \bar{w}_2, \ldots, \bar{w}_m\}$ in W_m

$$M = [[L(\bar{v}_1)]_w\, [L(\bar{v}_2)]_w\, \cdots\, [L(\bar{v}_n)]_w]_{m \times n}.$$

We get the same result if we apply wLv^{-1} to the vectors of the standard basis in $\mathbb{R}^n$, since we know that this expression is equal to the unknown M and that $M\bar{e}_i = \bar{c}_i$, $i = 1, 2, \ldots, n$ ($\bar{c}_i$ are the columns of M):

$$[wLv^{-1}]\bar{e}_i = [wL]\bar{v}_i = w[L(\bar{v}_i)] = [L(\bar{v}_i)]_w = \bar{c}_i, \; i = 1, 2, \ldots, n.$$

There is also a more practical formula for the elements a_{ij} of M. We may call it the *basic formula* for the matrix representation of the linmap L. We expand the images $L(\bar{v}_k), k = 1, 2, \ldots, n$, of the chosen basis $v = \{\bar{v}_1, \bar{v}_2, \ldots, \bar{v}_n\}$ in V_n in the basis $w = \{\bar{w}_1, \bar{w}_2, \ldots, \bar{w}_m\}$ in W_m

$$L(\bar{v}_k) = \sum_{i=1}^{m} a_{ik}\bar{w}_i, \; k = 1, 2, \ldots, n.$$

The order of indices in the elements of M corresponds to the fact that these expansion coefficients should form the columns of M, e.g., $L(\bar{v}_1) = \sum_{i=1}^{m} a_{i1}\bar{w}_i = a_{11}\bar{w}_1 + a_{21}\bar{w}_2 + \ldots + a_{m1}\bar{w}_m$, eventually gives

$$[L(\bar{v}_1)]_w = \begin{bmatrix} a_{11} \\ a_{21} \\ \vdots \\ a_{m1} \end{bmatrix} = \bar{c}_1.$$

We have already had this basic formula in Sect. 2.6.1 in the process of establishing the basis

$$\{E_{ij} \mid i = 1, 2, \ldots, m; \; j = 1, 2, \ldots, n\} \text{ for } \hat{L}(V_n, W_m).$$

The matrices which represent these basis vectors in $\hat{L}(V_n, W_m)$ are the $m \times n$ matrices

$$\{\mathscr{E}_{ij} \mid i = 1, 2, \ldots, m; \; j = 1, 2, \ldots, n\},$$

which form the *standard basis* in $\mathbb{R}_{m \times n}$. They are such matrices which have only one 1 as the (i, j) element and the rest of the elements are zeros.

In short, for the (p, k) element in $\mathscr{E}_{ij}$, we have

$$(\mathscr{E}_{ij})_{pk} = \delta_{ip}\delta_{jk}.$$

It is immediately obvious that the matrix $M = [a_{ij}]_{m \times n}$ can be uniquely expanded in this standard basis in $\mathbb{R}_{m \times n}$ as

$$M = \sum_{i=1}^{m}\sum_{j=1}^{n} a_{ij}\mathscr{E}_{ij}.$$

More explicitly,

$$\begin{bmatrix} a_{11} & a_{12} & \cdots & a_{1n} \\ a_{21} & a_{22} & \cdots & a_{2n} \\ \vdots & \vdots & \ddots & \vdots \\ a_{m1} & a_{m2} & \cdots & a_{mn} \end{bmatrix} = a_{11}\begin{bmatrix} 1 & 0 & \cdots & 0 \\ \vdots & \vdots & \ddots & \vdots \\ 0 & 0 & \cdots & 0 \end{bmatrix}_{m \times n} +$$

$$+ a_{12}\begin{bmatrix} 0 & 1 & \cdots & 0 \\ \cdots & \cdots & \cdots & \cdots \\ 0 & 0 & \cdots & 0 \end{bmatrix}_{m \times n} + \ldots + a_{mn}\begin{bmatrix} 0 & 0 & \cdots & 0 \\ \cdots & \cdots & \cdots & \cdots \\ 0 & 0 & \cdots & 1 \end{bmatrix}_{m \times n}.$$

Now, we can point out the *essence* of the $v - w$-isomorphism between $\hat{L}(V_n, W_m)$ and $\hat{L}(\mathbb{R}^n, \mathbb{R}^m)$:

$$L = \sum_{i=1}^{m}\sum_{j=1}^{n} a_{ij}E_{ij} \overset{v-w}{\leftrightarrow} M = \sum_{i=1}^{m}\sum_{j=1}^{n} a_{ij}\mathscr{E}_{ij},$$

where $\mathscr{E}_{ij}$ are the standard basis vectors in $\mathbb{R}_{m \times n}$, and E_{ij} are the elementary basis linmaps defined by the chosen bases $v = \{\bar{v}_1, \bar{v}_2, \ldots, \bar{v}_n\}$ in V_n and $w = \{\bar{w}_1, \bar{w}_2, \ldots, \bar{v}_m\}$ in W_m:

$$E_{ij}(\bar{v}_k) = \delta_{jk}\bar{w}_i = \sum_{p=1}^{m}\delta_{ip}\delta_{jk}\bar{w}_p, \quad i = 1, 2, \ldots, m; \; j, k = 1, 2, \ldots, n \text{ (see Sect. 2.6.1).}$$

2.7 An Example (Representation of a Linmap Which Acts between Vector Spaces of Polynomials)

Let P_4 and P_3 be the two vector spaces of polynomials with real coefficients of degree less than 4 and 3, respectively (see Sect. 1.14.3).

Let $D : P_4 \rightarrow P_3$ be the onto linmap (surjection), the *derivative*, defined by

$$D : p(x) \mapsto p'(x), \ p(x) \in P_4,$$

or, in more detail,

$$D(ax^3 + bx^2 + cx + d) = 3ax^2 + 2bx + c, \ a,b,c,d \in \mathbb{R}.$$

(1) Find the 3×4 matrix $\mathscr{D} = [D]_{v-w}$ representing the linmap D with respect to the bases $v = \{\bar{v}_1, \bar{v}_2, \bar{v}_3, \bar{v}_4\} = \{x^3 + x^2 + x + 1, \ x^2 + x + 1, \ x + 1, \ 1\}$ in P_4 and $w = \{\bar{w}_1, \bar{w}_2, \bar{w}_3\} = \{x^2 + x + 1, \ x + 1, \ 1\}$ in P_3;
(2) Represent $p(x) \in P_4$ and $p'(x) \in P_3$ in the given bases and show that $\mathscr{D}$ maps the representative column of $p(x)$ onto that of $p'(x)$;
(3) Find the kernel of D and the *inverse image* (which is a coset of ker D) of an arbitrary $q(x) \in \operatorname{ran} D = P_3$.

To answer this set of three questions, we shall first apply D to all four basis vectors in P_4 and then expand the obtained images in the basis of P_3. The expansion coefficients form the columns of a 3×4 matrix $\mathscr{D}$.

(1)

$$\begin{aligned}
D(\bar{v}_1) &= D(x^3 + x^2 + x + 1) = 3x^2 + 2x + 1 = \\
&= 3(x^2 + x + 1) - 1 \cdot (x + 1) - 1 \cdot 1 = 3\bar{w}_1 - \bar{w}_2 - \bar{w}_3; \\
D(\bar{v}_2) &= D(x^2 + x + 1) = 2x + 1 = \\
&= 0 \cdot (x^2 + x + 1) + 2(x + 1) - 1 \cdot 1 = 0 \cdot \bar{w}_1 + 2\bar{w}_2 - \bar{w}_3; \\
D(\bar{v}_3) &= D(x + 1) = 1 = \\
&= 0 \cdot (x^2 + x + 1) + 0 \cdot (x + 1) + 1 \cdot 1 = 0 \cdot \bar{w}_1 + 0 \cdot \bar{w}_2 + \bar{w}_3, \\
D(\bar{v}_4) &= D(1) = 0 = \\
&= 0 \cdot (x^2 + x + 1) + 0 \cdot (x + 1) + 0 \cdot 1 = 0 \cdot \bar{w}_1 + 0 \cdot \bar{w}_2 + 0 \cdots \bar{w}_3,
\end{aligned}$$

so that $\mathscr{D} = \begin{bmatrix} 3 & 0 & 0 & 0 \\ -1 & 2 & 0 & 0 \\ -1 & -1 & 1 & 0 \end{bmatrix}$.

(2)

$$\begin{aligned}
p(x) &= ax^3 + bx^2 + cx + d = \\
&= a(x^3 + x^2 + x + 1) + (b - a)(x^2 + x + 1) + (c - b)(x + 1) + (d - c) \cdot 1 = \\
&= a\bar{v}_1 + (b - a)\bar{v}_2 + (c - b)\bar{v}_3 + (d - c)\bar{v}_4, \ \text{so that}
\end{aligned}$$

$$[p(x)]_v = \begin{bmatrix} a \\ b-a \\ c-b \\ d-c \end{bmatrix}.$$

Furthermore,

$$
\begin{aligned}
Dp(x) = p'(x) &= 3ax^2 + 2bx + c = \\
&= 3a(x^2 + x + 1) + (2b - 3a)(x + 1) + (c - 2b) \cdot 1 = \\
&= 3a\bar{w}_1 + (2b - 3a)\bar{w}_2 + (c - 2b)\bar{w}_3, \quad \text{so that}
\end{aligned}
$$

$$[p'(x)]_w = \begin{bmatrix} 3a \\ 2b - 3a \\ c - 2b \end{bmatrix}.$$

Finally,

$$\mathscr{D}[p(x)]_v = [p'(x)]_w = [Dp(x)]_w, \quad \text{or}$$

$$
\begin{bmatrix} 3 & 0 & 0 & 0 \\ -1 & 2 & 0 & 0 \\ -1 & -1 & 1 & 0 \end{bmatrix}
\begin{bmatrix} a \\ b-a \\ c-b \\ d-c \end{bmatrix}
$$

$$
= \begin{bmatrix} 3a \\ -a + 2b - 2a \\ -a - b + a + c - b \end{bmatrix}
= \begin{bmatrix} 3a \\ 2b - 3a \\ c - 2b \end{bmatrix}.
$$

(3) To find the kernel of $\mathscr{D} : \mathbb{R}^4 \to \mathbb{R}^3$, we have to solve the following homogeneous system of linear equations using the representative matrix $\mathscr{D}$

$$\mathscr{D} \begin{bmatrix} x_1 \\ x_2 \\ x_3 \\ x_4 \end{bmatrix} = \begin{bmatrix} 0 \\ 0 \\ 0 \end{bmatrix},$$

this is the zero vector in $\mathbb{R}^3$ that represents the zero vector in P_3.

The augmented matrix of this system can be brought to the unique GJ reduced form in a few steps using elementary operations (see Sect. 2.13):

$$
\begin{bmatrix} 3 & 0 & 0 & 0 & | & 0 \\ -1 & 2 & 0 & 0 & | & 0 \\ -1 & -1 & 1 & 0 & | & 0 \end{bmatrix} \begin{matrix} R_1/3 \\ \, \\ \, \end{matrix}
\sim
\begin{bmatrix} 1 & 0 & 0 & 0 & | & 0 \\ -1 & 2 & 0 & 0 & | & 0 \\ -1 & -1 & 1 & 0 & | & 0 \end{bmatrix} \begin{matrix} \, \\ R_2 + R_1 \\ R_3 + R_1 \end{matrix}
\sim
$$

$$
\sim
\begin{bmatrix} 1 & 0 & 0 & 0 & | & 0 \\ 0 & 2 & 0 & 0 & | & 0 \\ 0 & -1 & 1 & 0 & | & 0 \end{bmatrix} \begin{matrix} \, \\ R_2/2 \\ \, \end{matrix}
\sim
\begin{bmatrix} 1 & 0 & 0 & 0 & | & 0 \\ 0 & 1 & 0 & 0 & | & 0 \\ 0 & -1 & 1 & 0 & | & 0 \end{bmatrix} \begin{matrix} \, \\ \, \\ R_3 + R_2 \end{matrix}
\sim
$$

$$\sim \begin{bmatrix} 1 & 0 & 0 & 0 & | & 0 \\ 0 & 1 & 0 & 0 & | & 0 \\ 0 & 0 & 1 & 0 & | & 0 \end{bmatrix}.$$

For the GJ modified form, we add one row of zeros and continue the diagonal of 1s with (-1) (see Sect. 2.17):
$$\begin{bmatrix} 1 & 0 & 0 & 0 & | & 0 \\ 0 & 1 & 0 & 0 & | & 0 \\ 0 & 0 & 1 & 0 & | & 0 \\ 0 & 0 & 0 & -1 & | & 0 \end{bmatrix}.$$

The solution of the system (i.e., ker $\mathscr{D}$) can be read immediately in the vector form in $\mathbb{R}^4$, since the fourth column is the unique basis vector in ker $\mathscr{D}$:

$$\text{ker } \mathscr{D} = \text{LIN} \left(\begin{bmatrix} 0 \\ 0 \\ 0 \\ -1 \end{bmatrix} \right) = \left\{ \begin{bmatrix} 0 \\ 0 \\ 0 \\ k \end{bmatrix} \middle| k \in \mathbb{R} \right\}.$$

Remembering that $[ax^3 + bx^2 + cx + d]_v = \begin{bmatrix} a \\ b-a \\ c-b \\ d-c \end{bmatrix}$, we see that $[k]_v = \begin{bmatrix} 0 \\ 0 \\ 0 \\ k \end{bmatrix}$,

so that going back to P_4, we have ker $D = \{k \mid k \in \mathbb{R}\}$.

Summary To find the kernel of the operator $A : V_n \to W_m$, we first choose bases v and w in V_n and W_m, respectively, and then using the $v-w$-isomorphism calculate the representing matrix $\mathscr{A} : \mathbb{R}^n \to \mathbb{R}^m$. Solving the homogeneous linear system $\mathscr{A}\vec{\mathscr{X}} = \bar{0}_m$, $\vec{\mathscr{X}} \in \mathbb{R}^n$ by the GJ modified method (see Sect. 2.17), we get the unique basis vectors of ker $\mathscr{A}$, so that ker $\mathscr{A}$ is their *linear span*. Going back by the inverse v-isomorphism, we get ker A as a subspace in V_n.

The unique preimage in P_4 of $q(x) = a'x^2 + b'x + c' \in P_3$ can be calculated in the following manner: Firstly, we find the w-representation of $q(x)$ in $\mathbb{R}^3$ $[q(x)]_w = \begin{bmatrix} a' \\ b'-a' \\ c'-b' \end{bmatrix}$, since $a'x^2 + b'x + c' = a'(x^2 + x + 1) + (b' - a')(x+1) + (c' - b') \cdot 1$.

Secondly, we expand it in terms of the three linearly independent columns of $\mathscr{D}$:

$$\begin{bmatrix} a' \\ b'-a' \\ c'-b' \end{bmatrix} = A \begin{bmatrix} 3 \\ -1 \\ -1 \end{bmatrix} + B \begin{bmatrix} 0 \\ 2 \\ -1 \end{bmatrix} + C \begin{bmatrix} 0 \\ 0 \\ 1 \end{bmatrix} \Rightarrow A = \frac{a'}{3}, B = \frac{b'}{2} - \frac{a'}{3}, C = c' - \frac{b'}{2}.$$

The *canonical expansion* of $[q(x)]_w$ in terms of *all* columns of $\mathscr{D}$ gives its unique preimage (see Sect. 2.8) in $\mathbb{R}^4$

$$[q(x)]_v^{pre} = \begin{bmatrix} a'/3 \\ b'/2 - a'/3 \\ c' - b'/2 \\ 0 \end{bmatrix}.$$

(Verify: $\mathscr{D}[q(x)]_v \overset{pre}{=} [q(x)]_w$).

Thirdly, making use of the inverse v-isomorphism, we find the unique D-preimage $\overset{pre}{q}(x)$ in P_4 of $q(x) \in P_3$

$$\overset{pre}{q}(x) = \frac{a'}{3}(x^3 + x^2 + x + 1) + (b'/2 - a'/3)(x^2 + x + 1) + (c' - b'/2)(x+1) + 0 \cdot 1 =$$

$$= \frac{a'}{3}x^3 + \frac{b'}{2}x^2 + c'x + c' = \int (a'x^2 + b'x + c')dx + c'$$

(This is one particular antiderivative).

Thus, the D-inverse image of $q(x) \in P_3$ is the indefinite integral [the set of all antiderivatives, which is a coset in P_4 consisting of ker D plus the unique preimage of $q(x)$ as the coset representative]:

$$\mathrm{Inv}[q(x)] = \underbrace{\mathrm{Inv}[a'x^2 + b'x + c']}_{\text{a coset in } P_4} = \underbrace{\frac{a'}{3}x^3 + \frac{b'}{2}x^2 + c'x + c'}_{\text{the coset representative}} + \underbrace{\{k \mid k \in \mathbb{R}\}}_{\text{ker } D}$$

$$(\text{or, in short}) \ = \int (a'x^2 + b'x + c')dx + k, \ k \in \mathbb{R}.$$

An illustration of the three-step procedure for finding the D-inverse image of $q(x) \in P_3$

2.8 Systems of Linear Equations (Linear Systems)

Many problems in Linear Algebra and its applications are solved by finding solutions of some appropriate system of linear equations. In the previous Sect. 2.7 we had an example. For this reason, we shall now perform a thorough study of systems of linear equations and their solutions. We shall try to emphasize the *real meaning* of this problem and consequently to suggest some *minor modifications* in the usual methods for finding solutions. We shall base this study almost entirely on our previous experience with linear maps, in particular those with matrix representation.

A system of linear simultaneous equations (a linear system) is given as

$$a_{11}x_1 + a_{12}x_2 + \cdots + a_{1n}x_n = b_1$$
$$a_{21}x_1 + a_{22}x_2 + \cdots + a_{2n}x_n = b_2$$
$$\dots\dots\dots\dots\dots\dots\dots\dots\dots\dots\dots\dots\dots$$
$$a_{m1}x_1 + a_{m2}x_2 + \cdots + a_{mn}x_n = b_m.$$

There are m linear equations with n unknowns $\{x_1, x_2, \ldots, x_n\}$. The coefficients $\{a_{ij} \mid i = 1, 2, \ldots, m; \ j = 1, 2, \ldots, n\}$ form an $m \times n$ matrix $A = [a_{ij}]_{m \times n}$ called the *coefficient matrix*. The unknowns can be arranged as an $n \times 1$ *matrix-column*

$$\bar{\mathscr{X}} = [x_1 \, x_2 \ldots x_n]^T,$$

which is a vector in $\mathbb{R}^n$.

The *free coefficients* $\{b_1, b_2, \ldots, b_m\}$ can be arranged as the $m \times 1$ *free vector* $\bar{b} = [b_1 \, b_2 \ldots b_m]^T$ in $\mathbb{R}^m$.

This linear system can be interpreted as the matrix product of the $m \times n$ coefficient matrix A with the $n \times 1$ *unknown vector* $\bar{\mathscr{X}}$, which gives the $m \times 1$ *free vector* $\bar{b}$ as the result:

$$\boxed{A\bar{\mathscr{X}} = \bar{b}}.$$

With this new interpretation, the problem of finding all possible solutions of the system can be reinterpreted as finding the *inverse image* (if it exists) of the given free vector $\bar{b} \in \mathbb{R}^m$ with respect to the matrix linmap $A : \mathbb{R}^n \to \mathbb{R}^m$.

From our previous investigations about linmaps, it follows that any solutions of the system will exist if the free vector $\bar{b}$ is one of the images of this map. In other words, if $\bar{b}$ is a vector from the range ran A.

In this case and only in this case, all solutions will form a coset in $\mathbb{R}^n$. This coset is the sum of the ker A and a particular, unique, preimage $\bar{b} \in \mathbb{R}^n$ of $\bar{b}$, which preimage serves as the representative of the coset.

As the first step in the attempt to find all solutions, we shall look more carefully into the range ran A. We already know that it is a subspace of $\mathbb{R}^m$, which contains the images of all vectors from $\mathbb{R}^n$ by the linmap $A : \mathbb{R}^n \to \mathbb{R}^m$.

We shall show now that the ran A is *spanned* by n column-vectors $\{\bar{c}_1, \bar{c}_2, \ldots, \bar{c}_n\}$ of A. In fact, we shall prove that the image $A\bar{\mathscr{X}}$ of any $\bar{\mathscr{X}} = [x_1 \, x_2 \ldots x_n]^T \in \mathbb{R}^n$ is a linear combination of the columns of the matrix A with expansion coefficients equal to the components of $\bar{\mathscr{X}}$:

$$A\bar{\mathscr{X}} = \begin{bmatrix} a_{11} & a_{12} & \cdots & a_{1n} \\ a_{21} & a_{22} & \cdots & a_{2n} \\ \cdots & \vdots & \ddots & \vdots \\ a_{m1} & a_{m2} & \cdots & a_{mn} \end{bmatrix} \begin{bmatrix} x_1 \\ x_2 \\ \vdots \\ x_n \end{bmatrix} = \begin{bmatrix} a_{11}x_1 + a_{12}x_2 + \cdots + a_{1n}x_n \\ a_{21}x_1 + a_{22}x_2 + \cdots + a_{2n}x_n \\ \cdots\cdots\cdots\cdots\cdots\cdots \\ a_{m1}x_1 + a_{m2}x_2 + \cdots + a_{mn}x_n \end{bmatrix} =$$

$$= x_1 \begin{bmatrix} a_{11} \\ a_{21} \\ \vdots \\ a_{m1} \end{bmatrix} + x_2 \begin{bmatrix} a_{12} \\ a_{22} \\ \vdots \\ a_{m2} \end{bmatrix} + \cdots + x_n \begin{bmatrix} a_{1n} \\ a_{2n} \\ \vdots \\ a_{mn} \end{bmatrix} = x_1\bar{c}_1 + x_2\bar{c}_2 + \cdots + x_n\bar{c}_n.$$

We call this the *canonical expansion* of $A\bar{\mathscr{X}} \in \operatorname{ran} A$.

Conclusion,

$$\operatorname{ran} A = \operatorname{LIN}(\{\bar{c}_1, \bar{c}_2, \ldots \bar{c}_n\}).$$

For this reason, the subspace $\operatorname{ran} A$ is often called the *column space* of A. Its dimension $r \leq n$ is equal to the maximal number of linearly independent columns of A. This dimension r is referred to as the *rank* of A.

These r linearly independent columns of A form the *natural basis* in $\operatorname{ran} A$. Note, we can always assume that the first r columns in A are linearly independent. If not, then the expansion $A\bar{\mathscr{X}} = x_1\bar{c}_1 + x_2\bar{c}_2 + \cdots + x_n\bar{c}_n$ enables us (since the terms commute) to consider the equivalent product $A_R\bar{\mathscr{X}}_R = A\bar{\mathscr{X}}$ in which A_R is a rearrangement of matrix A so that the first r columns are linearly independent, and $\bar{\mathscr{X}}_R$ has its components rearranged accordingly.

Thus, every vector $\bar{b} = [b_1\, b_2\, \ldots\, b_m]^T$ from $\operatorname{ran} A$ can be uniquely expressed as a linear combination of all the n columns of A, so that the first r terms are the unique expansion of $\bar{b} \in \operatorname{ran} A$ in the natural basis of $\operatorname{ran} A$, and the rest of the $(n-r)$ terms (with linearly *dependent* columns of A) have all coefficients equal to 0:

$$\bar{b} = b_1'\bar{c}_1 + b_2'\bar{c}_2 + \cdots + b_r'\bar{c}_r + 0 \cdot \bar{c}_{r+1} + \cdots + 0 \cdot \bar{c}_n.$$

We have already defined this kind of expansion of vectors from $\operatorname{ran} A$ in the from of linear combinations of all columns of A as their *canonical expansion*. This expansion results in the vector $\bar{b}' \in \mathbb{R}^n$, whose components are the above expansion coefficients

$$\boxed{\bar{b}' = [\underbrace{b_1'\, b_2'\, \ldots b_r'\, 0\, 0 \ldots 0}_{n \text{ components}}]^T}$$

Note that $\bar{b}'$ has the same expansion coefficients in the standard basis $\{\bar{e}_1, \bar{e}_2, \ldots, \bar{e}_n\}$ in $\mathbb{R}^n$ as $\bar{b}$ has in its canonical expansion in terms of all n columns of the matrix A.

It is a fascinating fact that $\bar{b}'$ is the *unique preimage* of $\bar{b}$ in the matrix map $A : \mathbb{R}^n \to \mathbb{R}^m$:

$$A\bar{b}' = b_1'\bar{c}_1 + b_2'\bar{c}_2 + \cdots + b_r'\bar{c}_r + 0 \cdot \bar{c}_{r+1} + \cdots + 0 \cdots \bar{c}_n = \bar{b}.$$

This $\bar{b}'$ is obviously the unique representative of the coset, which is the inverse image of $\bar{b}$:

$$\boxed{\mathrm{Inv}(\bar{b}) = \bar{b}' + \ker A}.$$

There is now the problem of how to find the (possibly unique) basis in $\ker A$. As we have already proved the columns of A span the range $\mathrm{ran}\,A$:

$$\mathrm{ran}\,A = \mathrm{LIN}(\{\bar{c}_1, \bar{c}_2, \ldots, \bar{c}_n\}).$$

This means that we can expand uniquely every column $\bar{c}_i \in \mathrm{ran}A$, $i = 1, 2, \ldots, n$ in terms of all columns (its canonical expansion): in front of r linearly independent columns (which form the natural basis in $\mathrm{ran}\,A$) we have the unique basis coefficients

$$a_1^{(i)}, a_2^{(i)}, \ldots, a_r^{(i)},$$

and in front of $(n - r)$ linearly dependent columns we have just zeros

$$\bar{c}_i = a_1^{(i)} \bar{c}_1 + a_2^{(i)} \bar{c}_2 + \cdots + a_r^{(i)} \bar{c}_r + 0 \cdot \bar{c}_{r+1} + \cdots + 0 \cdot \bar{c}_n, \ i = 1, 2, \ldots, n.$$

For linearly independent columns $\{\bar{c}_1, \bar{c}_2, \ldots, \bar{c}_r\}$, this expansion is extremely simple. Arranging their expansion coefficients as vectors in $\mathbb{R}^n$, we get

$$\bar{c}_1 \rightarrow \begin{bmatrix} 1 \\ 0 \\ \vdots \\ 0 \end{bmatrix}_{n \times 1} = \bar{e}_1, \ \bar{c}_2 \rightarrow \begin{bmatrix} 0 \\ 1 \\ \vdots \\ 0 \end{bmatrix}_{n \times 1} = \bar{e}_2, \ \ldots, \bar{c}_r \rightarrow \left.\begin{bmatrix} 0 \\ 0 \\ \vdots \\ 1 \\ \vdots \\ 0 \end{bmatrix}\right\}_{n \times 1}^{\textstyle r} = \bar{e}_r.$$

These columns are the first r vectors $\{\bar{e}_1, \bar{e}_2, \ldots, \bar{e}_r\}$ in the standard basis in $\mathbb{R}^n$ of n vectors, and they are the unique preimages of the linearly independent columns $\{\bar{c}_1, \bar{c}_2, \ldots, \bar{c}_r\}$ of A : $A\bar{e}_1 = \bar{c}_1, A\bar{e}_2 = \bar{c}_2, \ldots, A\bar{e}_r = \bar{c}_r$. Therefore, $\mathrm{LIN}(\{\bar{e}_1, \bar{e}_2, \ldots, \bar{e}_r\})$ is the r-dimensional subspace of $\mathbb{R}^n$, which consists of the unique preimages of all vectors from $\mathrm{ran}\,A$ and it is isomorphic to $\mathrm{ran}\,A$.

Conclusion: The set of r linearly independent vectors $\{\bar{e}_1, \bar{e}_2, \ldots, \bar{e}_r\}$ in $\mathbb{R}_n$ spans the unique *preimage subspace* $\mathrm{pre}\,A$ in $\mathbb{R}^n$, which is *isomorphic* to the r-dimensional $\mathrm{ran}\,A$ (a subspace in $\mathbb{R}^m$):

$$\mathrm{pre}\,A \overset{c}{\cong} \mathrm{ran}\,A,$$

where c stands for the *canonical expansion* of vectors from $\mathrm{ran}\,A$ in terms of all columns of A. Naturally, A maps every vector from *pre A* onto its unique image in $\mathrm{ran}\,A$.

This isomorphism is established as follows:

$$\mathrm{ran}\,A \ni \bar{b} = b_1' \bar{c}_1 + b_2' \bar{c}_2 + \cdots + b_r' \bar{c}_r + 0 \cdot \bar{c}_{r+1} + \cdots + 0 \cdot \bar{c}_n \overset{c}{\leftrightarrow} \bar{b}'$$

$$= \begin{bmatrix} b'_1 \\ b'_2 \\ \vdots \\ b'_r \\ 0 \\ \vdots \\ 0 \end{bmatrix} \left.\vphantom{\begin{matrix}0\\\vdots\\0\end{matrix}}\right\} n-r \qquad \in \text{pre } A.$$

$$A\bar{b}' = \bar{b} = [b_1 \, b_2 \, \ldots \, b_m]^T.$$

The vector $\bar{b}$ belongs to ran A if and only if the last $(n-r)$ expansion coefficients are 0.

As far as the linearly dependent columns of A $\bar{c}_{r+1}, \bar{c}_{r+2}, \ldots, \bar{c}_n$ are concerned, we can also apply the unique canonical expansion to all of them:

$$\bar{c}_i = a_1^{(i)} \bar{c}_1 + a_2^{(i)} \bar{c}_2 + \ldots + a_r^{(i)} \bar{c}_r + 0 \cdot \bar{c}_{r+1} + \cdots + 0 \cdot \bar{c}_n, \; i = r+1, r+2, \ldots, n.$$

A most simple modification of this expansion solves completely the problem of finding the $(n-r)$ unique basis vectors for ker A in $\mathbb{R}^n$:
Taking $\bar{c}_i$, $i = r+1, r+2, \ldots, n = r+(n-r)$, across the equality sign, we get $\bar{0}_m$ on the left hand side and an extra $(-1)\bar{c}_i$ on the right hand side

$$\bar{0}_m = a_1^{(i)} \bar{c}_1 + a_2^{(i)} \bar{c}_2 + \cdots + a_r^{(i)} \bar{c}_r + 0 \cdot \bar{c}_{r+1} + \cdots + (-1)\bar{c}_i + \cdots + 0 \cdot \bar{c}_n,$$
$$i = r+1, r+2, \ldots, n.$$

In this way, we get $(n-r)$ expansions of $\bar{0}_m$ and the corresponding $(n-r)$ vectors in $\mathbb{R}^n$:

$$\bar{k}_1 = \begin{bmatrix} a_1^{(r+1)} \\ a_2^{(r+1)} \\ \vdots \\ a_r^{(r+1)} \\ -1 \\ 0 \\ \vdots \\ 0 \end{bmatrix}_{n \times 1}, \; \bar{k}_2 = \begin{bmatrix} a_1^{(r+2)} \\ a_2^{(r+2)} \\ \vdots \\ a_r^{(r+2)} \\ 0 \\ -1 \\ \vdots \\ 0 \end{bmatrix}_{n \times 1}, \; \ldots, \bar{k}_{n-r} = \begin{bmatrix} a_1^{(n)} \\ a_2^{(n)} \\ \vdots \\ a_r^{(n)} \\ 0 \\ 0 \\ \vdots \\ -1 \end{bmatrix}_{n \times 1}.$$

These $(n-r)$ vectors are obviously *linearly independent* since in the last $(n-r)$ entries they have all zeros, except one (-1) which is in $(n-r)$ different positions. If we want to verify their linear independence, we make a linear combination

$$\sum_{i=r+1}^{n} x_i \bar{k}_i = \bar{0}_n,$$

and see immediately that all x_i must be zero. They all belong to $\ker A$, since the multiplication with A of all of them gives $\bar{0}_m$:

$$A\bar{k}_j = A \begin{bmatrix} a_1^{(r+j)} \\ a_2^{(r+j)} \\ \vdots \\ a_r^{(r+j)} \\ 0 \\ \vdots \\ 0 \\ -1 \leftarrow r+j \text{ position} \\ 0 \\ \vdots \\ 0 \end{bmatrix}_{n\times 1} = A \begin{bmatrix} a_1^{(r+j)} \\ a_2^{(r+j)} \\ \vdots \\ a_r^{(r+j)} \\ 0 \\ 0 \\ \vdots \\ 0 \end{bmatrix}_{n\times 1} - A \begin{bmatrix} 0 \\ 0 \\ \vdots \\ 0 \\ 0 \\ \vdots \\ 1 \\ \vdots \\ 0 \end{bmatrix}_{n\times 1} \left. \begin{matrix} \\ \\ \\ \end{matrix} \right\} r$$

with $r+j$ position marked.

$$= \bar{c}_{r+j} - \bar{c}_{r+j} = \bar{0}_m \quad j = 1, 2, \ldots, n-r.$$

Thus, the vectors $\{\bar{k}_1, \bar{k}_2, \ldots, \bar{k}_{n-r}\}$ from $\mathbb{R}^n$ are the *unique* basis in the kernel of A, whose dimension is $(n-r)$ as follows from the dimension theorem (see Sect. 2.4).

Theorem (on the general solution of a linear system) The general solution of the system of linear equations

$$A\mathscr{X} = \bar{b}, \quad \text{where}$$
$$A = [a_{ij}]_{m\times n}, \quad \mathscr{X} \in \mathbb{R}^n, \quad \bar{b} \in \mathbb{R}^m,$$

is the *coset* $\boxed{\text{Inv}(\bar{b}) = b' + \ker A}$ determined by the *unique representation* $\bar{b}'$, which is obtained by the canonical expansion of $\bar{b}$ in terms of all the columns of A (the unique preimage of $\bar{b}$ in $\mathbb{R}^n$), and the *unique basis* $\{\bar{k}_1, \bar{k}_2, \ldots, \bar{k}_{n-r}\}$ of $\ker A$ obtained by the slightly modified canonical expansion of the $(n-r)$ linearly dependent columns of A:

$$\text{Inv}(\bar{b}) = \bar{b}' + \text{LIN}(\{\bar{k}_1, \bar{k}_2, \ldots, \bar{k}_{n-r}\}) = \bar{b}' + \ker A.$$

Proof Obtained in the previous text. Δ

Comments

1. The free vector $\bar{b}$ in the system of linear equations $A\mathscr{X} = \bar{b}$ is a nonzero vector from $\mathbb{R}^m$ (this is called a *nonhomogeneous system*). The system has a solution if and only if $\bar{b}$ belongs to the range $\text{ran}\,A$, a subspace of $\mathbb{R}^m$ (this is called the *column-space criterion*).
   Looking at $\bar{b}$ itself one cannot tell whether it is from $\text{ran}\,A$. The practical criterion for the existence of solutions is if and only if $\bar{b}$ can be canonically expanded. In

that case, the last $(n - r)$ expansion coefficient are all zeros. A system which has solutions is called *consistent*.

2. If $\bar{b} = \bar{0}_m$, then we have a *homogeneous* system of linear equations. Since $\bar{0}_m$ always belongs to the subspace ran A, such a system is *necessarily consistent*. The canonical expansion of $\bar{0}_m$ has all the coefficients zero, so that the corresponding particular unique solution is the zero vector $\bar{0}_n$. If the matrix A has no linearly dependent columns, i.e., if the rank r of A is equal to n, the kernel is trivial, it consists only of $\bar{0}_n$. Consequently, $\mathrm{Inv}(\bar{0}_m) = \bar{0}_n$ (trivial solution).

If, on the other hand, there are $(n - r)$ linearly dependent columns of A, then their modified canonical expansions give immediately $(n - r)$ basis vectors for the kernel ker A.

So, the general solution of a homogeneous linear system is the kernel ker A itself – trivial or nontrivial, depending on the rank r of $A - r = n$ or $r < n$, respectively.

Remember that the kernel is also a coset $\bar{0}_n + \ker A = \ker A$, the only difference being that it is a subspace of $\mathbb{R}^n$ as well, while the other cosets are never subspaces.

3. As we have already shown the canonical expansions of vectors from ran A induce an isomorphism, denoted by c, between ran A and pre A (the subspace of $\mathbb{R}^n$ consisting of all unique preimages of vectors from ran A):

$$\mathrm{ran}\,A \overset{c}{\cong} \mathrm{pre}\,A, \quad \text{where } \bar{b} \overset{c}{\leftrightarrow} \bar{b}'.$$

At the same time, we have another isomorphism c' between ran A and the quotient space $\mathbb{R}/\ker A$ of all cosets in $\mathbb{R}^n$: ran $A \overset{c'}{\cong} \mathbb{R}^n/\ker A$, where $\bar{b} \overset{c'}{\longleftrightarrow} \bar{b}' + \ker A = \mathrm{Inv}(\bar{b})$.

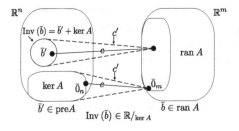

4. We say that the vector space V_n is a *direct sum* $V_n = V_{n_1} + V_{n_2}$ (see Sect. 3.4.1) of two of its subspaces V_{n_1} and V_{n_2} if every vector $\bar{x} \in V_n$ can be uniquely written as the sum of two components, one $\bar{x}_1$ from V_{n_1} and the other $\bar{x}_2$ from V_{n_2}:

$$\bar{x} = \bar{x}_1 + \bar{x}_2, \quad \bar{x} \in V_{n_1}, \quad \bar{x}_2 \in V_{n_2}.$$

From this definition, it follows immediately that V_{n_1} and V_{n_2} have the trivial intersection

$$V_{n_1} \cap V_{n_2} = \{\bar{0}_n\},$$

and that their dimensions add up to $n : n_1 + n_2 = n$.

It is almost obvious that $\mathbb{R}^n$ is the direct sum of two of its subspaces

$$\mathbb{R}^n = \operatorname{pre} A + \ker A.$$

To prove this, we have to remember that $\mathbb{R}^n$ is partitioned into the cosets of $\ker A$. Representatives of these cosets are vectors from $\operatorname{pre} A$, so that every $\bar{\mathscr{X}} \in \mathbb{R}^n$ is the unique sum

$$\bar{\mathscr{X}} = \bar{\mathscr{X}}_1 + \bar{\mathscr{X}}_2, \ \bar{\mathscr{X}}_1 \in \operatorname{pre} A, \ \bar{\mathscr{X}}_2 \in \ker A.$$

The dimension of $\operatorname{pre} A$ is r and that of $\ker A$ is $(n-r)$, which add up to n, the dimension of $\mathbb{R}^n$. $\quad \Delta$

To anticipate somewhat, we can say that the GJM reduced form of A (see Sect. 2.17) will explicitly exhibit the two unique bases of the above direct decomposition of $\mathbb{R}^n$: one set of the first r vectors from the standard basis in $\mathbb{R}^n$, as the basis of $\operatorname{pre} A$, and another set of $(n-r)$ vectors from $\mathbb{R}^n$, as the basis of $\ker A$.

2.9 The Four Tasks

By now, we have discovered that the canonical expansions of $(n-r)+1$ linearly dependent columns of the augmented matrix $[A \mid \bar{b}]$ of the consistent linear system $A\bar{\mathscr{X}} = \bar{b}$, where $A = [a_{ij}]_{m \times n}$, $\bar{\mathscr{X}} = [x_i]_{n \times 1}$, $\bar{b} = [b_i]_{m \times 1}$, provide the *general solution* of this consistent linear system in coset form:

$$\operatorname{Inv}(\bar{b}) = \bar{b}' + \ker A,$$

so that the coset representative $\bar{b}'$ and the basis $\{\bar{k}_1, \bar{k}_2, \ldots, \bar{k}_{n-r}\}$ of $\ker A$ are uniquely determined.

The afore-mentioned canonical expansions are immediately obtained in the GJ modified reduced row-echelon form of $[A \mid \bar{b}]$ as will be shown in Sect. 2.17.

To prepare the ground for this claim, we shall have to perform a number of tasks beforehand.

1. Firstly, we shall prove that the number of linearly independent (lind) rows and the number of lind columns of the matrix A is the same, regardless of how many rows m and how many columns n there are in the matrix A. We call this common number the *rank r* of A (see Sect. 2.10).
2. Secondly, we have to introduce three elementary row operations (eros), which can be performed on the augmented matrix $[A \mid \bar{b}]$ of $A\bar{\mathscr{X}} = \bar{b}$, so that one gets reduced equivalent linear systems with the same solution set. The same effect arises from multiplication of A from the left with one of three $m \times m$ elementary

matrices (see Sect. 2.12). Using this alternative method, we can derive a number of proofs in a much simpler way.

3. Furthermore, the original unique GJ reduced row-echelon form $[A \,|\bar{b}]_{\text{red}}$ of $[A \,|\, \bar{b}]$ will be defined and obtained by means of eros. Its amazing properties are discussed in detail (see Sects. 2.13 and 2.15).

The standard method for solving $A\vec{\mathscr{X}} = \bar{b}$ consists in writing the linear system corresponding to this reduced *GJ* form of $[A \,|\, \bar{b}]$ and deciding which unknowns have no solution, so that they must be taken as *free parameters*, which can have any real value. Finally, one expresses the basic unknowns in terms of these parameters and the coefficients of $[A \,|\, \bar{b}]_{\text{red}}$. The basic unknowns are those which multiply lind columns of A in the canonical expansion

$$A\vec{\mathscr{X}} = \bar{c}_1 x_1 + \bar{c}_2 x_2 + \cdots + \bar{c}_n x_n = \bar{b}.$$

Consequently, the nonbasic ones (declared as parameters) are those unknowns which multiply the linearly dependent columns of A in the above canonical expansion (see Sect. 2.16).

4. Transfer to the GJM form of $[A \,|\, \bar{b}]_{\text{red}}$ is the final simple step that immediately produces the general unique coset solution of the linear system $A\vec{\mathscr{X}} = \bar{b}$.

This GJM form of $[A \,|\, \bar{b}]_{\text{red}}$ gives explicitly the canonical expansions of all columns of the original $[A \,|\, \bar{b}]$; so providing the bases for the direct decomposition $\mathbb{R}^n = \text{pre}\,A \oplus \ker A$, as well as $\bar{b}'$, the unique coset representative for the coset solution $\text{Inv}(\bar{b}) = \bar{b}' + \ker A$ (see Sect. 2.17).

2.10 The Column Space and the Row Space

Task 1. Proof of the statement that there is the same number of lind columns and lind rows in every $m \times n$ matrix A. In other words, the column space of A has the same dimension as its row space. This dimension is called the *rank r* of A.

We consider the matrix $A = [a_{ij}]_{m \times n} = \begin{bmatrix} a_{11} & a_{12} & \cdots & a_{1n} \\ a_{21} & a_{22} & \cdots & a_{2n} \\ \vdots & \vdots & \ddots & \vdots \\ a_{m1} & a_{m2} & \cdots & a_{mn} \end{bmatrix}$ as the column of m

row-vectors with n components

$$A = \begin{bmatrix} \bar{r}_1 \\ \bar{r}_2 \\ \vdots \\ \bar{r}_m \end{bmatrix} = \begin{bmatrix} a_{11} & a_{12} & \cdots & a_{1n} \\ a_{21} & a_{22} & \cdots & a_{2n} \\ \vdots & \vdots & \ddots & \vdots \\ a_{m1} & a_{m2} & \cdots & a_{mn} \end{bmatrix}.$$

These vectors span the row space of A. Let us assume that there are s lind vectors among these rows, so that the dimension of the row space of A is $s \leq n$. We can always choose a basis in the row space, e.g., $B_{RS} = \{\bar{b}_1, \bar{b}_2, \ldots, \bar{b}_s\}$, where

$$\bar{b}_j = [b_{j1}\, b_{j2} \ldots b_{jn}], \quad j = 1, 2, \ldots, s,$$

and expand every row of A uniquely in this basis:

$$\bar{r}_1 = \sum_{j=1}^{s} k_{1j}\bar{b}_j, \; \bar{r}_2 = \sum_{j=1}^{s} k_{2j}\bar{b}_j, \ldots, \bar{r}_m = \sum_{j=1}^{s} k_{mj}\bar{b}_j, \quad \text{or, in short,}$$

$$\boxed{\bar{r}_i = \sum_{j=1}^{s} k_{ij}\bar{b}_j, \quad i = 1, 2, \ldots, m.}$$

Writing each row in terms of its components, we get

$$[a_{i1}\, a_{i2} \ldots a_{in}] = \sum_{j=1}^{s} k_{ij}[b_{j1}\, b_{j2} \ldots b_{jn}], \quad i = 1, 2, \ldots, m.$$

Such row-vector expansions give immediately unique expansions for each component in them

$$a_{ip} = \sum_{j=1}^{s} k_{ij} b_{jp}, \quad i = 1, 2, \ldots, m; \; p = 1, 2, \ldots, n.$$

We can now arrange these components as n columns of the matrix A:

$$\bar{c}_p = \begin{bmatrix} a_{1p} \\ a_{2p} \\ \vdots \\ a_{mp} \end{bmatrix} = b_{1p} \begin{bmatrix} k_{11} \\ k_{21} \\ \vdots \\ k_{m1} \end{bmatrix} + b_{2p} \begin{bmatrix} k_{12} \\ k_{22} \\ \vdots \\ k_{m2} \end{bmatrix} + \cdots + b_{sp} \begin{bmatrix} k_{1s} \\ k_{2s} \\ \vdots \\ k_{ms} \end{bmatrix}, \quad p = 1, 2, \ldots, n.$$

We conclude that every column $\bar{c}_p$, $p = 1, 2, \ldots, n$ of the matrix A can be expressed as a unique linear combination of s column vectors from $\mathbb{R}^m$:

$$\bar{k}'_1 = \begin{bmatrix} k_{11} \\ k_{21} \\ \vdots \\ k_{m1} \end{bmatrix}, \; \bar{k}'_2 = \begin{bmatrix} k_{12} \\ k_{22} \\ \vdots \\ k_{m2} \end{bmatrix}, \; \ldots \; \bar{k}'_s = \begin{bmatrix} k_{1s} \\ k_{2s} \\ \vdots \\ k_{ms} \end{bmatrix}.$$

Thus, these s column-vectors $B_{cs} = \{\bar{k}'_1, \bar{k}'_2, \ldots, \bar{k}'_s\}$ form a basis in the column space. This means that the dimension of the column space is the same as the dimension s of the row space. But, we have already defined the dimension of the column space as the rank r of the matrix A. Therefore, the final conclusion is that $s = r$. $\triangle$

It is a rather surprising conclusion since some $m \times n$ matrices can be very elongated rectangles. Nevertheless, r is always equal to s. Notice that r is less than or at most equal to the smaller of m and n ($r \le \min(m,n)$). For instance, a 2×6 matrix (2 rows and 6 columns) can have at most two lind columns ($\overset{\text{rank}}{r} \le 2$). If such a matrix is the coefficient matrix of the linear system $A\vec{x} = \vec{b}$, then we can be sure that out of six unknowns at most two can be basic ones. The rest of the unknowns are simply free parameters.

2.11 Two Examples of Linear Dependence of Columns and Rows of a Matrix

1. Consider the 3×5 matrix

$$\begin{bmatrix} 1 & -2 & 1 & 3 & 2 \\ 1 & 1 & -1 & 4 & 3 \\ 2 & 5 & -2 & 9 & 8 \end{bmatrix}.$$

It can have at most three lind columns since it has only three rows. Let us start with the first three columns to see if they are lind. To do this we shall form a linear combination of them and substantiate that this linear combination can produce the zero vector in $\mathbb{R}^3$ only with zero coefficients.

$$a\begin{bmatrix} 1 \\ 1 \\ 2 \end{bmatrix} + b\begin{bmatrix} -2 \\ 1 \\ 5 \end{bmatrix} + c\begin{bmatrix} 1 \\ -1 \\ -2 \end{bmatrix} = \begin{bmatrix} 0 \\ 0 \\ 0 \end{bmatrix} \Rightarrow$$

$$\left.\begin{matrix} (-1)\cdot/a - 2b + c = 0 \\ (-2)\cdot/a + b - c = 0 \\ 2a + 5b - 2c = 0 \end{matrix}\right\} \Rightarrow \left.\begin{matrix} 3b - 2c = 0 \\ 3b = 0 \end{matrix}\right\} \Rightarrow b = 0,\ c = 0,\ a = 0.$$

Thus, $r = 3$.

Now, we can be sure that the third and fourth columns are linearly dependent on the first three. Let us verify this.

$$\begin{bmatrix} 3 \\ 4 \\ 9 \end{bmatrix} = x\begin{bmatrix} 1 \\ 1 \\ 2 \end{bmatrix} + y\begin{bmatrix} -2 \\ 1 \\ 5 \end{bmatrix} + z\begin{bmatrix} 1 \\ -1 \\ -2 \end{bmatrix} \Rightarrow \begin{cases} x - 2y + z = 3/\cdot(-1) \\ x + y - z = 4/\cdot(-2) \\ 2x + 5y - 2z = 9 \end{cases} \Rightarrow$$

$$\left.\begin{matrix} 3y - 2z = 1 \\ 3y = 1 \end{matrix}\right\} \Rightarrow y = 1/3,\ z = 0,\ x = 11/3.$$

Similarly, $\begin{bmatrix} 2 \\ 3 \\ 8 \end{bmatrix} = x'\begin{bmatrix} 1 \\ 1 \\ 2 \end{bmatrix} = y'\begin{bmatrix} -2 \\ 1 \\ 5 \end{bmatrix} + z'\begin{bmatrix} 1 \\ -1 \\ -2 \end{bmatrix} \Rightarrow x' = 17/6,\ y' = 2/3,$

$z' = 1/2$.

We now know that the three rows of A are lind ($r = 3$). There is no need to verify this.

2. Consider now the 4×5 matrix

$$
B = \begin{bmatrix} 1 & -1 & 1 & -1 & 4 \\ 1 & 1 & 2 & 3 & 8 \\ 2 & 4 & 5 & 10 & 20 \\ 2 & -4 & 1 & -6 & 4 \\ \bar{c}_1 & \bar{c}_2 & \bar{c}_3 & \bar{c}_4 & \bar{c}_5 \end{bmatrix} \begin{matrix} \bar{r}_1 \\ \bar{r}_2 \\ \bar{r}_3 \\ \bar{r}_4 \\ \end{matrix} .
$$

It is simple to verify that only the first two columns are lind. The other three are linearly dependent.

A.

$$
a\begin{bmatrix} 1 \\ 1 \\ 2 \\ 2 \end{bmatrix} + b\begin{bmatrix} -1 \\ 1 \\ 4 \\ -4 \end{bmatrix} = \begin{bmatrix} 0 \\ 0 \\ 0 \\ 0 \end{bmatrix} \Rightarrow \left. \begin{array}{r} a - b = 0 \\ a + b = 0 \\ 2a + 4b = 0 \\ 2a - 4b = 0 \end{array} \right\}
$$

$a = 0, \ b = 0.$
linearly dependent equations
$Eq.3 = (-1)Eq.1 + 3Eq.2,$
$Eq.4 = 3Eq.1 - Eq.2$

$\Rightarrow \bar{c}_1$ and $\bar{c}_2$ form a basis W the column space.

B.

$$
\begin{bmatrix} 1 \\ 2 \\ 5 \\ 1 \end{bmatrix} = c_1\begin{bmatrix} 1 \\ 1 \\ 2 \\ 2 \end{bmatrix} + c_2\begin{bmatrix} -1 \\ 1 \\ 4 \\ -4 \end{bmatrix} \Rightarrow \left. \begin{array}{r} c_1 - c_2 = 1 \\ c_1 + c_2 = 2 \\ 2c_1 + 4c_2 = 5 \\ 2c_1 - 4c_2 = 1 \end{array} \right\}
$$

$c_1 = \frac{3}{2}, \ c_2 = \frac{1}{2}$
linearly dependent
equations
$Eq.3 = (-1)Eq.1 + 3Eq.2$
$Eq.4 = 3Eq.1 - Eq.2$

$$\Rightarrow \bar{c}_3 = \frac{3}{2}\bar{c}_1 + \frac{1}{2}\bar{c}_2$$

C.

$$
\begin{bmatrix} -1 \\ 3 \\ 10 \\ -6 \end{bmatrix} = d_1\begin{bmatrix} 1 \\ 1 \\ 2 \\ 2 \end{bmatrix} + d_2\begin{bmatrix} -1 \\ 1 \\ 4 \\ -4 \end{bmatrix} \Rightarrow \left. \begin{array}{r} d_1 - d_2 = -1 \\ d_1 + d_2 = 3 \\ 2d_2 + 4d_2 = 10 \\ 2d_2 - 4d_2 = -6 \end{array} \right\}
$$

$d_1 = 1, \ d_2 = 2$
linearly dependent
equations
(the same linear
dependence
as above)

$$\Rightarrow \bar{c}_4 = \bar{c}_1 + 2\bar{c}_2$$

D.

$$
\begin{bmatrix} 4 \\ 8 \\ 20 \\ 4 \end{bmatrix} = e_1\begin{bmatrix} 1 \\ 1 \\ 2 \\ 2 \end{bmatrix} + e_2\begin{bmatrix} -1 \\ 1 \\ 4 \\ -4 \end{bmatrix} \Rightarrow \left. \begin{array}{r} e_1 - e_2 = 4 \\ e_1 + e_2 = 8 \\ 2e_1 + 4e_2 = 20 \\ 2e_1 - 4e_2 = 4 \end{array} \right\}
$$

$e_1 = 6, \ e_2 = 2$
linearly dependent
equations
(the same linear
dependence
as above)

$$\Rightarrow \bar{c}_5 = 6\bar{c}_1 + 2\bar{c}_2$$

One expects that the first two rows are lind. Indeed,

$$a_1[1\ -1\ 1\ -1\ 4] + b_1[1\ 1\ 2\ 3\ 8] = [0\,0\,0\,0\,0] \quad \text{giving five linear equations}$$

$$\left.\begin{array}{l} a_1 + b_1 = 0 \\ -a_1 + b_1 = 0 \end{array}\right\} \Rightarrow a_1 = 0,\ b_1 = 0 \Rightarrow \bar{r}_1 \text{ and } \bar{r}_2 \text{ form a basis in the row space.}$$

$$\left.\begin{array}{l} +a_1 + 2b_1 = 0 \\ -a_1 + 3b_1 = 0 \\ 4a_1 + 8b_1 = 0 \end{array}\right\} \begin{array}{l} \text{linearly dependent} \\ \text{equations (no} \\ \text{new information).} \end{array} \quad \begin{array}{l} Eq.3 = \frac{3}{2}Eq.1 + \frac{1}{2}Eq.2 \\ Eq.4 = Eq.1 + 2Eq.2 \\ Eq.5 = 6Eq.1 + 2Eq.2 \end{array}$$

The third and fourth rows are linearly dependent:

$$\bar{r}_3 = (-1)\bar{r}_1 + 3\bar{r}_2, \quad \bar{r}_4 = 3\bar{r}_1 - \bar{r}_2.$$

The last two linear dependence relations of rows can be read immediately from the linear dependence of the third and fourth equations in B, C, D. above.

Illustration Now is a very good opportunity to illustrate the method of finding the general solution for the linear system $A\mathcal{X} = \bar{b}$ in the coset form by means of *canonical expansions* of the linearly dependent columns of the augmented matrix $[A \mid \bar{b}]$ of the system.

Let our matrix B be the augmented matrix $B = [A \mid \bar{b}]$ of the linear system of four equations with four unknowns

$$x_1 - x_2 + x_3 - x_4 = 4$$
$$x_1 + x_2 + 2x_3 + 3x_4 = 8$$
$$2x_1 + 4x_2 + 5x_3 + 10x_4 = 20$$
$$2x_1 - 4x_2 + x_3 - 6x_4 = 4.$$

Having already found the linear dependency relations for the columns of B, we can now write straightforwardly the general solution in the coset form

$$\begin{bmatrix} x_1 \\ x_2 \\ x_3 \\ x_4 \end{bmatrix} = \begin{bmatrix} 6 \\ 2 \\ 0 \\ 0 \end{bmatrix} + \text{LIN}\left(\begin{bmatrix} 3/2 \\ 1/2 \\ -1 \\ 0 \end{bmatrix}, \begin{bmatrix} 1 \\ 2 \\ 0 \\ -1 \end{bmatrix}\right) = \bar{b}' + \ker A.$$

Remember that the unique coset representative $\bar{b}' = \begin{bmatrix} 6 \\ 2 \\ 0 \\ 0 \end{bmatrix}$ consists of the canonical

expansion coefficients of $\bar{b} = \begin{bmatrix} 4 \\ 8 \\ 20 \\ 4 \end{bmatrix}$: $\bar{b} = 6\bar{c}_1 + 2\bar{c}_2 + 0 \cdot \bar{c}_3 + 0 \cdot \bar{c}_3$ in terms of all columns of A. One can easily verify that $A\bar{b}' = \bar{b}$:

$$\begin{bmatrix} 1 & -1 & 1 & -1 \\ 1 & 1 & 2 & 3 \\ 2 & 4 & 5 & 10 \\ 2 & -4 & 1 & -6 \end{bmatrix} \begin{bmatrix} 6 \\ 2 \\ 0 \\ 0 \end{bmatrix} = \begin{bmatrix} 6-2 \\ 6+2 \\ 12+8 \\ 12-8 \end{bmatrix} = \begin{bmatrix} 4 \\ 8 \\ 20 \\ 4 \end{bmatrix}.$$

The unique basis vectors for ker A

$$\bar{k}_1 = \begin{bmatrix} 3/2 \\ 1/2 \\ -1 \\ 0 \end{bmatrix} \text{ and } \bar{k}_2 = \begin{bmatrix} 1 \\ 2 \\ 0 \\ -1 \end{bmatrix} \text{ consist of modifed}$$

canonical expansion coefficients of the two linearly dependent columns $\bar{c}_3$ and $\bar{c}_4$ of A:

$$\bar{c}_3 = \frac{3}{2}\bar{c}_1 + \frac{1}{2}\bar{c}_2 + 0 \cdot \bar{c}_3 + 0 \cdot \bar{c}_4 \overset{\text{modification}}{\Rightarrow} \bar{0}_4 = \frac{3}{2}\bar{c}_1 + \frac{1}{2}\bar{c}_2 - \bar{c}_3 + 0 \cdot \bar{c}_4,$$

$$\bar{c}_4 = \bar{c}_1 + 2\bar{c}_2 + 0 \cdot \bar{c}_3 + 0\bar{c}_4 \overset{\text{modification}}{\Rightarrow} \bar{0}_4 = \bar{c}_1 + 2\bar{c}_2 + 0 \cdot \bar{c}_3 - \bar{c}_4.$$

One can easily verify that $A\bar{k}_1 = \bar{0}_4$, $A\bar{k}_2 = \bar{0}_4$, and that $\bar{k}_1$ and $\bar{k}_2$ are lind. So, they both belong to ker A, they are obviously linearly independent , and there are two of them as required by the dimension theorem

$$\dim \ker A = \dim \mathbb{R}^4 - \dim \mathrm{ran}\, A = 4 - 2 = 2.$$

Therefore, ker $A = \mathrm{LIN}(\bar{k}_1, \bar{k}_2) = \{s\bar{k}_1 + t\bar{k}_2 \mid s,t \in \mathbb{R}\}$.

2.12 Elementary Row Operations (Eros) and Elementary Matrices

Task 2. The last example presents a simple illustration of how to obtain the general solution of a consistent linear system $A\bar{\mathcal{X}} = \bar{b}$ in the unique coset form by means of the *canonical expansions* of linearly dependent columns in the augmented matrix $[A \mid \bar{b}]$.

There is a more sophisticated and much faster method to achieve the same result by means of the Gauss-Jordan modified (GJM) procedure. To perform this procedure, we need elementary row operations (eros).

2.12.1 Eros

First, we have to remember that every $m \times n$ linear system $A\bar{\mathcal{X}} = \bar{b}$ is obviously in a one-to-one correspondence with its $m \times (n+1)$ augmented matrix $[A \mid \bar{b}]$. Making use of the matrix $[A \mid \bar{b}]$ instead of the system of linear equations $A\mathscr{X} = \bar{b}$ is of course much more convenient, especially because with $[A \mid \bar{b}]$ one can apply different operations from matrix algebra.

The basic idea is to transform $[A \mid \bar{b}]$ into the canonical (simplest) form called the GJ *reduced row-echelon form*, so that the corresponding linear system has exactly the same general solution as $A\mathscr{X} = \bar{b}$. For the purpose of getting the GJ form, we use only three elementary row operations (eros) on $[A \mid \bar{b}]$, each of which preserves the general solution.

The first type of ero is an interchange of any two rows: $\bar{r}_i \leftrightarrow \bar{r}_j$ in $[A \mid \bar{b}]$.

The second type of ero consists of multiplying a row in $[A \mid \bar{b}]$ with a nonzero real number $\lambda \neq 1 : \bar{r}_i \rightarrow \lambda \bar{r}_i$, $\lambda \neq 0, 1$.

The third type of ero is performed by adding to a row another row multiplied by some $\lambda \neq 0 : \bar{r}_i \rightarrow \bar{r}_i + \lambda \bar{r}_j \, i \neq j$.

To prove that each type of ero produces an equivalent linear system (the linear system which has the same solution), we shall briefly investigate the effects of the above three types of ero considered as operations on the equations of the system $A\mathscr{X} = \bar{b}$.

1. Obviously, the interchange of any two equations in the system will not alter the solutions.
2. If the i-th equation of the system is multiplied by λ, we get

$$\lambda a_{i1}x_1 + \lambda a_{i2}x_2 + \cdots + \lambda a_{in}x_n = \lambda b_i.$$

Since $\lambda \neq 0$, we immediately see that if the set $\{c_1, c_2, \ldots, c_n\}$ is a solution for this new equation:

$$\lambda a_{i1}c_1 + \lambda a_{i2}c_2 + \cdots + \lambda a_{in}c_n = \lambda b_i,$$

then dividing by $\lambda \neq 0$, this set is also a solution for the original equation:

$$a_{i1}c_1 + a_{i2}c_2 + \cdots + a_{in}c_n = b_i$$

and vice versa. The remaining equations are not changed.
3. The i-th equation $a_{i1}x_1 + a_{i2}x_2 + \cdots + a_{in}x_n = b_i$ is changed into

$$(a_{i1} + \lambda a_{j1})x_1 + (a_{i2} + \lambda a_{j2})x_2 + \cdots + (a_{in} + \lambda a_{jn})x_n = b_i + \lambda b_j. \quad (*)$$

If the set $\{c_1, c_2, \ldots, c_n\}$ is a solution for the system before this operation is performed, it means that both

$$a_{i1}c_1 + a_{i2}c_2 + \cdots + a_{in}c_n = b_i \quad (**)$$

and

$$a_{j1}c_1 + a_{j2}c_2 + \cdots + a_{jn}c_n = b_j (***)$$

are satisfied.

Multiplying $(***)$ with λ and adding $(**)$, one sees that the new equation $(*)$ has this set as its solution:

$$(a_{i1} + \lambda a_{j1})c_1 + (a_{i2} + \lambda a_{j2})c_2 + \cdots + (a_{in} + \lambda a_{jn})c_n = b_i + \lambda b_j. \quad (****)$$

Consequently, the set $\{c_1, c_2, \ldots, c_n\}$ is a solution to the new system, since other equations are unchanged.

If, on the other hand, the set $\{c_1, c_2, \ldots, c_n\}$ is a solution to the new system, it implies that $(***)$ and $(****)$ are satisfied. Multiplying $(***)$ by λ and subtracting from $(****)$, we get $(**)$, so this set is a solution to the old system as well. Δ

For each ero, there is the inverse one:

1. For the interchange of rows $(\bar{r}_i \leftrightarrow \bar{r}_j)$, the inverse ero is the same interchange.
2. For the multiplication of a row with $\lambda \neq 1$ $(\bar{r}_i \rightarrow \lambda \bar{r}_i)$, the inverse ero is the multiplication with $1/\lambda$.
3. For the addition of a row multiplied with λ to another row $(\bar{r}_i \rightarrow \bar{r}_i + \lambda \bar{r}_j)$, the inverse ero is the subtraction of the same row multiplied with λ.

2.12.2 Elementary Matrices

For every ero, there is a square matrix $m \times m$ that produces the same effect as that ero has on any $m \times n$ matrix A by left multiplication (often referred as *premultiplication*). These matrices are called the *elementary matrices*.

For example, for $m = 3$, if one wants to interchange rows 1 and 2 $(\bar{r}_1 \leftrightarrow \bar{r}_2)$ on a 3×3 matrix $A = [a_{ij}]_{3 \times 3}$, one uses the elementary matrix $\begin{bmatrix} 0 & 1 & 0 \\ 1 & 0 & 0 \\ 0 & 0 & 1 \end{bmatrix}$:

$$\begin{bmatrix} 0 & 1 & 0 \\ 1 & 0 & 0 \\ 0 & 0 & 1 \end{bmatrix} \begin{bmatrix} a_{11} & a_{12} & a_{13} \\ a_{21} & a_{22} & a_{23} \\ a_{31} & a_{32} & a_{33} \end{bmatrix} = \begin{bmatrix} a_{21} & a_{22} & a_{23} \\ a_{11} & a_{12} & a_{13} \\ a_{31} & a_{32} & a_{33} \end{bmatrix}.$$

To multiply the second row with 3 $(\bar{r}_2 \rightarrow 3\bar{r}_2)$, one uses

$$\begin{bmatrix} 1 & 0 & 0 \\ 0 & 3 & 0 \\ 0 & 0 & 1 \end{bmatrix} \begin{bmatrix} a_{11} & a_{12} & a_{13} \\ a_{21} & a_{22} & a_{23} \\ a_{31} & a_{32} & a_{33} \end{bmatrix} = \begin{bmatrix} a_{11} & a_{12} & a_{13} \\ 3a_{21} & 3a_{22} & 3a_{23} \\ a_{31} & a_{32} & a_{33} \end{bmatrix}.$$

To add two times row 1 to row 3 $(\bar{r}_3 \rightarrow \bar{r}_3 + 2\bar{r}_1)$, one uses

$$\begin{bmatrix} 1 & 0 & 0 \\ 0 & 1 & 0 \\ 2 & 0 & 1 \end{bmatrix} \begin{bmatrix} a_{11} & a_{12} & a_{13} \\ a_{21} & a_{22} & a_{23} \\ a_{31} & a_{32} & a_{33} \end{bmatrix} = \begin{bmatrix} a_{11} & a_{12} & a_{13} \\ a_{21} & a_{22} & a_{23} \\ a_{31} + 2a_{11} & a_{32} + 2a_{12} & a_{33} + 2a_{13} \end{bmatrix}.$$

Another, more simple definition of elementary matrices is that they are $m \times m$ square matrices obtained from the $m \times m$ identity matrix I_m when the corresponding eros are applied to this matrix. The above examples are nice illustrations of this basic definition: when $\bar{r}_1 \leftrightarrow \bar{r}_2$ in $\begin{bmatrix} 1 & 0 & 0 \\ 0 & 1 & 0 \\ 0 & 0 & 1 \end{bmatrix}$, and also, $\bar{r}_2 \rightarrow 3\bar{r}_2$ and $\bar{r}_3 \rightarrow \bar{r}_3 + 2\bar{r}_1$ are performed in the same identity matrix I_3.

Obviously, every elementary matrix has its inverse that corresponds to the inverse ero. In the above examples, the inverse matrices are

the inverse
matrix

$$\begin{bmatrix} 0 & 1 & 0 \\ 1 & 0 & 0 \\ 0 & 0 & 1 \end{bmatrix} \begin{bmatrix} 0 & 1 & 0 \\ 1 & 0 & 0 \\ 0 & 0 & 1 \end{bmatrix} = \begin{bmatrix} 1 & 0 & 0 \\ 0 & 1 & 0 \\ 0 & 0 & 1 \end{bmatrix},$$

the inverse
matrix

$$\begin{bmatrix} 1 & 0 & 0 \\ 0 & 3 & 0 \\ 0 & 0 & 1 \end{bmatrix} \begin{bmatrix} 1 & 0 & 0 \\ 0 & 1/3 & 0 \\ 0 & 0 & 1 \end{bmatrix} = \begin{bmatrix} 1 & 0 & 0 \\ 0 & 1 & 0 \\ 0 & 0 & 1 \end{bmatrix},$$

the inverse
matrix

$$\begin{bmatrix} 1 & 0 & 0 \\ 0 & 1 & 0 \\ 2 & 0 & 1 \end{bmatrix} \begin{bmatrix} 1 & 0 & 0 \\ 0 & 1 & 0 \\ -2 & 0 & 1 \end{bmatrix} = \begin{bmatrix} 1 & 0 & 0 \\ 0 & 1 & 0 \\ 2-2 & 0 & 1 \end{bmatrix} = \begin{bmatrix} 1 & 0 & 0 \\ 0 & 1 & 0 \\ 0 & 0 & 1 \end{bmatrix}.$$

Another method for proving that all elementary matrices have an inverse is to calculate their determinants, which turn out to be always different from zero. In our examples

$$\det \begin{bmatrix} 0 & 1 & 0 \\ 1 & 0 & 0 \\ 0 & 0 & 1 \end{bmatrix} = -1, \quad \det \begin{bmatrix} 1 & 0 & 0 \\ 0 & 3 & 0 \\ 0 & 0 & 1 \end{bmatrix} = 3, \quad \det \begin{bmatrix} 1 & 0 & 0 \\ 0 & 1 & 0 \\ 2 & 0 & 1 \end{bmatrix} = 1.$$

Also, all three rows and all three columns in these matrices are lind, so their rank is 3. Remember, an $m \times m$ matrix has an inverse if its rank is m.

The general form of the elementary matrices can be easily obtained by making use of the standard basis matrices in $\mathbb{R}_{m \times m}$

$$\{ \mathcal{E}_{ij} \mid i, j = 1, 2, \ldots, m \}$$

where the (p,q) element in $\mathcal{E}_{ij}$ is

$$(\mathcal{E}_{ij})_{pq} = \delta_{ip} \delta_{jq}$$

(it has all $m^2 - 1$ entries 0, except $e_{ij} = 1$). Note: $\det I_m = 1$, $\det \mathcal{E}_{ii} = 1$, $\det \mathcal{E}_{ij} = 0$

1. The elementary $m \times m$ matrix F_{ij} for the interchange of the i-th and j-th row $(\bar{r}_i \leftrightarrow \bar{r}_j)$ is

$$F_{ij} = I_m - \mathscr{E}_{ii} + \mathscr{E}_{ji} - \mathscr{E}_{jj} + \mathscr{E}_{ij}.$$

Its determinant is (-1).

2. The elementary $m \times m$ matrix $F_i(\lambda)$ for the multiplication of the i-th row with a non-zero $\lambda \neq 1$ $(\bar{r}_i \rightarrow \lambda \bar{r}_i)$ is

$$F_i(\lambda) = I_m + (\lambda - 1)\mathscr{E}_{ii}.$$

Its determinant is equal to λ.

3. The elementary matrix $F_{ij}(\lambda)$ for adding λ times row j to row i $(\bar{r}_i \rightarrow \bar{r}_i + \lambda \bar{r}_j)$ is

$$F_{ij}(\lambda) = I_m + \lambda \mathscr{E}_{ij}.$$

Its determinant is equal to 1.

At this point, we shall introduce the notion of *row equivalence*. We say that A is row equivalent to B (written $A \sim B$) if there is a sequence of eros that transforms A into B.

It is quite simple to verify that this is a genuine equivalence relation since it is obviously reflexive (every matrix is row equivalent to itself by multiplication of any row by 1); it is naturally symmetric (since if B is obtained from A by a sequence of eros, then A can be obtained from B by the sequence of inverse eros performed in the opposite order); it is also transitive (since if $A \sim B$ and $B \sim C$, then $A \sim C$ by performing on A the set of eros to get B, and then continuing with the set of eros that change B into C).

2.13 The GJ Form of a Matrix

Task 3. A matrix A is in *row-echelon form* if it satisfies the following two conditions:

1. All rows that consist entirely of zeros (zero rows) are grouped together at the bottom of the matrix.
2. The first nonzero number in any nonzero row is a 1 (called *the leading one*, and it appears to the right of the leading 1 in the preceding row).

    As an important consequence of these requirements, the matrix A in row-echelon form has all zeros below each leading 1.
    A matrix A is said to be in GJ *reduced* row-echelon form A_{red} if it also satisfies the third condition:

3. It has only zeros above, as well as below, each leading 1.

    The GJ standard method consists of using elementary row operations (eros) on the augmented matrix $[A \mid \bar{b}]$ of the given linear system $A\mathscr{X} = \bar{b}$ to bring the coefficient matrix A into reduced row-echelon form A_{red}. Then the general solution of the system can be obtained by inspection if the system is consistent, i.e., if all zero rows in A_{red} continue to be zero rows in $[A \,|\bar{b}]_{red}$. It should be emphasized that for linear systems that have no unique solution, some unknowns (those that multiply linearly dependent columns in the canonical expansion of $A\bar{\mathscr{X}}$) must be first declared as free parameters (see Sect. 2.16).

We shall now give a short instruction on how one can achieve reduced row-echelon form of the given matrix:

(A) If the first column of the matrix contains only zeros, we forget about it and go to the second column. We continue in this manner until the left column of the remaining matrix has at least one nonzero number.

(B) We use row interchange, if necessary, to obtain a nonzero number a at the top of that column.

(C) We multiply the first row with $1/a$ in order to obtain the leading 1.

(D) We add suitable multiples of the top row to other rows so that all numbers below the leading 1 become zeros.

(E) Now, we can forget the first row to obtain an even smaller matrix. We go back to step (A) and repeat the procedure with this smaller matrix until the row-echelon form is obtained.

(F) Beginning with the last nonzero row and working upward, we add suitable multiples of each row to the rows above to introduce zeros above the leading 1s.

Let us again consider as an illustration the 4×5 matrix B (the second example in Sect. 2.11):

$$B = \begin{bmatrix} 1 & -1 & 1 & -1 & 4 \\ 1 & 1 & 2 & 3 & 8 \\ 2 & 4 & 5 & 10 & 20 \\ 2 & -4 & 1 & -6 & 4 \end{bmatrix}.$$

Since at the top of column 1 we already have the leading 1, we start with step (D).

$$B = \begin{bmatrix} 1 & -1 & 1 & -1 & 4 \\ 1 & 1 & 2 & 3 & 8 \\ 2 & 4 & 5 & 10 & 20 \\ 2 & -4 & 1 & -6 & 4 \end{bmatrix} \begin{matrix} \\ \bar{r}_2 - \bar{r}_1 \\ \bar{r}_3 - 2\bar{r}_1 \\ \bar{r}_4 - 2\bar{r}_1 \end{matrix} \sim \begin{bmatrix} 1 & -1 & 1 & -1 & 4 \\ 0 & 2 & 1 & 4 & 4 \\ 0 & 6 & 3 & 12 & 12 \\ 0 & -2 & -1 & -4 & -4 \end{bmatrix} \begin{matrix} \\ \frac{1}{2}\bar{r}_2 \\ \bar{r}_3 - 3\bar{r}_2 \\ \bar{r}_4 + \bar{r}_2 \end{matrix} \sim$$

$$\sim \begin{bmatrix} 1 & -1 & 1 & -1 & 4 \\ 0 & 1 & 1/2 & 2 & 2 \\ 0 & 0 & 0 & 0 & 0 \\ 0 & 0 & 0 & 0 & 0 \end{bmatrix} \begin{matrix} \bar{r}_1 + \bar{r}_2 \\ \\ \\ \end{matrix} \sim \begin{bmatrix} 1 & 0 & 3/2 & 1 & 6 \\ 0 & 1 & 1/2 & 2 & 2 \\ 0 & 0 & 0 & 0 & 0 \\ 0 & 0 & 0 & 0 & 0 \end{bmatrix} = B_{red}.$$

It is obvious that performing eros we do not change the row space, which is closed under linear operations like vector addition and scalar multiplication. The final effect is to separate lind rows $\bar{r}_1$ and $\bar{r}_2$ from those that are linearly dependent $\bar{r}_3 = -\bar{r}_1 + 3\bar{r}_2$ and $\bar{r}_4 = 3\bar{r}_1 - \bar{r}_2$, and to change the latter ones into zero rows.

This can be illustrated in detail in the above example. The effect of the performed eros on the third and fourth rows is to transform into zero rows by "undoing" these linear dependences:

$$\bar{r}_3 - 2\bar{r}_1 - 3(\bar{r}_2 - \bar{r}_1) = \bar{r}_3 + \bar{r}_1 - 3\bar{r}_2 = \bar{r}_3 - \bar{r}_3 = \bar{0},$$
$$\bar{r}_4 - 2\bar{r}_1 + (\bar{r}_2 - \bar{r}_1) = \bar{r}_4 - 3\bar{r}_1 + \bar{r}_2 = \bar{r}_4 - \bar{r}_4 = \bar{0}.$$

This is true in every Gauss reduction procedure: linearly dependent rows in the original matrix are transformed by eros into zero rows and as such placed at the bottom of the reduced row-echelon form.

It should be noticed that the column space of B is not preserved by performing these eros. But, the most remarkable property of the G-J reduced row-echelon form is that linear independence and linear dependence of the columns present in the first matrix B is exactly preserved among the columns of the reduced row-echelon form B_{red}: the first two columns are lind, the third one is a linear combination of these two with coefficients $3/2$ and $1/2$. The same is true for columns 4 and 5 (coefficients 1, 2 and 6, 2, respectively). The main difference is that in the reduced form these dependences are exposed explicitly.

2.14 An Example (Preservation of Linear Independence and Dependence in GJ Form)

Let us now consider the 3×4 matrix

$$\begin{bmatrix} 1 & -2 & 1 & 2 \\ 1 & 1 & -1 & 3 \\ 2 & 5 & -2 & 8 \end{bmatrix}.$$

The first three columns are lind, which can be checked by calculating the determinant of the 3×3 matrix, which they form

$$\begin{vmatrix} 1 & -2 & 1 \\ 1 & 1 & -1 \\ 2 & 5 & -2 \end{vmatrix} = -2 + 5 + 4 - 2 - 4 + 5 = 6 \neq 0.$$

(remember: in an $n \times n$ matrix with determinant different from zero all n columns and n rows are lind)

The fourth column must therefore be linearly dependent since such a matrix with three rows cannot possibly have more than three lind columns.

To find this dependence, we proceed in a pedestrian manner:

$$\begin{bmatrix} 2 \\ 3 \\ 8 \end{bmatrix} = x \begin{bmatrix} 1 \\ 1 \\ 2 \end{bmatrix} + y \begin{bmatrix} -2 \\ 1 \\ 5 \end{bmatrix} + z \begin{bmatrix} 1 \\ -1 \\ -2 \end{bmatrix} \Rightarrow \begin{matrix} x - 2y + z = 2 \\ x + y - z = 3 \\ 2x + 5y - 2z = 8 \end{matrix} \Rightarrow \begin{matrix} x = 17/6 \\ y = 2/3 \\ z = 1/2 \end{matrix}.$$

We express this linear dependence explicitly

$$C = \begin{bmatrix} 1 & -2 & 1 \\ 1 & 1 & -1 \\ 1 & -1 & -2 \end{bmatrix} \; 17/6 \begin{bmatrix} 1 \\ 1 \\ 2 \end{bmatrix} + 2/3 \begin{bmatrix} -2 \\ 1 \\ 5 \end{bmatrix} + 1/2 \begin{bmatrix} 1 \\ -1 \\ -2 \end{bmatrix}.$$

We shall now apply eros to obtain the *reduced* row-echelon form of C and observe that the first three columns and the same ones in the expansion of the fourth column $\bar{c}_4$ transform simultaneously: Thus, eros do not alter the formulas (linear combinations) that relate linearly dependent columns and lind ones:

$$\begin{bmatrix} 1 & -2 & 1 \\ 1 & 1 & -1 \\ 2 & 5 & -2 \end{bmatrix} \; 17/6 \begin{bmatrix} 1 \\ 1 \\ 2 \end{bmatrix} + 2/3 \begin{bmatrix} -2 \\ 1 \\ 5 \end{bmatrix} + 1/2 \begin{bmatrix} 1 \\ -1 \\ -2 \end{bmatrix} \quad \begin{matrix} \bar{r}_2 - \bar{r}_1 \\ \bar{r}_3 - 2\bar{r}_1 \end{matrix} \quad \sim$$

$$\sim \begin{bmatrix} 1 & -2 & 1 \\ 0 & 3 & -2 \\ 0 & 9 & -4 \end{bmatrix} \; 17/6 \begin{bmatrix} 1 \\ 0 \\ 0 \end{bmatrix} + 2/3 \begin{bmatrix} -2 \\ 3 \\ 9 \end{bmatrix} + 1/2 \begin{bmatrix} 1 \\ -2 \\ -4 \end{bmatrix} \quad \begin{matrix} \\ \bar{r}_3 - 3\bar{r}_2 \end{matrix} \quad \sim$$

$$\sim \begin{bmatrix} 1 & -2 & 1 \\ 0 & 3 & -2 \\ 0 & 0 & 2 \end{bmatrix} \; 17/6 \begin{bmatrix} 1 \\ 0 \\ 0 \end{bmatrix} + 2/3 \begin{bmatrix} -2 \\ 3 \\ 0 \end{bmatrix} + 1/2 \begin{bmatrix} 1 \\ -2 \\ +2 \end{bmatrix} \quad \begin{matrix} \\ \cdot 1/3 \\ \cdot 1/3 \end{matrix} \quad \sim$$

$$\sim \begin{bmatrix} 1 & -2 & 1 \\ 0 & 1 & -2/3 \\ 0 & 0 & 1 \end{bmatrix} \; 17/6 \begin{bmatrix} 1 \\ 0 \\ 0 \end{bmatrix} + 2/3 \begin{bmatrix} -2 \\ 1 \\ 0 \end{bmatrix} + 1/2 \begin{bmatrix} 1 \\ -2/3 \\ 1 \end{bmatrix} \quad \bar{r}_1 + 2\bar{r}_2 \quad \sim$$

$$\sim \begin{bmatrix} 1 & 0 & -1/3 \\ 0 & 1 & -2/3 \\ 0 & 0 & 1 \end{bmatrix} \; 17/6 \begin{bmatrix} 1 \\ 0 \\ 0 \end{bmatrix} + 2/3 \begin{bmatrix} 0 \\ 1 \\ 0 \end{bmatrix} + 1/2 \begin{bmatrix} -1/3 \\ -2/3 \\ 1 \end{bmatrix} \quad \begin{matrix} \bar{r}_1 + \frac{1}{3}\bar{r}_3 \\ \bar{r}_2 + \frac{2}{3}\bar{r}_3 \end{matrix} \quad \sim$$

$$\sim \begin{bmatrix} 1 & 0 & 0 \\ 0 & 1 & 0 \\ 0 & 0 & 1 \end{bmatrix} \; 17/6 \begin{bmatrix} 1 \\ 0 \\ 0 \end{bmatrix} + 2/3 \begin{bmatrix} 0 \\ 1 \\ 0 \end{bmatrix} + 1/2 \begin{bmatrix} 0 \\ 0 \\ 1 \end{bmatrix} =$$

$$= \begin{bmatrix} 1 & 0 & 0 & 17/6 \\ 0 & 1 & 0 & 2/3 \\ 0 & 0 & 1 & 1/3 \end{bmatrix}.$$

Obviously, this example can be considered as the proof since in any other $m \times n$ matrix the phenomenon will be the same: linear dependence and independence among columns of the first matrix of rank r will be preserved and explicitly demonstrated in its reduced row-echelon form – lind columns will become the first r vectors of the standard basis in $\mathbb{R}^m$, while $(n - r)$ linearly dependent columns will have as their entries the coefficients of their linear dependences, thus showing them explicitly. This is the *fundamental property* of the reduced row-echelon form.

2.15 The Existence of the Reduced Row-Echelon (GJ) Form for Every Matrix

We shall now demonstrate that every matrix can be brought by eros into reduced row-echelon form. Then, the above statement about exact preservation of linear dependence among columns proves that the reduced row-echelon form for every matrix is *unique*.

We actually want to prove that every matrix A can be transformed to a reduced row-echelon form A_{red} by multiplying on the left of A by a finite sequence of elementary matrices or, equivalently, by applying a finite sequence of eros to A. We do so by mathematical induction.

Theorem [on the existence of GJ form] If A is an $m \times n$ matrix, then there are $m \times m$ elementary matrices $F_1, F_2, \ldots, F_q$ such that

$$\boxed{(F_q \, F_{q-1} \, \cdots \, F_2 \, F_1)A = A_{\text{red}}},$$

where A_{red} is a reduced row-echelon (GJ) matrix.

Proof The proof is by induction on the number of rows of A. The plan then is to show that

(a) the theorem is true if A has one row and
(b) if the theorem is true for all matrices having k rows, then it is also true for all matrices having $k + 1$ rows. The principle of mathematical induction will then assure us that the theorem is true.

If A has 1 row, there are two possibilities: either A is the zero matrix, in which case A is already a reduced row-echelon matrix and $F_{11}A = A_{\text{red}}$, or A has a nonzero entry. In this case, let j be the first column (from left to right) of A that has a nonzero entry a_{1j}. Then, $F_1(1/a_{1j})A$ is a reduced row-echelon matrix and the theorem is true for $q = 1$.

Next, we assume that the theorem is true for all matrices having k rows. Let A be a matrix with $k + 1$ rows. If $A = \hat{0}$, then it is a reduced row-echelon matrix. If $A \neq \hat{0}$, then A has at least one nonzero entry. Let t be the first column that has at least one nonzero entry, and let a_{st} be one of nonzero entries in the column t

$$A = \begin{array}{c} \\ \\ \\ \\ s \\ \\ \\ \end{array} \left.\begin{bmatrix} 0 & \cdots & 0 & \overset{t}{*} & \cdots & * \\ 0 & \cdots & 0 & * & \cdots & * \\ \vdots & & \vdots & \vdots & & \vdots \\ 0 & \cdots & 0 & a_{st} & \cdots & * \\ \vdots & & \vdots & \vdots & & \vdots \\ 0 & \cdots & 0 & * & \cdots & * \end{bmatrix}\right\} k+1$$

The matrix $F_{1s}F_s(1/a_{st})A$ has all zeros in each of the first $(t-1)$ columns, and the entry in the first row and t-th column is 1

$$F_2F_1A = F_{1s}F_s(1/a_{st})A = \left.\begin{bmatrix} 0 & \cdots & 0 & \overset{t}{1} & * & \cdots & * \\ 0 & \cdots & 0 & * & * & \cdots & * \\ \vdots & & \vdots & \vdots & \vdots & & \vdots \\ 0 & \cdots & 0 & * & * & \cdots & * \end{bmatrix}\right\} k+1$$

The next step is to sweep out the t-th column by performing appropriate eros of the form $\bar{r}_i \to \bar{r}_j - a_{jt}\bar{r}_1$, $j = 2,3,\ldots,k+1$. The effect of these eros is obtained by multiplying on the left by the corresponding k elementary matrices $F_{j1}(-a_{jt})$, $j = 2,3,\ldots,k+1$. We obtain

$$B = F_{k+2}\,F_{k+1}\,F_k \ldots F_2\,F_1A = \left.\begin{bmatrix} 0 & \cdots & 0 & 1 & * \cdots * \\ 0 & \cdots & 0 & 0 & \\ \vdots & & \vdots & \vdots & \boxed{\quad C \quad} \\ 0 & \cdots & 0 & 0 & \end{bmatrix}\right\} k+1$$

where
$F_1 = F_s(1/a_st)$, $F_2 = F_{1s}$, $F_3 = F_{21}(-a_{2t})$, $F_4 = F_{31}(-a_{3t}), \ldots$, $F_{k+2} = F_{k+1\,1}(-a_{k+1\,t})$.

The matrix C has k rows, and by our induction hypothesis, C can be transformed to reduced row-echelon form by multiplying on the left by the appropriate elementary matrices of order k or, what amounts to the same thing, by performing the appropriate eros on C. Multiplying on the left of B by the elementary matrices of order $(k+1)$ corresponding to the eros that reduce the matrix C, we obtain

$$D = F_p \ldots F_2\,F_1\,A = \begin{bmatrix} 0 & \cdots & 0 & 1 & * \cdots * \\ 0 & \cdots & 0 & 0 & \\ \vdots & & \vdots & \vdots & \boxed{\begin{matrix} F \\ \text{reduced} \\ \text{r-e form} \end{matrix}} \\ 0 & \cdots & 0 & 0 & \end{bmatrix} \rightarrow \text{row-echelon form}$$

Next, the nonzero entries in the first row of D that lie above the leading 1s in the matrix F are changed to zeros by multiplying on the left by appropriate elementary matrices of the form $F_{1i}(-a_{1j})$, where j is the column number and i is the row number in D in which the leading 1 appears. Finally, we have

$$F_q \ldots F_2 F_1 A = A_{\text{red}},$$

a reduced row-echelon matrix. The theorem now follows from the principle of mathematical induction. $\triangle$

Note the reduced row-echelon matrix in the theorem is unique, as we proved previously, since A_{red} exposes explicitly linear independence and dependence of the corresponding columns in A. In other words, if we know the linear independent and dependent columns in A, we can write the unique A_{red} immediately.

An Important Remark If A is an *invertible* (nonsingular) $n \times n$ matrix, then it is equivalent to say that A can be reduced by eros to the identity matrix I_n. The reason for this is that all n columns of the invertible A are *lind*, so its reduced row-echelon form has all 1s on its diagonal.

We can also say that an invertible matrix A is a *product of elementary matrices*, since $F_q \ldots F_2 F_1 A = I_n$ implies $F_q \ldots F_2 F_1 = A^{-1}$ or

$$A = (F_q \ldots F_2 F_1)^{-1} = F_1^{-1} F_2^{-1} \ldots F_q^{-1}.$$

2.16 The Standard Method for Solving $A\bar{\mathscr{X}} = \bar{b}$

This standard method is called the GJ method, and it consists in bringing the corresponding *augmented matrix* $[A \mid \bar{b}]$ [which has m rows and $(n+1)$ columns] to reduced row-echelon form $[A_{\text{red}} \mid \bar{b}_{\text{red}}]$. Some solutions exist (we say that the system $A\bar{\mathscr{X}} = \bar{b}$ is consistent) if and only if $\bar{b} \in \text{ran } A$, which implies that $\bar{b}$ has its inverse image that is a subset (called a coset) in $\mathbb{R}^n$ as the general solution of the system.

A more sophisticated way to express the same condition is to say that the ranks of A and $[A \mid \bar{b}]$ must be the same, meaning again that $\bar{b}$ is linearly dependent on the columns of A that span ran A (the Kronecker–Capelli theorem). Practically, consistent systems are immediately recognizable if A_{red} and $[A_{\text{red}} \mid \bar{b}_{\text{red}}]$ have the same number of nonzero rows, i.e., the same number of lind rows, in fact *the same rank*. (Remember that A and A_{red} have always the same rank, since eros do not change the number of lind rows.)

On the other hand, for inconsistent systems, the rank of the $m \times n$ matrix A is r (r lind columns), then A_{red} has r nonzero rows. But since $\bar{b} \in \text{ran } A$, then $[A_{\text{red}} \mid \bar{b}_{\text{red}}]$ has $(r+1)$ nonzero rows since $\bar{b}$ is also lind and the rank of $[A \mid \bar{b}]$ is $(r+1)$.

Example of a 4×4 inconsistent system (a system with no solution):

$$\left.\begin{array}{l} x_1 - 2x_2 - 3x_3 + 2x_4 = -4 \\ -3x_1 + 7x_2 - x_3 + x_4 = -3 \\ 2x_1 - 5x_2 + 4x_3 - 3x_4 = 7 \\ -3x_1 + 6x_2 + 9x_3 - 6x_4 = -1 \end{array}\right\} \Rightarrow \left[\begin{array}{cccc|c} 1 & -2 & -3 & 2 & -4 \\ -3 & 7 & -1 & 1 & -3 \\ 2 & -5 & 4 & -3 & 7 \\ -3 & 6 & 9 & -6 & -1 \end{array}\right] \begin{array}{l} \bar{r}_2 + 3\bar{r}_1 \\ \bar{r}_3 - 2\bar{r}_1 \\ \bar{r}_4 + 3\bar{r}_1 \end{array} \sim$$

$$\sim \left[\begin{array}{cccc|c} 1 & -2 & -3 & 2 & -4 \\ 0 & 1 & -10 & 7 & -15 \\ 0 & -1 & 10 & -7 & 15 \\ 0 & 0 & 0 & 0 & -13 \end{array}\right] \begin{array}{l} \\ \\ \bar{r}_3 + \bar{r}_2 \\ :(-13) \end{array} \sim \left[\begin{array}{cccc|c} 1 & -2 & -3 & 2 & -4 \\ 0 & 1 & -10 & 7 & -15 \\ 0 & 0 & 0 & 0 & 0 \\ 0 & 0 & 0 & 0 & 1 \end{array}\right] \begin{array}{l} \bar{r}_1 + 2\bar{r}_2 \\ \\ \end{array} \sim$$

$$\sim \left[\begin{array}{cccc|c} 1 & 0 & -23 & 16 & -34 \\ 0 & 1 & -10 & 7 & -15 \\ 0 & 0 & 0 & 0 & 1 \\ 0 & 0 & 0 & 0 & 0 \end{array}\right] \Rightarrow$$

(the reduced form of A has two nonzero rows, whereas that of $[A \mid \bar{b}]$ has 3)

$$\begin{array}{rl} x_1 & - 23x_3 + 16x_4 = -34 \\ x_2 & - 10x_3 + 7x_4 = -15. \\ 0 \cdot x_1 + 0 \cdot x_2 & + 0 \cdot x_3 + 0 \cdot x_4 = 1 \end{array}$$

This system has no solution, since there are no x_1, x_2, x_3, x_4 that would satisfy the last equation.

One can see why the system is inconsistent, since the l.h.s. of the fourth equation is (-3) times the l.h.s. of the first one, while the rhs of the fourth equation is (-1) instead of 12, so that the first and fourth equations are in *contradiction*.

Obviously, the column $\bar{b}$ is not linearly dependent on the columns $\bar{c}_1, \bar{c}_2, \bar{c}_3, \bar{c}_4$ that span ran A (meaning $\bar{b} \notin$ ran A), since an attempt to verify this: $\bar{b} = x_1 \bar{c}_1 + x_2 \bar{c}_2 + x_3 \bar{c}_3 + x_4 \bar{c}_4$ is precisely our original system, which, as we have already seen, has no solution for the expansion coefficients x_1, x_2, x_3, x_4. Δ

2.16.1 When Does a Consistent System $A\bar{\mathscr{X}} = \bar{b}$ Have a Unique Solution?

When the system $A\bar{\mathscr{X}} = \bar{b}$, $A : \mathbb{R}^n \to \mathbb{R}^m$ is consistent, the question arises whether the general solution is only one vector in $\mathbb{R}^n$ (the so-called *unique* solution). This, in other words, means that the coset, which is the inverse image of $\bar{b}$, has only one vector, so that ker A is trivial: ker $A = \{\bar{0}_n\}$. From the dimension theorem $\dim(\ker A) + \dim(\text{ran } A) = \dim(\text{Dom } A) = n$, it follows that the kernel is trivial $[\dim(\ker A) = 0]$ when $r = $ rank of $A = \dim(\text{ran } A) = n$, i.e., when all n columns of A are lind $(r = n)$. This implies that the $m \times n$ reduced matrix A_{red} must have exactly n nonzero rows, implying $m \geq n$.

If $m > n$, we have $(m - n)$ zero rows in A_{red}, so that $(m - n)$ equations are linearly dependent, they carry no new information and can be ignored. In conclusion, a unique solution of $A\bar{\mathscr{X}} = \bar{b}$ (A is an $m \times n$ matrix with $m > n$) exists if all n columns are lind $(r = n)$, also n rows are lind, and $(m - n)$ rows are linearly dependent. In this case, A_{red} has at its top the $n \times n$ identity matrix I_n and the unique solution is the top n entries in $\bar{b}_{\mathrm{red}}$.

Example

$$
\left.\begin{array}{r}
x_1 - x_2 = 4 \\
x_1 + x_2 = 8 \\
2x_1 + 4x_2 = 20 \\
2x_1 - 4x_2 = 4
\end{array}\right\} \Rightarrow
\left[\begin{array}{rr|r}
1 & -1 & 4 \\
1 & 1 & 8 \\
2 & 4 & 20 \\
2 & -4 & 4
\end{array}\right]
\begin{array}{l}
\\ \bar{r}_2 - \bar{r}_1 \\ \bar{r}_3 - 2\bar{r}_1 \\ \bar{r}_4 - 2\bar{r}_1
\end{array} \sim
$$

$$
\sim
\left[\begin{array}{rr|r}
1 & -1 & 4 \\
0 & 2 & 4 \\
0 & 6 & 12 \\
0 & -2 & -4
\end{array}\right]
\begin{array}{l}
\\ \frac{1}{2}\bar{r}_2 \\ \bar{r}_3 - 3\bar{r}_2 \\ \bar{r}_4 + \bar{r}_2
\end{array} \sim
\left[\begin{array}{rr|r}
1 & -1 & 4 \\
0 & 1 & 2 \\
0 & 0 & 0 \\
0 & 0 & 0
\end{array}\right]
\begin{array}{l}
\bar{r}_1 + \bar{r}_2 \\ \\ \\
\end{array} \sim
$$

$$
\sim
\left[\begin{array}{rr|r}
1 & 0 & 6 \\
0 & 1 & 2 \\
0 & 0 & 0 \\
0 & 0 & 0
\end{array}\right]
\Rightarrow x_1 = 6,\ x_2 = 2.
$$

$m = 4,\ n = 2,\ \bar{c}_1$ and $\bar{c}_2$ are lind, since $a\bar{c}_1 + b\bar{c}_2 = \bar{0}_4 \Rightarrow a = b = 0$

$$
a\begin{bmatrix} 1 \\ 1 \\ 2 \\ 2 \end{bmatrix} + b\begin{bmatrix} -1 \\ 1 \\ 4 \\ -4 \end{bmatrix} = \begin{bmatrix} 0 \\ 0 \\ 0 \\ 0 \end{bmatrix}
\Rightarrow
\left.\begin{array}{r}
a - b = 0 \\
a + b = 0 \\
2a + 4b = 0 \\
2a - 4b = 0
\end{array}\right\}
a = 0,\quad b = 0.
$$

and $\bar{b}$ is linearly dependent on $\bar{c}_1$ and $\bar{c}_2$ with coefficients 6 and 2:

$$
6\begin{bmatrix} 1 \\ 1 \\ 2 \\ 2 \end{bmatrix} + 2\begin{bmatrix} -1 \\ 1 \\ 4 \\ -4 \end{bmatrix} = \begin{bmatrix} 6 \\ 6 \\ 12 \\ 12 \end{bmatrix} + \begin{bmatrix} -2 \\ 2 \\ 8 \\ -8 \end{bmatrix} = \begin{bmatrix} 4 \\ 8 \\ 20 \\ 4 \end{bmatrix} \quad \Delta
$$

If $m = n$, i.e., if the coefficient matrix A is square, then there are more efficient criteria for the unique solution of $A\bar{\mathscr{X}} = \bar{b}$. In this case, the $n \times n$ matrix A is a linmap $A : \mathbb{R}^n \to \mathbb{R}^n$. Since the uniqueness of the solution requires $\dim \ker A = 0 \Rightarrow \dim \mathrm{ran}\, A = n = \dim \mathbb{R}^n \Rightarrow \mathrm{ran}\, A = \mathbb{R}^n$, the map is obviously an onto linmap (the range $\mathrm{ran}\, A$ is equal to the co-domain $\mathbb{R}^n$). This uniqueness of the solution also means that A is a 1-1 linmap. In conclusion, A is a bijection. A linmap $A : \mathbb{R}^n \to \mathbb{R}^n$, which is also a bijection, is usually called an *automorphism* in $\mathbb{R}^n$.

This is equivalent to saying that the inverse matrix A^{-1} exists. For A^{-1} to exist, a necessary and sufficient condition is that the determinant of A is not zero: $\det A \neq 0$. A unique solution exists for every $\bar{b} \in \mathbb{R}^n$, since $\mathbb{R}^n = \operatorname{ran} A$.

All these conditions can be summarized:

Theorem (on the unique solution of $A\bar{\mathscr{X}} = \bar{b}$) A linear system $A\bar{\mathscr{X}} = \bar{b}$, $A : \mathbb{R}^n \to \mathbb{R}^n$, is consistent and has a unique solution for every $\bar{b} \in \mathbb{R}^n$ iff any of the following five equivalent conditions are satisfied:

1) The inverse matrix A^{-1} exists, so that $AA^{-1} = A^{-1}A = I_n$;
2) The determinant of A is different from zero, $\det A \neq 0$;
3) The reduced row-echelon form of $[A \mid \bar{b}]$ is $[I_n \mid \bar{b}_{\text{red}}]$, where $\bar{b}_{\text{red}}$ is that unique solution, $A\bar{b}_{\text{red}} = \bar{b}$;
4) All n columns and all n rows of A are lind;
5) The rank r of $A = \dim(\operatorname{ran} A) = n$, which is equivalent to the fact that the defect of $A = \dim(\ker A) = 0$. Δ

Example 3×3 linear system

$$-2x_1 + 3x_2 - x_3 = 1$$
$$x_1 + 2x_2 - x_3 = 4$$
$$-2x_1 - x_2 + x_3 = -3$$

As we have seen, there are several different approaches to find the solution when one is dealing with linear systems that have $n \times n$ coefficient matrices. These methods are basically only two: either to calculate A^{-1} (if it exists) or to apply the GJ reduction method. The latter is always the most straightforward and gives a direct answer as to whether the system is consistent and also what is the general solution (is it unique or with infinitely many solutions). Consequently, we shall first use the GJ method and then calculate A^{-1} (if GJ shows that there is a unique solution).

(A) *GJ method*

$$B = [A' \mid \bar{b}] = \begin{bmatrix} -2 & 3 & -1 & 1 \\ 1 & 2 & -1 & 4 \\ -2 & -1 & 1 & -3 \end{bmatrix} \sim \begin{bmatrix} 1 & 2 & -1 & 4 \\ -2 & 3 & -1 & 1 \\ -2 & -1 & 1 & -3 \end{bmatrix} \begin{matrix} \\ \bar{r}_2 + 2\bar{r}_1 \\ \bar{r}_3 + 2\bar{r}_1 \end{matrix} \sim$$

$$\sim \begin{bmatrix} 1 & 2 & -1 & 4 \\ 0 & 7 & -3 & 9 \\ 0 & 3 & -1 & 5 \end{bmatrix} \begin{matrix} \\ \frac{1}{7}\bar{r}_2 \\ \bar{r}_3 - \frac{3}{7}\bar{r}_2 \end{matrix} \sim \begin{bmatrix} 1 & 2 & -1 & 4 \\ 0 & 1 & -3/7 & 9/7 \\ 0 & 0 & 2/7 & 8/7 \end{bmatrix} \frac{7}{2}\bar{r}_3 \sim$$

$$\sim \begin{bmatrix} 1 & 2 & -1 & 4 \\ 0 & -1 & -3/7 & 9/7 \\ 0 & 0 & 1 & 4 \end{bmatrix} \begin{matrix} \bar{r}_1 + \bar{r}_3 \\ \bar{r}_2 + \frac{3}{7}\bar{r}_3 \end{matrix} \sim \begin{bmatrix} 1 & 2 & 0 & 8 \\ 0 & 1 & 0 & 3 \\ 0 & 0 & 1 & 4 \end{bmatrix} \bar{r}_1 - 2\bar{r}_2 \sim$$

$$\sim \begin{bmatrix} 1 & 0 & 0 & | & 2 \\ 0 & 1 & 0 & | & 3 \\ 0 & 0 & 1 & | & 4 \end{bmatrix} = [I_3 \, | \, \bar{b}_{\text{red}}].$$

Obviously, all columns of A are lind, since so are the columns in $A_{\text{red}} = I_3$. So, $r = 3$. The coefficients of $\bar{b}_{\text{red}}$ show explicitly how $\bar{b}$ is linearly dependent on the three lind columns of $A : \bar{b} = 2\bar{c}_1 + 3\bar{c}_2 + 4\bar{c}_3 = A\bar{b}_{\text{red}}$. So, $x_1 = 2, x_2 = 3, x_3 = 4$ is the unique solution of the system.

Remark 1: The vectors $\begin{bmatrix} -2 \\ 1 \\ -2 \end{bmatrix}$, $\begin{bmatrix} 3 \\ 2 \\ -1 \end{bmatrix}$, $\begin{bmatrix} -1 \\ -1 \\ 1 \end{bmatrix}$ (the columns of A) in the ex-ample obviously form a basis in $\mathbb{R}^3$ (they are lind), so this example illustrates another application of the GJ reduction method: how to find the components of a given column vector $\bar{b} \in \mathbb{R}^n$ in a basis $\{\bar{v}_1, \bar{v}_2, \ldots, \bar{v}_n\}$ of $\mathbb{R}^n$ [the representation of the column $\bar{b}$ (which is given in the standard basis in $\mathbb{R}^n$) by a column in another basis]. The solution is simply to form the augmented matrix $B = [\bar{v}_1 \, \bar{v}_2, \ldots \bar{v}_n \, | \, \bar{b}]$ and to apply the GJ method to obtain $B_{\text{red}} = [I_n \, | \, \bar{b}_{\text{red}}]$. The column $\bar{b}_{\text{red}}$ consists of the components of $\bar{b}$ in the given basis:

$$\bar{b} = b'_1 \bar{v}_1 + b'_2 \bar{v}_2 + \ldots + b'_n \bar{v}_n.$$

(Note that I_n appears in B_{red} since the $n \times n$ matrix $[\bar{v}_1 \, \bar{v}_2 \ldots \bar{v}_n]$ has all of its columns linearly independent).

(B) *Calculating* A^{-1}

For this approach, one must first verify that

$$\det A \neq 0.$$

This is because one has to be sure that all the columns of A are lind (i.e., that $r = 3$). To see this connection, one writes the lind condition for columns $a\bar{c}_1 + b\bar{c}_2 + c\bar{c}_3 = \bar{0}_3$, which turns out to be a homogeneous linear system with unknowns a, b, c:

$$a\begin{bmatrix} -2 \\ 1 \\ -2 \end{bmatrix} + b\begin{bmatrix} 3 \\ 2 \\ -1 \end{bmatrix} + c\begin{bmatrix} -1 \\ -1 \\ 1 \end{bmatrix} = \begin{bmatrix} 0 \\ 0 \\ 0 \end{bmatrix} \Rightarrow \begin{array}{c} -2a + 3b - c = 0 \\ a + 2b - c = 0 \\ -2a - b + c = 0 \end{array}.$$

Such a system has only the trivial solution $a = b = c = 0$ (the columns are lind) if and only if $\det A \neq 0$. So

$$\det A = \begin{vmatrix} -2 & 3 & -1 \\ 1 & 2 & -1 \\ -2 & -1 & 1 \end{vmatrix} = -2 \neq 0 \Rightarrow \text{all columns are lind} \Rightarrow r = 3.$$

One can see immediately that this homogeneous system for a, b, c is just the homogeneous version of our original system. So, the condition $(\det A \neq 0)$ that the homogeneous system has a (unique) trivial solution is naturally the same as the condition that any nonhomogeneous system with the same A has a unique solution, since the homogeneous system is just the special case when $\bar{b} = \bar{0}_3$.

The basic formula for A^{-1} is

$$A^{-1} = \frac{\text{adj}A}{\det A},$$

where adj A is the transposed matrix of cofactors of A (see the Appendix A.2(4) to Chap. 1):

$$A^{-1} = \frac{1}{-2}\left[\begin{array}{ccc} \begin{vmatrix} 2 & -1 \\ -1 & 1 \end{vmatrix} & -\begin{vmatrix} 1 & -1 \\ -2 & 1 \end{vmatrix} & \begin{vmatrix} 1 & 2 \\ -2 & -1 \end{vmatrix} \\ -\begin{vmatrix} 3 & -1 \\ -1 & 1 \end{vmatrix} & \begin{vmatrix} -2 & -1 \\ -2 & 1 \end{vmatrix} & -\begin{vmatrix} -2 & 3 \\ -2 & -1 \end{vmatrix} \\ \begin{vmatrix} 3 & -1 \\ 2 & -1 \end{vmatrix} & -\begin{vmatrix} -2 & -1 \\ 1 & -1 \end{vmatrix} & \begin{vmatrix} -2 & 3 \\ 1 & 2 \end{vmatrix} \end{array}\right]^T$$

$$= \frac{1}{-2}\begin{bmatrix} 1 & 1 & 3 \\ -2 & -4 & -8 \\ -1 & -3 & -7 \end{bmatrix}^T = \frac{1}{2}\begin{bmatrix} -1 & 2 & 1 \\ -1 & 4 & 3 \\ -3 & 8 & 7 \end{bmatrix}.$$

Verification of this result consists in showing $AA^{-1} = I_3$ (there is no need to show $A^{-1}A = I_3$, since it is implied by the former verification).

$$\frac{1}{2}\begin{bmatrix} -2 & 3 & -1 \\ 1 & 2 & -1 \\ -2 & -1 & 1 \end{bmatrix}\begin{bmatrix} -1 & 2 & 1 \\ -1 & 4 & 3 \\ -3 & 8 & 7 \end{bmatrix} = \frac{1}{2}\begin{bmatrix} 2-3+3 & -4+12-8 & -2+9-7 \\ -1-2+3 & 2+8-8 & 1+6-7 \\ 2+1-3 & -4-4+8 & -2-3+7 \end{bmatrix} =$$

$$= \frac{1}{2}\begin{bmatrix} 2 & 0 & 0 \\ 0 & 2 & 0 \\ 0 & 0 & 2 \end{bmatrix} = \begin{bmatrix} 1 & 0 & 0 \\ 0 & 1 & 0 \\ 0 & 0 & 1 \end{bmatrix}.$$

Now, we are sure that $A^{-1}\bar{b} = \bar{b}_{\text{red}}$ (the unique solution).

$$A^{-1}\bar{b} = \frac{1}{2}\begin{bmatrix} -1 & 2 & 1 \\ -1 & 4 & 3 \\ -3 & 8 & 7 \end{bmatrix}\begin{bmatrix} 1 \\ 4 \\ -3 \end{bmatrix} = \frac{1}{2}\begin{bmatrix} -1+8-3 \\ -1+16-9 \\ -3+32-21 \end{bmatrix} = \frac{1}{2}\begin{bmatrix} 4 \\ 6 \\ 8 \end{bmatrix} = \begin{bmatrix} 2 \\ 3 \\ 4 \end{bmatrix} = \bar{b}_{\text{red}},$$

which agrees with the result of the GJ method.

Remark 2: There are two more methods for calculating A^{-1} (as a matter of fact, the first of them gives directly $A^{-1}\bar{b}$).

(A) The first method is known as *Cramer's rule*, and it is a method that separates the basic expression $A^{-1}\bar{b} = \bar{b}_{red}$ into components of $\bar{b}_{red}$.

Let us derive Cramer's rule for a 3×3 coefficient matrix $A = \begin{bmatrix} a_{11} & a_{12} & a_{13} \\ a_{21} & a_{22} & a_{23} \\ a_{31} & a_{32} & a_{33} \end{bmatrix}$:

$$\bar{b}_{red} = \begin{bmatrix} b'_1 \\ b'_2 \\ b'_3 \end{bmatrix} = \frac{1}{\det A} \begin{bmatrix} c_{11} & c_{21} & c_{31} \\ c_{12} & c_{22} & c_{32} \\ c_{13} & c_{23} & c_{33} \end{bmatrix} \begin{bmatrix} b_1 \\ b_2 \\ b_3 \end{bmatrix}, \quad \text{where } c_{ij}, \ i,j = 1,2,3, \text{ are the cofactors of } a_{ij}.$$

This explicit expression implies the following expansion for

$$b'_1 = \frac{1}{\det A}(b_1 c_{11} + b_2 c_{21} + b_3 c_{31}) = \frac{\det \begin{bmatrix} b_1 & a_{12} & a_{13} \\ b_2 & a_{22} & a_{23} \\ b_3 & a_{32} & a_{33} \end{bmatrix}}{\det \begin{bmatrix} a_{11} & a_{12} & a_{13} \\ a_{21} & a_{22} & a_{23} \\ a_{31} & a_{32} & a_{33} \end{bmatrix}} = \frac{\det A'}{\det A},$$

since $(b_1 c_{11} + b_2 c_{21} + b_3 c_{31})$ is precisely the expansion of the upper determinant in terms of elements of its first column $\bar{b}$.

Similarly, $b'_2 = \frac{\det A''}{\det A}, b'_3 = \frac{\det A'''}{\det A}$, where in the matrices A'' and A''' the second and third columns of A, respectively, are replaced by $\bar{b}$.

So we do not get A^{-1}, but application of A^{-1} to the free vector $\bar{b}$. This method is very complicated even for 4×4 matrices, and for larger ones it is hardly applicable.

In our case,

$$b'_1 = \frac{\det A'}{\det A} = \frac{\begin{vmatrix} 1 & 3 & -1 \\ 4 & 2 & -1 \\ -3 & -1 & 1 \end{vmatrix}}{\begin{vmatrix} -2 & 3 & -1 \\ 1 & 2 & -1 \\ -2 & -1 & 1 \end{vmatrix}} = \frac{-4}{-2} = 2;$$

$$b'_2 = \frac{\det A''}{\det A} = \frac{-6}{-2} = 3, \ b'_3 = \frac{\det A'''}{\det A} = \frac{-8}{-2} = 4, \ \text{or } A^{-1}\bar{b} = \bar{b}' = \begin{bmatrix} 2 \\ 3 \\ 4 \end{bmatrix}.$$

(B) The second method for calculating A^{-1} directly consists in making use of eros. Consider any sequence of eros that transforms A into I_n. Then the same sequence of eros transforms I_n into A^{-1}. Namely, this sequence of eros has the same effect as premultiplication by a certain nonsingular matrix Q (which is the product $F_q \ldots F_2 F_1$ of the corresponding elementary matrices): $QA = I_n$. But this immediately implies that Q is the inverse of A, i.e., $Q = A^{-1}$ or $QI_n = A^{-1}$. So, we write a matrix

$$C = [A \mid I_3] = \begin{bmatrix} -2 & 3 & -1 & 1 & 0 & 0 \\ 1 & 2 & -1 & 0 & 1 & 0 \\ -2 & -1 & 1 & 0 & 0 & 1 \end{bmatrix}$$

and apply a sequence of eros to both blocks, which brings A to I_n (we have already done that) and I_n to A^{-1}:

$$C_{\text{red}} = \begin{bmatrix} 1 & 0 & 0 & -1/2 & 1 & 1/2 \\ 0 & 1 & 0 & -1/2 & 2 & 3/2 \\ 0 & 0 & 1 & -3/2 & 4 & 7/2 \end{bmatrix} = [I_3 \mid A^{-1}] = [QA \mid QI_3].$$

This result enables us to see the GJ method (for square matrices with determinant different from zero) in a concise from:

$$B = [A \mid \bar{b}] = [A \mid I_n \bar{b}] \quad \text{and}$$
$$B_{\text{red}} = [QA \mid QI_n \bar{b}] = [I_n \mid A^{-1}\bar{b}] = [I_n \mid \bar{b}_{\text{red}}].$$

So, if we want only $\bar{b}_{\text{red}}$ there is no need to calculate A^{-1} and then to apply it to $\bar{b}$, since the GJ method does it at one go.

2.16.2 When a Consistent System $A\bar{\mathcal{X}} = \bar{b}$ Has No Unique Solution

If the rank of an $m \times n$ matrix A is $r < n$, this means that $(n - r)$ columns of A are linearly dependent, implying that $(n - r)$ nonbasic unknowns (that multiply these columns in the canonical expansion $A\mathcal{X} = x_1\bar{c}_1 + x_2\bar{c}_2 + \cdots + x_n\bar{c}_n = \bar{b}$) have no specific solution and must be declared as *free parameters*. So, for $r < n$, there is no unique solution and this nonuniqueness consists in the choice of these free parameters.

Example Let us again consider the 4×4 linear system (Sect. 2.11, example 2)

$$\left.\begin{array}{l} x_1 - x_2 + x_3 - x_4 = 4 \\ x_1 + x_2 + 2x_3 + 3x_4 = 8 \\ 2x_1 + 4x_2 + 5x_3 + 10x_4 = 20 \\ 2x_1 - 4x_2 + x_3 - 6x_4 = 4 \end{array}\right\} \Rightarrow x_1 \begin{bmatrix} 1 \\ 1 \\ 2 \\ 2 \end{bmatrix} + x_2 \begin{bmatrix} -1 \\ 1 \\ 4 \\ -4 \end{bmatrix} + x_3 \begin{bmatrix} 1 \\ 2 \\ 5 \\ 1 \end{bmatrix} + x_4 \begin{bmatrix} -1 \\ 3 \\ 10 \\ -6 \end{bmatrix} = \begin{bmatrix} 4 \\ 8 \\ 20 \\ 4 \end{bmatrix}.$$

Its augmented matrix $B = [A \mid \bar{b}] = \begin{bmatrix} 1 & -1 & 1 & -1 & 4 \\ 1 & 1 & 2 & 3 & 8 \\ 2 & 4 & 5 & 10 & 20 \\ 2 & -4 & 1 & -6 & 4 \end{bmatrix}$ has the reduced row-echelon form

$$B_{\text{red}} = [A_{\text{red}} \mid \bar{b}_{\text{red}}] = \begin{bmatrix} 1 & 0 & \frac{3}{2} & 1 & 6 \\ 0 & 1 & \frac{1}{2} & 2 & 2 \\ 0 & 0 & 0 & 0 & 0 \\ 0 & 0 & 0 & 0 & 0 \end{bmatrix}$$

which shows that the first two columns $\bar{c}_1, \bar{c}_2$ in A are lind, so that the corresponding unknowns x_1 and x_2 are the *basic ones*. The original system is reduced to the simplest equivalent one

$$\begin{aligned} x_1 \quad + \tfrac{3}{2}x_3 + x_4 &= 6 \\ x_2 + \tfrac{1}{2}x_3 + 2x_4 &= 2. \end{aligned}$$

We must declare the two nonbasic unknowns x_3 and x_4, which are related to linearly dependent columns $\bar{c}_3$ and $\bar{c}_4$ in A

$$\begin{aligned} \bar{c}_3 &= \frac{3}{2}\bar{c}_1 + \frac{1}{2}\bar{c}_2 \\ \bar{c}_4 &= \bar{c}_1 + 2\bar{c}_2, \end{aligned}$$

as *free parameters* $s, t \in \mathbb{R}$, so that the general solution reads as

$$\begin{aligned} x_1 &= 6 - \frac{3}{2}s - t \\ x_2 &= 2 - \frac{1}{2}s - 2t \\ x_3 &= s \\ x_4 &= t, \quad s, t, \in \mathbb{R}. \end{aligned}$$

This general solution differs from the coset form solution (which exploits the canonical expansions of all linearly dependent columns in B) by the irrelevant replacement of $-s$ by s and $-t$ by t.

Obviously, when there are fewer equations than unknowns ($m < n$), the system cannot have a unique solution, since we have always $r \leq m$, so that with $m < n$ it leads to $r < n$. But a nonunique solution ($r < n$) can also occur if $m \geq n$.

Summary for consistent systems $\bar{b} \in \text{ran } A$

unique solution exists	no unique solution
	$m < n \Rightarrow r < n$
$m = n$ and $r = n = m$	$m = n$ and $r < n$
$m > n$ and $r = n < m$	$m > n$ and $r < n$.

2.17 The GJM Procedure – a New Approach to Solving Linear Systems with Nonunique Solutions

Task 4. The basic idea in this approach is to modify slightly the GJ approach, i.e., to modify the reduced form B_{red} of $B = [A \mid \bar{b}]$, where B is the augmented matrix of the linear system $A\bar{\mathscr{X}} = \bar{b}$ under consideration, in order to get the canonical expansion coefficients of $\bar{b}$ and also the modified canonical expansion coefficients of all linearly dependent columns in A.

In doing this, we get the unique coset representative $\bar{b}'$ and also the unique basis vectors $\{\bar{k}_1, \bar{k}_2, \ldots, \bar{k}_{n-r}\}$ of ker A, so that we have the general solution of the consistent ($\bar{b} \in$ ran A) system $A\bar{\mathscr{X}} = \bar{b}$ in the coset form:

$$\boxed{\text{Inv}(\bar{b}) = \bar{b}' + \ker A = \bar{b}' + \text{LIN}(\bar{k}_1, \bar{k}_2, \ldots, \bar{k}_{n-r})}.$$

2.17.1 Detailed Explanation

(A) *About* $\bar{b}'$

Remember that the unique canonical expansion of $\bar{b} \in$ ran A (dim ran $A = r$, the rank of A)

$$\bar{b} = b_1'\bar{c}_1 + b_2'\bar{c}_2 + \cdots + b_r'\bar{c}_r + 0 \cdot \bar{c}_{r+1} + \cdots + 0 \cdot \bar{c}_n$$

provides the vector $\bar{b}' \in \mathbb{R}^n$

$$\bar{b}' = [b_1' \, b_2' \ldots b_r' \, 0 \ldots 0]^T,$$

which is the very special preimage of $\bar{b}$

$$\boxed{A\bar{b}' = \bar{b}}.$$

In short, the matrix A maps the canonical expansion $\bar{b}'$ of $\bar{b}$ in terms of all columns of A [since $\bar{b} \in$ ran $A = \text{LIN}(\bar{c}_1, \bar{c}_2, \ldots, \bar{c}_n)$] onto its representation in the standard basis in $\mathbb{R}^m$: $\bar{b} = b_1\bar{e}_1 + b_2\bar{e}_2 + \ldots + b_m\bar{e}_m = [b_1 \, b_2 \ldots b_m]^T$, when $\bar{b}$ is considered as a vector in the wider space, codomain.

When ker A is trivial, ker $A = \{\bar{0}_n\}$, then $\bar{b}'$ is the unique solution of the system. This is equivalent to $r = n$, i.e., all the columns of A are lind and they form the natural basis in ran A, so that the canonical expansion of $\bar{b}$ is its unique expansion in that basis. In this case, there is an isomorphism between the domain $\mathbb{R}^n$ and the range ran A :

$$\boxed{\mathbb{R}^n \cong \text{ran } A}.$$

When ker A is not trivial ($r < n$), then $\bar{b}'$ is the unique coset representative. In fact, all the vectors from the coset $\bar{b}$ + ker A are preimages of $\bar{b}$, but only $\bar{b}'$ is entirely determined by $\bar{b}'$ itself.

(B) *About* ker A

The linearly dependent columns $\bar{c}_{r+1}, \bar{c}_{r+2}, \ldots, \bar{c}_n$ of A can also be canonically expanded in terms of all columns

$$\bar{c}_i = a_1^{(i)}\bar{c}_1 + a_2^{(i)}\bar{c}_2 + \cdots + a_r^{(i)}\bar{c}_r + 0\cdot\bar{c}_{r+1} + \cdots + 0\cdot\bar{c}_n, \ i = r+1, r+2, \ldots, n.$$

(Note: the coefficients $a_1^{(i)}, a_2^{(i)}, \ldots, a_r^{(i)}$ are the unique expansion coefficients in the basis of ran A, which make lind columns $\{\bar{c}_1, \bar{c}_2, \ldots, \bar{c}_r\}$).

Modifying the above expansions simply by taking $\bar{c}_i$ across the equality sign we get

$$\bar{0}_m = a_1^{(i)}\bar{c}_1 + a_2^{(i)}\bar{c}_2 + \cdots + a_r^{(i)}\bar{c}_r + 0\cdot\bar{c}_{r+1} + \cdots + (-1)\bar{c}_i + \cdots + 0\cdot\bar{c}_n,$$
$$i = r+1, r+2, \ldots, n = r+(n-r)$$

Making use of the canonical expansion $A\begin{bmatrix} x_1 \\ x_2 \\ \vdots \\ x_n \end{bmatrix} = x_1\bar{c}_1 + x_2\bar{c}_2 + \cdots + x_n\bar{c}_n$, we

see that the coefficients in these modified expansions form $(n-r)$ vectors in $\mathbb{R}^n$ $\{\bar{k}_1, \bar{k}_2, \ldots, \bar{k}_{n-r}\}$ that all belong to ker A:

$$A\bar{k}_j = A \begin{bmatrix} a_1^{(r+j)} \\ a_2^{(r+j)} \\ \vdots \\ a_r^{(r+j)} \\ \vdots \\ (-1) \\ \vdots \\ 0 \end{bmatrix}_{n\times 1} = \bar{0}_m, \qquad (r+j) \text{ position}$$

The number of these vectors is just the dimension of ker A (the dimension theorem: $\dim\ker A = \dim\mathbb{R}^n - \dim\operatorname{ran} A = n - r$). They are lind, which can be immediately seen since they all have in the last $(n-r)$ entries only zeros, except one (-1), which takes $(n-r)$ different positions. Writing the test for their linear independence

$$\sum_{j=1}^{n-r} a_j\bar{k}_j = \bar{0}_n$$

we see without further calculation that all a_j, $j = 1, 2, \ldots, n-r$ must be zero, since $(-1)\cdot a_j = 0$ implies $a_j = 0$.

So, the vectors $\{\bar{k}_1, \bar{k}_2, \ldots, \bar{k}_{n-r}\}$ form a basis in ker A, which is unique in the sense that these vectors are modified canonical expansions of $(n-r)$ linearly dependent columns of A.

(C) *The GJM procedure*

The final step in explaining the GJM procedure is almost obvious. Naturally, we discuss here only the cases of consistent linear systems when $r < n$ (there are $n - r$ linearly dependent columns in A), when solutions are not unique, but their multitude consists of all vectors of the nontrivial $\ker A$:

$$\boxed{\mathrm{Inv}(\bar{b}) = \bar{b}' + \ker A}$$

There are two possible cases.

(1) Case $m = n$ $(r < m = n)$

The reduced GJ form $B_{\mathrm{red}} = [A_{\mathrm{red}} \mid \bar{b}_{\mathrm{red}}]$ of the augmented matrix $B = [A \mid \bar{b}]$ of the consistent linear $n \times n$ system $A\mathscr{X} = \bar{b}$ has necessarily $(n - r)$ zero rows at the bottom. This $\bar{b}_{\mathrm{red}}$ is just the unique representative $\bar{b}'$ of the coset (i.e., the general solution of the system) $\mathrm{Inv}(\bar{b}) = \bar{b}' + \ker A$. The reason for this statement is the fact that the first r entries in $\bar{b}'$ are the unique expansion coefficients of $\bar{b}$ in the basis $\{\bar{c}_1, \bar{c}_2, \ldots, \bar{c}_r\}$ in $\mathrm{ran}\, A$. Remember the fundamental property of the reduced GJ form. The other $(n - r)$ entries are zeros. It follows that the entries of $\bar{b}'$ are the coefficients in the canonical expansion of $\bar{b}$ in terms of all columns of A:

$$\bar{b} = b_1'\bar{c}_1 + b_2'\bar{c}_2 + \cdots + b_r'\bar{c}_r + 0 \cdot \bar{c}_{r+1} + \cdots + 0 \cdot \bar{c}_n.$$

Thus, $\bar{b}'$ is from $\mathbb{R}^n$ and is the very special preimage of $\bar{b}$:

$$\boxed{A\bar{b}' = \bar{b}}.$$

As for the reduced forms in A_{red} of linearly dependent columns of A, they have the same features as in the above case of $\bar{b}'$: Their entries are the coefficients of the canonical expansions of $\bar{c}_{r+1}, \bar{c}_{r+2}, \ldots, \bar{c}_n$. To get $\bar{k}_1, \bar{k}_2, \ldots, \bar{k}_{n-r}$ from those reduced forms, we have only to change one 0 to (-1) in each of them: For $\bar{k}_1$ in the $(r+1)$-th, entry, for $\bar{k}_2$ in the $(r+2)$-th entry and so on. Graphically, this amounts to continuing the diagonal of 1s in the top left $r \times r$ identity submatrix of A_{red} by the diagonal of (-1)s in the bottom right $(n-r) \times (n-r)$ zero submatrix of A_{red}.

Example 17/1 For comparison, we shall again solve the previous 4×4 linear system, $(n = 4, m = 4)$ with $r = 2$ (see the standard method)

$$\left.\begin{array}{r} x_1 - x_2 + x_3 - x_4 = 4 \\ x_1 + x_2 + 2x_3 + 3x_4 = 8 \\ 2x_1 + 4x_2 + 5x_3 + 10x_4 = 20 \\ 2x_1 - 4x_2 + x_3 - 6x_4 = 4 \end{array}\right\} B = [A \mid \bar{b}] = \begin{bmatrix} 1 & -1 & 1 & -1 & 4 \\ 1 & 1 & 2 & 3 & 8 \\ 2 & 4 & 5 & 10 & 20 \\ 2 & -4 & 1 & -6 & 4 \end{bmatrix} \Rightarrow$$

$$\Rightarrow B_{\mathrm{red}} = \begin{bmatrix} 1 & 0 & \frac{3}{2} & 1 & 6 \\ 0 & 1 & \frac{1}{2} & 2 & 2 \\ 0 & 0 & 0 & 0 & 0 \\ 0 & 0 & 0 & 0 & 0 \end{bmatrix} \Rightarrow$$

$$\Rightarrow B_{GJM} = \begin{bmatrix} 1 & 0 & \frac{3}{2} & 1 & | & 6 \\ 0 & 1 & \frac{1}{2} & 2 & | & 2 \\ 0 & 0 & \mathbf{-1} & 0 & | & 0 \\ 0 & 0 & 0 & \mathbf{-1} & | & 0 \end{bmatrix} \quad \begin{array}{l} \text{continue} \\ \text{the diagonal} \\ \text{of 1s by } (-1)\text{s} \end{array}$$

(modifications are printed bold).

The general coset solution is obtained immediately:

$$\mathrm{Inv}(\bar{b}) = \begin{bmatrix} 6 \\ 2 \\ 0 \\ 0 \end{bmatrix} + \mathrm{LIN}\left(\begin{bmatrix} 3/2 \\ 1/2 \\ -1 \\ 0 \end{bmatrix}, \begin{bmatrix} 1 \\ 2 \\ 0 \\ -1 \end{bmatrix} \right) =$$

$$= \begin{bmatrix} 6 \\ 2 \\ 0 \\ 0 \end{bmatrix} + \left\{ s \begin{bmatrix} 3/2 \\ 1/2 \\ -1 \\ 0 \end{bmatrix} + t \begin{bmatrix} 1 \\ 2 \\ 0 \\ -1 \end{bmatrix}, \quad s, t \in \mathbb{R} \right\}$$

This solution differs from the standard one by an irrelevant change $s \to -s$ and $t \to -t$. But, the above coset solution is more natural from the linmap theory. Furthermore, the unique coset representative $\bar{b}'$ and the unique basis $\bar{k}_1, \bar{k}_2$ of $\ker A$ are in the form of the canonical expansion of all linearly dependent columns of $B = [A \mid \bar{b}]$, and they are obtained straightforwardly from B_{red}: the transition from B_{red} (the standard method) to B_{GJM} (the coset solution method) consists in continuation of the diagonal of 1s by the diagonal of (-1)s. Δ

(2) Case $m < n$ (fewer equations than unknowns). Since A is an $m \times n$ matrix, it follows that $B = [A \mid \bar{b}]$ is an $m \times (n+1)$ matrix.
We now have two subcases:

(2A) case $r = m$.

In this case, B_{red} will not have zero rows at the bottom (all m rows in A are lind). But, A_{red} is not an $n \times n$ matrix, its columns are $m \times 1$ vectors and one cannot fulfill the requirement that its $(n - r)$ linearly dependent columns provide (after slight modification) the basis $\bar{k}_1, \bar{k}_2, \ldots, \bar{k}_{n-r}$ for $\ker A$ (these vectors must be in $\mathbb{R}^n$). So, all columns in B_{red} (including $\bar{b}'$) must be $n \times 1$ vectors from $\mathbb{R}^n$. Thus, the GJ modification consists of two steps: we add $(n - m)$ zero rows at the bottom of B_{red} and then continue the diagonal of 1s in the top left submatrix with (-1)s, as in the previous case (1) $m = n$.

Example 17/2 Consider the 2×4 linear system ($m = 2, n = 4$) with $r = 2$.

$$\left. \begin{array}{l} x_1 - x_2 + x_3 - x_4 = 4 \\ x_1 + x_2 + 2x_3 + 3x_4 = 8 \end{array} \right\} \Rightarrow B = \begin{bmatrix} 1 & -1 & 1 & -1 & | & 4 \\ 1 & 1 & 2 & 3 & | & 8 \end{bmatrix} \Rightarrow$$

$$
B_\text{red} = \begin{array}{cc} A_\text{red} & \bar{b}_\text{red} \\ \begin{bmatrix} 1 & 0 & 3/2 & 1 & 6 \\ 0 & 1 & 1/2 & 2 & 2 \end{bmatrix} \end{array}
\Rightarrow B_\text{GJM} =
\begin{array}{cccc} A_\text{GJM} & \bar{k}_1 & \bar{k}_2 & \bar{b}' \\ \begin{bmatrix} 1 & 0 & 3/2 & 1 & 6 \\ 0 & 1 & 1/2 & 2 & 2 \\ 0 & 0 & -1 & 0 & 0 \\ 0 & 0 & 0 & -1 & 0 \end{bmatrix} \end{array}
$$

[two extra zero rows $(n - m = 2)$ and the diagonal of 1s continued with (-1)s. The modifications are printed in bold.] The general coset solution is

$$
\text{Inv}(\bar{b}) = \begin{bmatrix} 6 \\ 2 \\ 0 \\ 0 \end{bmatrix} + \left\{ s \begin{bmatrix} 3/2 \\ 1/2 \\ -1 \\ 0 \end{bmatrix} + t \begin{bmatrix} 1 \\ 2 \\ 0 \\ -1 \end{bmatrix}, s, t \in \mathbb{R} \right\}.
$$

Verification

$$
A\bar{b}' = \begin{bmatrix} 1 & -1 & 1 & -1 \\ 1 & 1 & 2 & 3 \end{bmatrix} \begin{bmatrix} 6 \\ 2 \\ 0 \\ 0 \end{bmatrix} = \begin{bmatrix} 4 \\ 8 \end{bmatrix} = \bar{b},
$$

$$
A\bar{k}_1 = \begin{bmatrix} 1 & -1 & 1 & -1 \\ 1 & 1 & 2 & 3 \end{bmatrix} \begin{bmatrix} 3/2 \\ 1/2 \\ -1 \\ 0 \end{bmatrix} = \begin{bmatrix} 0 \\ 0 \end{bmatrix} = \bar{0}_2,
$$

$$
A\bar{k}_2 = \begin{bmatrix} 1 & -1 & 1 & -1 \\ 1 & 1 & 2 & 3 \end{bmatrix} \begin{bmatrix} 1 \\ 2 \\ 0 \\ -1 \end{bmatrix} = \begin{bmatrix} 0 \\ 0 \end{bmatrix} = \bar{0}_2,
$$

$$
a_1 \bar{k}_1 + a_2 \bar{k}_2 = \bar{0}_4 \Rightarrow a_1 = a_2 = 0.
$$

[Note that the systems in examples 17/1 and 17/2 have the same solution, since the system in 17/1 has two more linearly dependent equations, which contain no more useful information about unknowns]. Δ

(2B) Case $m < n$, $r < m$.

In this case B_red will have $m - r$ zero rows at the bottom and again A_red will not be an $n \times n$ matrix, as required for coset solution, so one will add an extra $(n - m)$ zero rows at the bottom of B_red, and continue as above.

Example 17/3

$$
A : \mathbb{R}^4 \to \mathbb{R}^3 \ (n = 4, m = 3) \ \text{and} \ r = 2 \ (m < n, r < m)
$$

$$
\left. \begin{array}{r} x_1 - x_2 + x_3 + x_4 = 5 \\ x_1 + 2x_3 - x_4 = 3 \\ x_1 + x_2 + 3x_3 - 3x_4 = 1 \end{array} \right\} \Rightarrow B = \begin{bmatrix} 1 & -1 & 1 & 1 & 5 \\ 1 & 0 & 2 & -1 & 3 \\ 1 & 1 & 3 & -3 & 1 \end{bmatrix} \begin{array}{l} \\ r_2 - r_1 \sim \\ r_3 - r_1 \end{array}
$$

$$\sim \begin{bmatrix} 1 & -1 & 1 & 1 & | & 5 \\ 0 & 1 & 1 & -2 & | & -2 \\ 0 & 2 & 2 & -4 & | & 1 \end{bmatrix} \begin{matrix} \\ \\ r_3 - 2r_2 \end{matrix} \sim \begin{bmatrix} 1 & -1 & 1 & 1 & | & 5 \\ 0 & 1 & 1 & -2 & | & -2 \\ 0 & 0 & 0 & 0 & | & 0 \end{bmatrix} \begin{matrix} r_1 + r_2 \\ \sim \\ \\ \end{matrix}$$

$$\sim \begin{matrix} A_{\text{red}} & & & & b_{\text{red}} \\ \begin{bmatrix} 1 & 0 & 2 & -1 & | & 3 \\ 0 & 1 & 1 & -2 & | & -2 \\ 0 & 0 & 0 & 0 & | & 0 \end{bmatrix} \end{matrix} = B_{\text{red}} \Rightarrow B_{\text{GJM}} = \begin{matrix} A_{\text{GJM}} & \bar{k}_1 & \bar{k}_2 & b'_{\text{red}} \\ \begin{bmatrix} 1 & 0 & 2 & -1 & | & 3 \\ 0 & 1 & 1 & -2 & | & -2 \\ \mathbf{0} & \mathbf{0} & \mathbf{-1} & \mathbf{0} & | & \mathbf{0} \\ \mathbf{0} & \mathbf{0} & \mathbf{0} & \mathbf{-1} & | & \mathbf{0} \end{bmatrix} \end{matrix}$$

[one extra zero row $(n - m = 1)$, and continuation of the diagonal of 1s with (-1)s. The modifications are printed in bold.]

The general coset solution

$$\text{Inv}(\bar{b}) = \begin{bmatrix} 3 \\ -2 \\ 0 \\ 0 \end{bmatrix} + \left\{ s \begin{bmatrix} 2 \\ 1 \\ -1 \\ 0 \end{bmatrix} + t \begin{bmatrix} -1 \\ -2 \\ 0 \\ -1 \end{bmatrix}, s, t \in \mathbb{R} \right\} = \bar{b}' + \ker A$$

Verification

$$A\bar{b}' = \begin{bmatrix} 1 & -1 & 1 & 1 \\ 1 & 0 & 2 & 1 \\ 1 & 1 & 3 & -3 \end{bmatrix} \begin{bmatrix} 3 \\ -2 \\ 0 \\ 0 \end{bmatrix} = \begin{bmatrix} 5 \\ 3 \\ 1 \end{bmatrix} = \bar{b},$$

$$A\bar{k}_1 = \begin{bmatrix} 1 & -1 & 1 & 1 \\ 1 & 0 & 2 & 1 \\ 1 & 1 & 3 & -3 \end{bmatrix} \begin{bmatrix} 2 \\ 1 \\ -1 \\ 0 \end{bmatrix} = \begin{bmatrix} 0 \\ 0 \\ 0 \end{bmatrix} = \bar{0}_3,$$

$$A\bar{k}_2 = \begin{bmatrix} 1 & -1 & 1 & 1 \\ 1 & 0 & 2 & 1 \\ 1 & 1 & 3 & -3 \end{bmatrix} \begin{bmatrix} -1 \\ -2 \\ 0 \\ -1 \end{bmatrix} = \begin{bmatrix} 0 \\ 0 \\ 0 \end{bmatrix} = \bar{0}_3,$$

$\bar{k}_1$ and $\bar{k}_2$ are lind:

$$a_1\bar{k}_1 + a_2\bar{k}_2 = \bar{0}_4 \Rightarrow a_1 \begin{bmatrix} 2 \\ 1 \\ -1 \\ 0 \end{bmatrix} + a_2 \begin{bmatrix} -1 \\ -2 \\ 0 \\ -1 \end{bmatrix} = \begin{bmatrix} 0 \\ 0 \\ 0 \\ 0 \end{bmatrix} \Rightarrow$$

$$\Rightarrow \left. \begin{matrix} 2a_1 - a_2 = 0 \\ a_1 - 2a_2 = 0 \\ -a_1 \qquad = 0 \\ - a_2 = 0 \end{matrix} \right\} \Rightarrow a_1 = 0, a_2 = 0. \; \Delta$$

2.18 Summary of Methods for Solving Systems of Linear Equations

A system of m simultaneous linear equations with n unknowns

$$\begin{array}{l} a_{11}x_1 + a_{12}x_2 + \cdots + a_{1n}x_n = b_1 \\ a_{21}x_1 + a_{22}x_2 + \cdots + a_{2n}x_n = b_2 \\ \cdots\cdots\cdots\cdots\cdots\cdots\cdots\cdots\cdots \\ a_{m1}x_1 + a_{m2}x_2 + \cdots + a_{mn}x_n = b_1 \end{array}$$

can be expressed in the matrix form

$$\begin{bmatrix} a_{11} & a_{12} & \cdots & a_{1n} \\ a_{21} & a_{22} & \cdots & a_{2n} \\ \vdots & \vdots & \ddots & \vdots \\ a_{m1} & a_{m2} & \cdots & a_{mn} \end{bmatrix} \begin{bmatrix} x_1 \\ x_2 \\ \vdots \\ x_n \end{bmatrix} = \begin{bmatrix} b_1 \\ b_2 \\ \vdots \\ b_n \end{bmatrix},$$

or, in short,

$$\boxed{A\bar{\mathscr{X}} = \bar{b}},$$

where the $m \times n$ real coefficient matrix $A = [a_{ij}]_{m \times n}$ is multiplied with the column-matrix (vector) of unknowns $\bar{\mathscr{X}} = [x_1 x_2 \ldots x_n]^T \in \mathbb{R}^n$ to get the free vector $\bar{b} = [b_1 b_2 \ldots b_m]^T \in \mathbb{R}^m$ with known entries.

The matrix A is a linmap $A : \mathbb{R}^n \to \mathbb{R}^m$ which defines two important subspaces, one is the kernel ker A in $\mathbb{R}^n$, which consists of all vectors from $\mathbb{R}^n$ that are mapped by A onto the zero vector $\bar{0}_m$ in $\mathbb{R}^m$. The second subspace is the range ran A in $\mathbb{R}^m$, which consists of all images by A of vectors from $\mathbb{R}^n$. (As a map, A takes every vector from $\mathbb{R}^n$ onto its unique image in $\mathbb{R}^m$.)

Since A is a map, it defines a *bijection* between ran A and the set of all inverse images, which set partitions $\mathbb{R}^n$.

Furthermore, since A is a linear map, every one of these inverse images in $\mathbb{R}^n$ is a coset, i.e., a linear translation of ker A: for $\bar{b} \in$ ran A the inverse image is the coset

$$\boxed{\text{Inv}(\bar{b}) = \bar{b}' + \ker A},$$

where $\bar{b}'$ is the unique coset representative entirely determined by $\bar{b}$. The set of all cosets $\mathbb{R}^n / \ker A$ forms a vector space isomorphic to ran A:

$$\boxed{\mathbb{R}^n / \ker A \cong \text{ran } A}.$$

Since $A\bar{\mathscr{X}} = \bar{b}$ can be expressed as $A\bar{\mathscr{X}} = x_1\bar{c} + x_2\bar{c}_2 + \cdots + x_n\bar{c}_n = \bar{b}$, where $\bar{c}_1, \bar{c}_2, \ldots, \bar{c}_n$ are the columns of A, (we call this expression the *canonical expansion* of $A\bar{\mathscr{X}}$) we see that the columns of A span the subspace ran A:

$$\boxed{\text{ran } A = \text{LIN}(\bar{c}_1, \bar{c}_2, \dots, \bar{c}_n)}.$$

Some columns $\bar{c}_1, \bar{c}_2, \dots, \bar{c}_r$ ($r \le n$, r is the rank of A) are linearly independent, and they form the *natural basis* in ran A.

If $r = n$ (for $n \le m$), this means that all columns are lind, and if the system is consistent ($\bar{b} \in$ ran A), then there is only one expansion of that $\bar{b}$ in terms of this basis. The coefficients of this unique expansion form the vector $\bar{b}'$, which is the only solution of the system. Remember that $r = n$ implies that ker $A = \{\bar{0}_n\}$, the kernel is trivial. The most natural way to get this $\bar{b}'$ is to reduce the augmented matrix $B = [A \mid \bar{b}]$ to its unique reduced row-echelon form (using eros)

$$\boxed{B_{\text{red}} = [I_n \mid \bar{b}_{\text{red}}] = [I_n \mid \bar{b}']}$$

(If $m > n$, then B_{red} consists of the top n rows). Remember that for the case $m = n$, the test for $r = n$ is $\det A \ne 0$, and then A_{red} is I_n.

When the $n \times n$ system is homogeneous, i.e., $\bar{b} = \bar{0}_n$, the only (trivial) solution is $\bar{b}' = \bar{0}_n$.

If $r < n$, then there are $(n - r)$ linearly dependent columns in A. Their modified reduced forms will provide the unique basis for the $(n - r)$-dimensional ker A, which in turn defines all the multitude of solutions of the system.

The GJM procedure, which is relevant only for $r < n$ cases, can be formulated for all these cases in one sentence

Only if $m < n$ add at the bottom of B_{red} $(n - m)$ zero rows, and in every $r < n$ case put $(n - r)$ numbers (-1) instead of 0 on the continuation of 1s on the main diagonal of the $r \times r$ left upper identity matrix (see examples 17/1, 17/2, 17/3).

Now, $\bar{b}'$ is the last column of the modified B_{red} (the unique coset representative), and the $(n - r)$ columns in front of $\bar{b}'$ in the modified B_{red} are the unique basis vectors $\bar{k}_1, \bar{k}_2, \dots, \bar{k}_{n-r}$ of ker A. The general solution of the consistent system ($\bar{b} \in$ ran A) is

$$\boxed{\text{Inv}(\bar{b}) = \bar{b}' + \text{LIN}(\bar{k}_1, \bar{k}_2, \dots, \bar{k}_{n-r})}.$$

When the $m \times n$ system is homogeneous, i.e.,

$$\bar{b} = \bar{0}_m,$$

the solution always exists since $\bar{0}_m \in$ ran A, and the general solution is simply

$$\boxed{\text{ker } A = \text{LIN}(\bar{k}_1, \bar{k}_2, \dots, \bar{k}_{n-r})}$$

since $\bar{b}' = \bar{0}_n$.

Therefore, in the coset solution approach, the homogeneous systems are treated in the same manner as nonhomogeneous ones:

$$\boxed{\mathrm{Inv}(\bar{0}_m) = \bar{0}_n + \ker A}\,,$$

so that everything depends on whether the kernel is trivial, $\ker A = \{\bar{0}_n\} \Leftrightarrow r = n \Leftrightarrow$ the unique solution $\bar{0}_n$, or nontrivial

$$\boxed{\ker A = \mathrm{LIN}(\bar{k}_1, \bar{k}_2, \ldots, \bar{k}_{n-r})}\,,$$

which is then the general solution.

Chapter 3
Inner-Product Vector Spaces (Euclidean and Unitary Spaces)

Applications in physics of vector spaces $V_n(\mathbb{R})$ or $V_n(\mathbb{C})$ is insignificant, since in them we cannot define measurable quantities like lengths or angles.

3.1 Euclidean Spaces E_n

In $\mathbb{R}^3$, we were able to do this by making use of the dot product: if $\bar{x} = [x_1 x_2 x_3]^T$ and $\bar{y} = [y_1 y_2 y_3]^T$ are two vectors (matrix-columns of three real numbers) in $\mathbb{R}^3$, then their dot product is defined as

$$\bar{x} \cdot \bar{y} = x_1 y_1 + x_2 y_2 + x_3 y_3 = \sum_{i=1}^{3} x_i y_i.$$

As a matter of fact, the dot product of two geometric vectors (arrows) $\vec{x}$ and $\vec{y}$ was originally introduced, as it is done in physics, as the product of their lengths with the cosine of the smaller angle α between them:

$$\vec{x} \cdot \vec{y} = |\vec{x}| |\vec{y}| \cos\alpha \text{ (the original definition).}$$

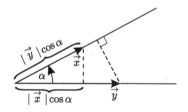

It is the product of the length $|\vec{x}|$ of $\vec{x}$ with the (positive or negative) length of projection $|\vec{y}| \cos\alpha$ of $\vec{y}$ along the line of $\vec{x}$ or the other way round $|\vec{y}|$ times the length of projection $|\vec{x}|\cos\alpha$. Remember that in physics the work done by a force $\vec{F}$ producing a *displacement* $\vec{d}$ is

$$\vec{F}\cdot\vec{d}=|\vec{F}||\vec{d}|\cos\alpha.$$

The four most important properties of the dot product

1. $\vec{x}\cdot\vec{y}=\vec{y}\cdot\vec{x}$—symmetry or commutative property (obvious);
2. $(\vec{x_1}+\vec{x_2})\cdot\vec{y}=\vec{x_1}\cdot\vec{y}+\vec{x_2}\cdot\vec{y}$—distributive property with regard to the vector addition in the first factor (the projection of $\vec{x_1}+\vec{x_2}$ along the line of $\vec{y}$ is the sum of the projections of $\vec{x_1}$ and $\vec{x_2}$, see the figure);

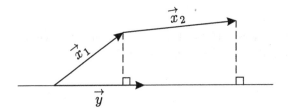

Due to the symmetry, it is also valid for the second factor $\vec{x}\cdot(\vec{y_1}+\vec{y_2})=\vec{x}\cdot\vec{y_1}+\vec{x}\cdot\vec{y_2}$.

3. $a(\vec{x}\cdot\vec{y})=(a\vec{x})\cdot\vec{y}=\vec{x}\cdot(a\vec{y})$, $a\in\mathbb{R}$—associative property with respect to multiplication with scalars

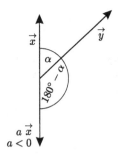

[for $a>0$ obvious, and for $a<0$ it can be proved as follows

$$(a\vec{x})\cdot\vec{y}=|a\vec{x}||\vec{y}|\cos(180°-\alpha)=|a||\vec{x}||\vec{y}|(-\cos\alpha)=$$
$$=-a|\vec{x}||\vec{y}|(-\cos\alpha)=a|\vec{x}||\vec{y}|\cos\alpha=a(\vec{x}\cdot\vec{y})];$$

4. $(\vec{x}\cdot\vec{x})=|\vec{x}|^2>0$ if $\vec{x}\neq0$; $\vec{x}\cdot\vec{x}=|\vec{x}|^2=0$ iff $\vec{x}=0$—positive definite property [only the zero vector $\vec{0}$ has zero length, others have positive lengths, $|\vec{x}|^2=0\Leftrightarrow|\vec{x}|=0\Leftrightarrow\vec{x}=0$];

Properties 2 and 3 can be generalized obviously by saying that the dot product is linear in both factors (i.e., bilinear)

$$\left(\sum_{i=1}^n a_i \, \vec{x_i}\right) \cdot \left(\sum_{j=1}^n b_j \, \vec{y_j}\right) = \sum_{i=1}^m \sum_{j=1}^n a_i b_j \, \vec{x_i} \cdot \vec{y_j} \, .$$

Note: The dot product of three vectors $\vec{x}, \vec{y}, \vec{z}$ does not exist.

The length (or norm) of vector $\vec{x}$ is immediately obtained as $| \, \vec{x} \, | = (\vec{x} \cdot \vec{x})^{1/2}$. The angle α between two vectors $\vec{x}$ and $\vec{y}$ is obviously defined as

$$\alpha = \arccos \frac{\vec{x} \cdot \vec{y}}{| \, \vec{x} \, || \, \vec{y} \, |} \, .$$

Two nonzero vectors are orthogonal (*perpendicular*) if their dot product is zero $\vec{x} \cdot \vec{y} = 0 \Leftrightarrow \alpha = 0°$.

When we choose a basis of orthogonal unit vectors $\{\vec{e_1}, \vec{e_2}, \vec{e_3}\}$ in the vector space of geometric vectors—$\vec{e_i} \cdot \vec{e_j} = \delta_{ij}$—then we establish an isomorphism with the space $\mathbb{R}^3$

$$\vec{x} = x_1 \, \vec{e_1} + x_2 \, \vec{e_2} + x_3 \, \vec{e_3} \Rightarrow \vec{x}$$

is represented by $\begin{bmatrix} x_1 \\ x_2 \\ x_3 \end{bmatrix}$.

The dot product of two vectors $\vec{x}$ and $\vec{y}$ is now

$$\vec{x} \cdot \vec{y} = (x_1 \, \vec{e_1} + x_2 \, \vec{e_2} + x_3 \, \vec{e_3}) \cdot (y_1 \, \vec{e_1} + y_2 \, \vec{e_2} + y_3 \, \vec{e_3}) =$$

$$= \left(\sum_{i=1}^3 x_i \, \vec{e_i}\right) \cdot \left(\sum_{j=1}^3 y_j \vec{e_j}\right) = \sum_{i,j=1}^3 x_i y_i \, \vec{e_i} \cdot \vec{e_j} =$$

$$= \sum_{i=1}^3 x_i y_i = x_1 y_1 + x_2 y_2 + x_3 y_3$$

(the *coordinate* definition)

In another basis of orthogonal unit vectors $\{\vec{e_1'}, \vec{e_2'}, \vec{e_3'}\}$ this dot product is $x_1' y_1' + x_2' y_2' + x_3' y_3'$, but its value remains unchanged, since these bases are connected by a 3×3 orthogonal matrix A (see Sect. 4.4), which is characterized by $A^{-1} = A^T$ or $A^T A = I_3$. In this case, the column of $\begin{bmatrix} x_1 \\ x_2 \\ x_3 \end{bmatrix}$ is changed into $A \begin{bmatrix} x_1 \\ x_2 \\ x_3 \end{bmatrix} = \begin{bmatrix} x_1' \\ x_2' \\ x_3' \end{bmatrix}$ in the new basis (see Chap. 4). In matrix notation, the dot product $\vec{x} \cdot \vec{y}$ in the old basis is $[x_1 x_2 x_3] \begin{bmatrix} y_1 \\ y_2 \\ y_3 \end{bmatrix}$, and in the new one $[x_1' x_2' x_3'] \begin{bmatrix} y_1' \\ y_2' \\ y_3' \end{bmatrix} = [x_1 x_2 x_3] A^T A \begin{bmatrix} y_1 \\ y_2 \\ y_3 \end{bmatrix}$, so it

remains unchanged. In the theory of tensors, every quantity which does not change when the coordinate system (basis of orthogonal unit vectors) changes is called a *scalar*, so it is usual (as physicists do) to call the dot product the *scalar product* of two vectors in $\mathbb{R}^3$.

Notice that the above four properties of the dot product can be proved directly from the coordinate definition.

1. It is obvious;
2.

$$(\vec{x_1} + \vec{x_2}) \cdot \vec{y} = (x_{11} + x_{21})y_1 + (x_{12} + x_{22})y_2 + (x_{13} + x_{23})y_3 =$$
$$= (x_{11}y_1 + x_{12}y_2 + x_{13}y_3) + (x_{21}y_1 + x_{22}y_2 + x_{23}y_3) = \vec{x_1} \cdot \vec{y} + \vec{x_2}\vec{y_2};$$

3. $a(\vec{x} \cdot \vec{y}) = a(x_1 y_1 + x_2 y_2 + x_3 y_3) = (ax_1)y_1 + (ax_2)y_2 + (ax_3)y_3 = (a\vec{x}) \cdot \vec{y};$
4. $\vec{x} \cdot \vec{x} = x_1^2 + x_2^2 + x_3^2 > 0$, $\vec{x} \cdot \vec{x} = 0$ iff $x_1 = x_2 = x_3 = 0$ or $\vec{x} = 0$.

The vector spaces $\mathbb{R}^n$ for $n > 3$ have no geometrical interpretation. Nevertheless, they are essential for many problems in mathematics (e.g., for systems of linear equations), in physics (e.g., $n = 4$ for some formulations of the special theory of relativity), as well as in economics (e.g., linear economic models).

The generalization of the dot product in $\mathbb{R}^3$ to $\mathbb{R}^n$, $n > 3$, is straightforward. We need only to extend the summation in the coordinate definition from 3 to n: if $\bar{x}, \bar{y} \in \mathbb{R}^n$, then the dot product is $\bar{x} \cdot \bar{y} = \sum_{i=1}^{n} x_i y_i$, where $\bar{x} = [x_1 x_2 \ldots x_n]^T$ and $\bar{y} = [y_1 y_2 \ldots y_n]^T$. The dot product of two vectors $\bar{x}$ and $\bar{y}$ in $\mathbb{R}^n$ can be expressed as a matrix product of the transposed $\bar{x}$ with the column of $\bar{y}$: $\bar{x} \cdot \bar{y} = \bar{x}^T \bar{y}$. This notation has a number of practical and useful applications.

There are several changes in terminology and notation that we are dealing with in $\mathbb{R}^n$.

The product $\sum_{i=1}^{n} x_i y_i$ is no longer called the dot product nor is it denoted as $\bar{x} \cdot \bar{y}$. It is called the *inner product* and denoted as $(\bar{x}, \bar{y})$. It is important not to confuse this notation with that for the ordered pair of vectors. For that reason, we shall further denote an ordered pair of vectors $\bar{x}$ and $\bar{y}$ as $[\bar{x}, \bar{y}]$.

The other difference in $\mathbb{R}^n$ compared with $\mathbb{R}^3$ is that the magnitude of the vector $\bar{x}$ is called its norm, and it is denoted as $\|\bar{x}\|$, but it is defined analogously as

$$\|\bar{x}\| = (\bar{x}, \bar{x})^{1/2} = \sqrt{x_1^2 + x_2^2 + \cdots + x_n^2}.$$

Note that in the theory of tensor multiplication of vector spaces this product $\sum_{i=1}^{n} x_i y_i$ is an example of a tensor product of matrix-column spaces with a subsequent contraction (which amounts to equalizing two indices and summing over the common one). Such tensor products are called *inner products*.

It can be seen, without much effort, that the inner product in $\mathbb{R}^n$ has all four characteristic properties of the dot product in $\mathbb{R}^3$: commutativity, distributivity with respect to the addition of vectors, associativity with respect to the multiplication of vectors with numbers, and finally, positive definiteness. One can generalize the second and third of these properties by saying that this inner product is *bilinear*.

As far as other real vector spaces $V_n(\mathbb{R})$ are concerned, the inner product in them can be defined axiomatically by taking the above properties of this product in $\mathbb{R}^n$ as postulates:

The inner product in a real vector space $V_n(\mathbb{R})$ can be every scalar function on the Descartes square $V_n(\mathbb{R}) \times V_n(\mathbb{R})$, i.e., any function $V_n(\mathbb{R}) \times V_n(\mathbb{R}) \to \mathbb{R}$ which maps every ordered pair $[\bar{x}, \bar{y}]$ of vectors from $V_n(\mathbb{R})$ onto a unique real number $(\bar{x}, \bar{y})$, provided that it is

1. *commutative* (symmetric)—$(\bar{x}, \bar{y}) = (\bar{y}, \bar{x})$;
2. *linear* in the first factor—$(a\bar{x}_1 + b\bar{x}_2, \bar{y}) = a(\bar{x}_1, \bar{y}) + b(\bar{x}_2, \bar{y})$;
3. *positive definite*—$(\bar{x}, \bar{x}) > 0$ for $\bar{x} \neq \bar{0}$ and $(\bar{x}, \bar{x}) = 0$ iff $\bar{x} = \bar{0}$.

Combining the first and second axioms, we can say that every inner product in $V_n(\mathbb{R})$ is *bilinear*:

$$\left(\sum_{i=1}^{p} a_i \bar{x}_i, \ \sum_{j=1}^{q} b_j \bar{y}_j \right) = \sum_{i=1}^{p} \sum_{j=1}^{q} a_i b_j (\bar{x}_i, \bar{y}_j).$$

A real vector space $V_n(\mathbb{R})$ with this kind of inner product is called a *Euclidean space* E_n.

Remark Some authors reserve this term for $\mathbb{R}^n$, while other real spaces of this kind are called *real inner-product vector spaces*.

There are two possible ways to define different inner products in the same $V_n(\mathbb{R})$ and $\mathbb{R}^n$.

1. Any basis $v = \{\bar{v}_1, \bar{v}_2, \dots, \bar{v}_n\}$ in $V_n(\mathbb{R})$ defines an inner product if we expand vectors from $V_n(\mathbb{R})$ in that basis $\bar{x} = \sum_{i=1}^{n} x_i \bar{v}_i, \bar{x} \to \mathscr{X} = [x_1, x_2, \dots, x_n]^T$ (note the differences in notation for the vector $\bar{x}$ and its representing column $\mathscr{X}$) of two vectors $\bar{x}, \bar{y} \in V_n(\mathbb{R})$ as

$$(\bar{x}, \bar{y})_v = \sum_{i=1}^{n} x_i y_i = \chi^T x.$$

If we replace this basis v by a new one $v' = \{\bar{v}'_1, \bar{v}'_2, \dots, \bar{v}'_n\}$ by means of an $n \times n$ replacement matrix $\mathbb{R}$,

$$\begin{bmatrix} \bar{v}'_1 \\ \bar{v}'_2 \\ \vdots \\ \bar{v}'_n \end{bmatrix} = \mathscr{R} \begin{bmatrix} \bar{v}_1 \\ \bar{v}_2 \\ \vdots \\ \bar{v}_n \end{bmatrix} \quad \text{or} \quad \bar{v}_i = \sum_{j=1}^{n} r_{ij} \vec{v}_j, \quad i = 1, 2, \dots, n,$$

then the representing column $\mathscr{X}$ of $\bar{x}$ will change by the contragredient matrix $(\mathscr{R}^T)^{-1}$:

$$\mathscr{X}' = \begin{bmatrix} \bar{x}_1' \\ \bar{x}_2' \\ \vdots \\ \bar{x}_n' \end{bmatrix} = (\mathscr{R}^T)^{-1} \begin{bmatrix} \bar{x}_1 \\ \bar{x}_2 \\ \vdots \\ \bar{x}_n \end{bmatrix} = (\mathscr{R}^T)^{-1} \mathscr{X}$$

(see Sect. 4.3). So $(\bar{x},\bar{y})_{v'} = \sum_{i=1}^n x_i' y_i' = (\mathscr{X}')^T \mathscr{Y}' = \mathscr{X} \mathscr{R}^{-1} (\mathscr{R}^T)^{-1} \mathscr{Y} = \mathscr{X}^T (\mathscr{R}^T \mathscr{R})^{-1} \mathscr{Y}$. This will be the same number as $(\bar{x},\bar{y})_v$ if $\mathscr{R}^T \mathscr{R} = I_n$ (or $\mathscr{R}^{-1} = \mathscr{R}^T$). Matrices of this kind are known as *orthogonal matrices*.

In conclusion, we can say that the basis v and all other bases in $V_n(R)$ obtained from v by orthogonal replacement matrices define one inner product in $V_n(\mathbb{R})$ (all three axioms are obviously satisfied).

Orthogonal $n \times n$ matrices $\mathscr{R}^{-1} = \mathscr{R}^T$ form a group $O(n)$. If we now define a relation in the set of all bases in $V_n(\mathbb{R})$ such that two bases are related if they can be obtained from each other by an orthogonal matrix (and its inverse), then this is obviously an *equivalence relation* [reflexive, symmetric, and transitive since $O(n)$ is a group]. Thus, all bases in $V_n(\mathbb{R})$ are partitioned into equivalence classes of orthogonally equivalent bases and every such class defines one inner product in $V_n(\mathbb{R})$. Since the vectors from v are represented in the same v by the standard basis in $\mathbb{R}^n$

$$\bar{e}_p = [\delta_{1p} \, \delta_{2p} \ldots \delta_{np}]^T, \quad p = 1,2,\ldots,n,$$

it follows immediately that all vectors from v have unit norm and that the inner product $(\bar{x},\bar{y})_v$ between different basis vectors is zero, so they are orthogonal. Such a basis of orthogonal unit vectors is usually called an *orthonormal (ON) basis*.

Thus, each class of orthogonally equivalent bases in $V_n(\mathbb{R})$ defines one inner product in $V_n(\mathbb{R})$, and all bases from the class (and only they) are orthonormal in that inner product.

2. We have the standard inner product in $\mathbb{R}^n$,

$$(\bar{x},\bar{y}) = \bar{x}^T \bar{y},$$

defined by the class of orthogonally equivalent bases which class is represented by the standard basis $\bar{e} = [\delta_{1p}, \delta_{2p},\ldots,\delta_{np}]^T$, $p = 1,2,\ldots,n$, in $\mathbb{R}^n$. Other inner products in $\mathbb{R}^n$ can be formulated if we choose any $n \times n$ *positive definite* real *symmetric* matrix $\mathscr{A}$ [$\mathscr{A}$ is symmetric if $\mathscr{A}^T = \mathscr{A}$ and it is positive definite if $(\bar{x},\mathscr{A}\bar{x}) > 0$ for all $\bar{x} \neq \bar{0}_n$ and it is zero iff $\bar{x} = \bar{0}_n$], and define a new inner product in $\mathbb{R}^n$

$$(\bar{x},\bar{y})_{\mathscr{A}} = \bar{x}^T \mathscr{A} \bar{y}.$$

To verify that it is an inner product, we have to prove only that it is symmetric (commutative). Indeed,

$$(\bar{y},\bar{x})_{\mathscr{A}} = \bar{y}^T \mathscr{A} \bar{x} = [y_1 \ y_2 \ldots y_n] \begin{bmatrix} a_{11}x_1 + a_{12}x_2 + \cdots + a_{1n}x_n \\ a_{21}x_1 + a_{22}x_2 + \cdots + a_{2n}x_n \\ \cdots\cdots\cdots\cdots\cdots\cdots\cdots \\ a_{n1}x_1 + a_{n2}x_2 + \cdots + a_{nn}x_n \end{bmatrix} =$$

$$= a_{11}x_1y_1 + a_{12}x_2y_1 + \cdots + a_{1n}x_ny_1 +$$
$$+ a_{21}x_1y_2 + a_{22}x_2y_2 + \cdots + a_{2n}x_ny_2 + \cdots$$
$$\cdots + a_{n1}x_1y_n + a_{n2}x_2y_n + \cdots + a_{nn}x_ny_n,$$

and because the matrix $\mathscr{A}$ is symmetric $(\mathscr{A}^T = \mathscr{A})$, it is symmetric with respect to the main diagonal $a_{ij} = a_{ji}$, $i \ne j$, $i,j = 1,2,\ldots,n$, so that the above result is equal to

$$(\bar{y},\bar{x})_{\mathscr{A}} = \bar{x}^T \mathscr{A} \bar{y} = [x_1 \ x_2 \ldots x_n] \begin{bmatrix} a_{11}y_1 + a_{12}y_2 + \cdots + y_{1n}x_n \\ a_{21}y_1 + a_{22}y_2 + \cdots + y_{2n}x_n \\ \cdots\cdots\cdots\cdots\cdots\cdots\cdots \\ a_{n1}y_1 + a_{n2}y_2 + \cdots + a_{nn}x_n \end{bmatrix} =$$

$$= a_{11}x_1y_1 + a_{12}x_1y_2 + \cdots + a_{1n}x_1y_n +$$
$$+ a_{21}x_2y_1 + a_{22}x_2y_2 + \cdots + a_{2n}x_2y_n + \cdots$$
$$\cdots + a_{n1}x_ny_1 + a_{n2}x_ny_2 + \cdots + a_{nn}x_ny_n.$$

Thus, $(\bar{y},\bar{x})_{\mathscr{A}} = (\bar{x},\bar{y})_{\mathscr{A}}$.

The other two axioms are obviously satisfied due to the linear properties of matrix multiplication: $(aA + bB)C = aAC + bBC$, where a,b are numbers, and A,B,C matrices, i.e.,

$$(a\bar{x}_1 + b\bar{x}_2, \bar{y})_{\mathscr{A}} = (a\bar{x}_1^T + b\bar{x}_2^T)\mathscr{A}\bar{y} = a\bar{x}_1^T\mathscr{A}\bar{y} + b\bar{x}_2^T\mathscr{A}\bar{y} = a(\bar{x}_1,\bar{y})_{\mathscr{A}} + b(\bar{x}_2,\bar{y})_{\mathscr{A}}$$

as well as due to the positive definiteness of the matrix $\mathscr{A}$:

$$(\bar{x},\bar{x})_{\mathscr{A}} = (\bar{x},\mathscr{A}\bar{x}) > 0 \text{ for } \bar{x} \ne \bar{0}_n \text{ and it is zero iff } \bar{x} = \bar{0}_n.$$

Examples of inner products in real vector spaces of matrices and polynomials.

a) In the vector space $\mathbb{R}_{m \times n}$ of real $m \times n$ matrices, we have the standard inner product:

$$\text{for } A,B \in \mathbb{R}_{m \times n}, \quad (A,B) = \text{tr}(A^T B) = \sum_{i=1}^{m} \sum_{j=1}^{n} a_{ij}b_{ij}.$$

It is

1. *symmetric* (commutative): $(B,A) = \text{tr}(B^T A) = \text{tr}(B^T A)^T = \text{tr}(A^T B) = (AB)$
   [since the trace (the sum of diagonal elements) is invariant under transposition $\text{tr}A^T = \text{tr}A$];

2. *linear*: $(A+B,C) = \text{tr}[(A+B)^T C] = \text{tr}(A^T C + B^T C) = \text{tr}(A^T C) + \text{tr}(B^T C) = (A,C) + (B,C)$ (since $(A+B)^T = A^T + B^T$ and $\text{tr}(A+B) = \text{tr}A + \text{tr}B$); $(aA,B) = \text{tr}[(aA)^T B] = \text{tr}(aA^T B) = a\,\text{tr}(A^T B) = a(A,B)$ [since $\text{tr}(aA) = a\,\text{tr}A$];

3. *positive definite*: $(A,A) = \text{tr}(A^T A) = \sum_{i=1}^{m} \sum_{j=1}^{n} a_{ij}^2 > 0$ for $A \neq 0_{m\times n}$ and it is zero iff $A = 0_{m\times n} = [0]_{m\times n}$.

b) In the infinite dimensional vector space $P_{[a,b]}(\mathbb{R})$ of real polynomials $p(x)$ defined on a closed interval $[a,b]$, the standard inner product is

$$(p(x),q(x)) = \int_a^b p(x)q(x)dx.$$

The three axioms are obviously satisfied.

b') In $P_{[a,b]}(\mathbb{R})$ one can define an inner product with the weight

$$(p(x),q(x))_\rho = \int_a^b \rho(x)p(x)q(x)dx,$$

where the weight $\rho(x)$ is a nonnegative ($\rho(x) \geq 0$) and continuous function on the interval (a,b) (it cannot be zero on the whole interval).

One may notice that both a and b are natural generalizations of the standard inner product in $\mathbb{R}^n$

$$(\bar{x},\bar{y}) = \bar{x}^T \cdot \bar{y} = \sum_{i=1}^{n} x_i y_i.$$

The first case is a generalization from one-column matrices with one index ($\mathbb{R}_{n\times 1}$) to the general type of matrices with two indices ($\mathbb{R}_{m\times n}$)

$$(A,B) = \sum_{i=1}^{m} \sum_{j=1}^{n} a_{ij}b_{ij}.$$

The second case is a generalization from a variable with the discrete index $[x_1\, x_2 \ldots x_n]^T$ to a variable with the continuous index $p(x)$, $x \in [a,b]$, so that the summation is replaced with an integral

$$(p(x),q(x)) = \int_a^b p(x)q(x)dx.$$

3.2 Unitary Spaces U_n (or Complex Inner-product Vector Spaces)

The definition of an inner product in the complex vector space $\mathbb{C}^n$ of matrix-columns with n complex numbers is not so straightforward. If we define it as the standard inner product in $\mathbb{R}^n$:

$$(\bar{x}, \bar{y}) = \sum_{i=1}^{n} x_i y_i, \quad x_i, y_i \in \mathbb{C} \text{ for all } i,$$

we immediately see that the norm

$$||\bar{x}|| = (\bar{x}, \bar{x})^{1/2} = \left(\sum_{i=1}^{n} x_i^2\right)^{1/2}$$

is not, in this case, a real number, since the sum of squares of complex numbers $\sum_{i=1}^{n} x_i^2$ can be a negative number. However, for applications in Quantum Mechanics, it is necessary that the norm is a real number, since the probabilities of measurement in QM are defined as norms of some vectors.

For this reason, the standard inner product in $\mathbb{C}^n$ is defined as

$$\boxed{(\bar{x}, \bar{y}) = (\bar{x}^*)^T \bar{y} = \sum_{i=1}^{n} x_i^* y_i}$$

(the asterisk * denotes the complex conjugation [1, 2])

With this inner product the norm of vector $\bar{x}$

$$||\bar{x}|| = (\bar{x}, \bar{x})^{1/2} = \left(\sum_{i=1}^{n} x_i^* x_i\right)^{1/2} = \left(\sum_{i=1}^{n} |x_i|^2\right)^{1/2}$$

is always a positive real number. (It is zero iff $\bar{x} = \bar{0}$.)

Note. In mathematical literature, it is more usual to define the inner product [3, 1] in $\mathbb{C}^n$ as

$$(\bar{x}, \bar{y}) = \sum_{i=1}^{n} x_i y_i^* = \bar{x}^T \bar{y}^*.$$

The reason for our definition lies in Quantum Mechanics which is mathematically based on complex vector spaces and which uses the so-called Dirac notation that requires the above "physical" definition.

It is usual and very practical to introduce one notation [1] for both the transposition T and complex conjugation $*$ of a matrix $(A^*)^T = A^\dagger$, where $\dagger$ is a *dagger* and this combined operation is called the adjoining of A ($A^\dagger$ is *adjoint of A*). (Some authors [3] denote the adjoint of A as A^* or A^H or call it [2] the *Hermitian adjoint*).

It should be distinguished from adj$A$ which is the *classical adjoint* [3] and represents the transposed matrix of the cofactors of a square matrix A.

Our inner product in $\mathbb{C}^n$ $(\bar{x}, \bar{y}) = \bar{x}^\dagger \bar{y}$ has the following three obvious properties:

1. It is skew (Hermitian)-symmetric $(\bar{x}, \bar{y}) = (\bar{y}, \bar{x})^*$;   [1]
2. It is linear in the second factor

$$(\bar{z}, a\bar{x} + b\bar{y}) = a(\bar{z}, \bar{x}) + b(\bar{z}, \bar{y});$$

3. It is positive definite (strictly positive)

$$(\bar{x},\bar{x}) > 0 \text{ for } \bar{x} \neq \bar{0} \text{ and } (\bar{x},\bar{x}) = 0 \text{ iff } \bar{x} = \bar{0}.$$

From properties 1 and 2, it follows that this inner product is *antilinear* [1] (skew-linear) or *conjugate linear* [3] in the first factor

$$(a\bar{x}+\bar{b},\bar{z}) = (\bar{z},a\bar{x}+b\bar{y})^* = [a(\bar{z},\bar{x})+b(\bar{z},\bar{y})]^* = a^*(\bar{x},\bar{z})+b^*(\bar{y},\bar{z}).$$

Being antilinear in the first factor and linear in the second one, we say that this inner product is *conjugate bilinear*:

$$\left(\sum_{i=1}^{m} a_i\bar{x}_i, \sum_{j=1}^{n} b_j\bar{y}_j\right) = \sum_{i,j} a_i^* b_j (\bar{x}_i,\bar{y}_j).$$

When we want to define an inner product in an arbitrary complex vector space $V(\mathbb{C})$, we can use the above three properties as postulates:

Definition An inner product in a complex vector space $V(\mathbb{C})$ is a complex scalar function on $V(\mathbb{C}) \times V(\mathbb{C})$, i.e., $V(\mathbb{C}) \times V(\mathbb{C}) \to \mathbb{C}$, which associates to each ordered pair $[\bar{x},\bar{y}]$ of vectors from $V(\mathbb{C})$ a complex number (a complex scalar) $(\bar{x},\bar{y})$, which has the following three properties:

1. $(\bar{x},\bar{y}) = (\bar{y},\bar{x})^*$—*skew (or Hermitian or conjugate) symmetry;*
2. $(\bar{z},a\bar{x}+b\bar{y}) = a(\bar{z},\bar{x})+b(\bar{z},\bar{y})$—*linearity in the second factor;*
3. $(\bar{x},\bar{x}) > 0$ for $\bar{x} = \bar{0}$ *and* $(\bar{x},\bar{x}) = 0$ iff $\bar{x} = \bar{0}$—*positive definiteness.*

(This inner product is obviously antilinear [2] in the first factor $(a\bar{x}+b\bar{y},\bar{z}) = a^*(\bar{x},\bar{z})+b^*(\bar{y},\bar{z})$.)

Together with 2, this means that it is *conjugate bilinear* [4]:

$$\left(\sum_{i=1}^{m} a_i\bar{x}_i, \sum_{j=1}^{n} b_j y_j\right) = \sum_{i,j} a_i^* b_j (\bar{x}_i,\bar{y}_i).$$

As in $V_n(\mathbb{R})$, we can define an inner prodcut in $V_n(\mathbb{C})$ by choosing any basis $v = \{\bar{v}_1,\bar{v}_2,\ldots,\bar{v}_n\}$ and expanding vectors from $V_n(\mathbb{C})$ in that basis:

$$\bar{x} = \sum_{i=1}^{n} x_i\bar{v}_i \text{ and } \bar{y} = \sum_{i=1}^{n} y_i\bar{v}_i.$$

Furthermore, we define $(\bar{x},\bar{y})_v = \sum_{i=1}^{n} x_i^* y_i = x^\dagger y$, $x = \begin{bmatrix} x_1 \\ x_2 \\ \vdots \\ x_n \end{bmatrix}$, $y = \begin{bmatrix} y_1 \\ y_2 \\ \vdots \\ y_n \end{bmatrix}$ as their

inner product induced by the basis v in the standard form.

In this inner product, the basis v becomes orthonormal [2, 3, 4] $(\bar{v}_i, \bar{v}_j)_v = \delta_{ij}$ since

$$\bar{v}_i = \sum_{k=1}^{n} \delta_{ik}\bar{v}_k \text{ and } \bar{v}_j = \sum_{k=1}^{n} \delta_{jk}\bar{v}_k, \text{ so}$$

$$(\bar{v}_i, \bar{v}_j)_v = \sum_{k=1}^{n} \delta_{ik}\delta_{jk} = \delta_{ij}.$$

Also orthonormal are all bases in $V_n(\mathbb{C})$ which are obtained from v by unitary replacement matrices $\mathscr{R}$, i.e., such that $\mathscr{R}^{-1} = \mathscr{R}^\dagger$.

To prove this, let us consider changing the basis $v = \{\bar{v}_1, \bar{v}_2, \ldots, \bar{v}_n\}$ in $V_n(\mathbb{C})$ to another basis $v' = \{\bar{v}'_1, \bar{v}'_2, \ldots, \bar{v}'_n\}$ by the invertible replacement matrix $\mathscr{R}$:

$$\begin{bmatrix} \bar{v}'_1 \\ \bar{v}'_2 \\ \vdots \\ \bar{v}'_n \end{bmatrix} = \mathscr{R} \begin{bmatrix} \bar{v}_1 \\ \bar{v}_2 \\ \vdots \\ \bar{v}_n \end{bmatrix} \text{ or } \bar{v}'_i = \sum_{j=1}^{n} r_{ij}\bar{v}_j.$$

Then the representing columns x' and x of a vector $\bar{x} \in V_n(\mathbb{C})$ in these two bases are connected by the so-called *contragredient* matrix [4] $(\mathscr{R}^{-1})^T$:

$$x' = \begin{bmatrix} x'_1 \\ x'_2 \\ \vdots \\ x'_n \end{bmatrix} = (\mathscr{R})^T \begin{bmatrix} x_1 \\ x_2 \\ \vdots \\ x_n \end{bmatrix} = (\mathscr{R}^{-1})^T x, \text{ (see Sect. 4.3.1)}$$

The inner product of the vectors $\bar{x}$ and $\bar{y}$ is defined in the first basis v as $(\bar{x}, \bar{y})_v = \sum_{i=1}^{n} x_i^* y_i = x^\dagger y$, where x and y are the representing columns of $\bar{x}$ and $\bar{y}$ in the first basis v. If we use the same definition with the second basis v', we get

$$(\bar{x}, \bar{y})_{v'} = (x')^\dagger y' = (\mathscr{R}^{-1^T} x)^\dagger (\mathscr{R}^{-1^T} y) = x^\dagger (\mathscr{R}^{-1^*})(\mathscr{R}^{-1^T}) y.$$

This will be the same number if $(\mathscr{R}^*)^{-1}(\mathscr{R}^T)^{-1} = I_n$ or $(\mathscr{R}^T \mathscr{R}^*)^{-1} = I_n$ or $\mathscr{R}^T \mathscr{R}^* = I_n$ or $\mathscr{R}^* = (\mathscr{R}^T)^{-1}$ or $(\mathscr{R}^{-1})^T = \mathscr{R}^*$ or $\mathscr{R}^{-1} = \mathscr{R}^{*T} = \mathscr{R}^\dagger$. Thus, when the replacement matrix $\mathscr{R}$ is unitary $(\mathscr{R}^{-1} = \mathscr{R}^\dagger)$, the definition of the inner product in these two bases will be the same.

All bases in $V_n(\mathbb{C})$ can be partitioned into *disjoint classes* of unitary equivalent bases. The reason for this is that unitary $n \times n$ matrices form the group $U(n)$ which defines an equivalence relation $\sim$ in the set of all bases:

$$\begin{bmatrix} \bar{v}_1 \\ \bar{v}_2 \\ \vdots \\ \bar{v}_n \end{bmatrix} \sim \begin{bmatrix} \bar{v}'_1 \\ \bar{v}'_2 \\ \vdots \\ \bar{v}'_n \end{bmatrix} \text{ if } \begin{bmatrix} \bar{v}'_1 \\ \bar{v}'_2 \\ \vdots \\ \bar{v}'_n \end{bmatrix} = \mathscr{R} \begin{bmatrix} \bar{v}_1 \\ \bar{v}_2 \\ \vdots \\ \bar{v}_n \end{bmatrix}, \quad \mathscr{R} \in U(n).$$

This relation is *reflexive, symmetric, and transitive (RST)*, which follows from the basic axioms of group theory—the existence of identity, also inversion and group multiplication are closed operations. Every such equivalence class defines one inner product in $V_n(\mathbb{C})$, and for that inner product the bases from the class (and only they) are orthonormal.

Also, there are no more inner products than those defined by classes of unitary equivalent bases. Namely, an arbitrary inner product in $V_n(\mathbb{C})$ defines such a class of orthonormal (unitary equivalent) bases, so there is a bijection between all inner products in $V_n(\mathbb{C})$ and all such classes. To see this in more detail, take an arbitrary inner product in $V_n(\mathbb{C})$ and choose any basis in $V_n(\mathbb{C})$. Then, find by the *Gramm-Schmidt orthonormalization process* (Sect. 3.3) its corresponding orthonormal basis $\{\bar{v}_1, \bar{v}_2, \ldots, \bar{v}_n\}$.

Our first task is to find the expansion coefficents of any vector $\bar{x} \in V_n(\mathbb{C})$ in this basis $\bar{x} = \sum_{j=1}^{n} x_j \bar{v}_j$. Multiplying this expansion from the left by $\bar{v}_i$, $i = 1, 2, \ldots, n$, we get

$$(\bar{v}_i, \bar{x}) = \sum_{j=1}^{n} x_j (\bar{v}_i, \bar{v}_j) = \sum_{j=1}^{n} x_j \delta_{ij} = x_i, \text{ so}$$

$$\boxed{\bar{x} = \sum_{i=1}^{n} (\bar{v}_i, \bar{x}) \bar{v}_i}.$$

These expansion coefficients $(\bar{v}_i, \bar{x})$, $i = 1, 2, \ldots, n$, are called the *Fourier* coefficients of $\bar{x}$ in this ON basis.

The inner product of two vectors $\bar{x}$ and $\bar{y}$ is calculated as

$$(\bar{x}, \bar{y}) = \left(\sum_{i=1}^{n} (\bar{v}_i, \bar{x}) \bar{v}_i, \sum_{j=1}^{n} (\bar{v}_j, \bar{y}) \bar{v}_j \right) = \sum_{i=1}^{n} \sum_{j=1}^{n} (\bar{v}_i, \bar{x})^* (\bar{v}_j, \bar{y}) (\bar{v}_i, \bar{v}_j) =$$

$$= \sum_{i,j=1}^{n} (\bar{v}_i, \bar{x})^* (\bar{v}_j, \bar{y}) \delta_{ij} = \sum_{i=1}^{n} (\bar{v}_i, \bar{x})^* (\bar{v}_i, \bar{y}) =$$

$$= \boxed{\sum_{i=1}^{n} (\bar{x}, \bar{v}_i)(\bar{v}_i, \bar{y})}.$$

It is called *Parseval's identity* [4] and can be written as $\boxed{(\bar{x}, \bar{y}) = x^\dagger y}$, where $x = \begin{bmatrix} (\bar{v}_1, \bar{x}) \\ (\bar{v}_2, \bar{x}) \\ \vdots \\ (\bar{v}_n, \bar{x}) \end{bmatrix}$ and similarly for y. So, relative to this ON basis $\{\bar{v}_1, \bar{v}_2, \ldots, \bar{v}_n\}$ and all other ON bases from its unitary equivalence class, the given inner product takes on the standard form, i.e., it is defined by them.

A complex vector space $V_n(\mathbb{C})$ with the above inner product (see the definition) is called a *unitary space* [4, 1] U_n (sometimes such a space is called a *complex inner-product vector space*, while $\mathbb{C}^n$ is called a *complex Euclidean space*).

We have already defined the standard inner product in $\mathbb{C}^n$ as $(\bar{x},\bar{y}) = \bar{x}^\dagger \bar{y}$, $\bar{x}, \bar{y} \in \mathbb{C}^n$. Now, we see that it is defined by the class of bases in $\mathbb{C}^n$ unitary equivalent to the usual (standard) basis $\{\bar{e}_1, \bar{e}_2, \ldots, \bar{e}_n\}$, where

$$\bar{e}_p = [\delta_{1p}\,\delta_{2p}\ldots\delta_{np}]^T, \quad p = 1, 2, \ldots, n.$$

Examples of inner products in other unitary spaces.

A) In the space $\mathbb{C}_{m \times n}$ of complex $m \times n$ matrices, the standard inner product is defined as $(A, B \in \mathbb{C}_{m \times n})$

$$\boxed{(A,B) = \mathrm{tr}(A^\dagger B) = \sum_{i=1}^m \sum_{j=1}^n a_{ij}^* b_{ij}}.$$

We can easily verify that this inner product satisfies the 3 axioms in the definition:

1. $(B,A) = \sum_{i=1}^m \sum_{j=1}^n b_{ij}^* a_{ij} = (\sum_{i=1}^m \sum_{j=1}^n a_{ij}^* b_{ij})^* = (A,B)^*$—*skew symmetry*;
2. $(A+B,C) = \mathrm{tr}[(A+B)^\dagger C] = \mathrm{tr}(A^\dagger C) + \mathrm{tr}(B^\dagger C) = (A,C) + (B,C)$, $(aA,B) = \mathrm{tr}[(aA)^\dagger B] = \mathrm{tr}(a^* A^\dagger B) = a^* \mathrm{tr}(A^\dagger B) = a^*(A,B)$—*antilinearity* in the first factor;
3. $(A,A) = \mathrm{tr}(A^\dagger A) = \sum_{i=1}^m \sum_{j=1}^n |a_{ij}|^2 > 0$ if $A \neq \hat{0}$, and equal to 0 iff $A = \hat{0}$ —*positive definiteness*.

B) If $x(t)$ and $y(t)$ are polynomials in the vector space $P(\mathbb{C})$ of complex polynomials on the real variable $t \in [a,b]$, then their inner product is defined as

$$(x(t),y(t)) = \int_a^b x^*(t)y(t)dt.$$

In both of the above examples, the inner products are natural generalizations of the standard inner product in $\mathbb{C}^n$—from one to two indices and from a discrete to a continuous variable, respectively.

3.3 Orthonormal Bases and the Gram-Schmidt Procedure for Orthonormalization of Bases

Definition Two nonzero vectors $\bar{x}$ and $\bar{y}$ in an inner product (real or complex) vector space are orthogonal $\bar{x} \perp \bar{y}$ iff their inner product is zero: $(\bar{x},\bar{y}) = 0 \Leftrightarrow \bar{x} \perp \bar{y}$. (see Sect. 3.1)

For two orthogonal vectors, we can easily prove the Pythagorean theorem:

$$||\bar{x}+\bar{y}||^2 = (\bar{x}+\bar{y},\bar{x}+\bar{y}) = ||\bar{x}||^2 + (\bar{x},\bar{y}) + (\bar{y},\bar{x}) + ||\bar{y}||^2 = ||\bar{x}||^2 + ||\bar{y}||^2.$$

Similarly, for an orthogonal set of vectors $\{\bar{x}_1, \bar{x}_2, \ldots, \bar{x}_k\}$

$$||\bar{x}_1 + \bar{x}_2 + \cdots + \bar{x}_k||^2 = ||\bar{x}||^2 + ||\bar{x}_2||^2 + \cdots + ||\bar{x}_k||^2,$$

where $(\bar{x}_i, \bar{x}_j) = 0$ for $i \neq j$.

A familiar illustration in $\mathbb{R}^2$ is

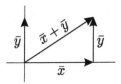

It is obvious that the zero vector $\bar{0}$ in any inner product vector space is orthogonal to any other vector $\bar{x}$ from that space: $(\bar{x}, \bar{0}) = 0$, but it is very useful in many proofs that the zero vector is the only one with this property.

Definition The set of vectors $\{\bar{x}_1, \bar{x}_2, \ldots, \bar{x}_k\}$ in any inner product vector space is called an *Ortho Normal (ON)* set if

$$(\bar{x}_i, \bar{x}_j) = \delta_{ij} = \begin{cases} 1 \text{ for } i = j \\ 0 \text{ for } i = j \end{cases}, i, j = 1, 2, \ldots, k.$$

In other words, this set of vectors is orthonormal if each vector from the set is of unit norm and at the same time is orthogonal to every other vector from the set.

It is a very important property of any orthonormal set of vectors that they are also linearly independent. This can be demonstrated easily: Consider an orthonormal (ON) set of vectors $\{\bar{x}_1, \bar{x}_2, \ldots, \bar{x}_k\}$. We have to show that this set is linearly independent, i.e., to show that a linear combination of these vectors $\sum_{i=1}^{k} a_i \bar{x}_i$ can be equal to the zero vector $\bar{0}$ only if all the coefficients a_i, $i = 1, 2, \ldots, k$ are zero— $\sum_{i=1}^{k} a_i \bar{x}_i = \bar{0} \Rightarrow$ all $a_i = 0$. Indeed, we shall multiply the above equality from the left with any $\bar{x}_j$, $j = 1, 2, \ldots, k$ and get on one side

$$(\bar{x}_j, \sum_{i=1}^{k} a_i \bar{x}_i) = \sum_{i=1}^{k} a_i (\bar{x}_j, \bar{x}_i) = \sum_{i=1}^{k} a_i \delta_{ij} = a_j,$$

while the other side gives $(\bar{x}_j, \bar{0})$, so $a_j = 0$, $i = 1, 2, \ldots, k$. Thus, every orthonormal ordered set of vectors which also spans its vector space is an orthonormal (ON) basis.

If $\{\bar{x}_1, \bar{x}_2, \ldots, \bar{x}_k\}$ is an ON set of vectors in a complex inner product vector space and if $\bar{x}$ is any vector from that space, then Bessel's inequality is valid:

$$\sum_{i=1}^{k} |(\bar{x}_i, \bar{x})|^2 \leq ||\bar{x}||^2.$$

Proof

$$0 \leq (\bar{x} - \sum_{i=1}^{k}(\bar{x}_i,\bar{x})\bar{x}_i, \bar{x} - \sum_{j=1}^{k}(\bar{x}_j,\bar{x})\bar{x}_j) =$$

$$= ||\bar{x}||^2 - \sum_{j=1}^{k}(\bar{x}_j,\bar{x})(\bar{x},\bar{x}_j) - \sum_{i=1}^{k}(\bar{x}_i,\bar{x})^*(\bar{x}_i,\bar{x}) + \sum_{i,j=1}^{k}(\bar{x}_i,\bar{x})^*(\bar{x}_j,\bar{x})(\bar{x}_i,\bar{x}_j) =$$

$$= ||\bar{x}||^2 - \sum_{j=1}^{k}|(\bar{x}_j,\bar{x})|^2 - \sum_{i=1}^{k}|(\bar{x}_i,\bar{x})|^2 + \sum_{j=1}^{k}|(\bar{x}_j,\bar{x})|^2 = ||\bar{x}||^2 - \sum_{i=1}^{k}|(\bar{x}_i,\bar{x})|^2.$$

In a real inner product vector space this inequality looks simpler

$$\sum_{i=1}^{k}(\bar{x}_i,\bar{x})^2 \leq ||\bar{x}||^2,$$

and the proof is analogous.

We shall be using only orthonormal (ON) bases because it is much easier to get useful results with them. We shall show below that every basis can be replaced with an equivalent ON basis.

But, first, let us list some useful formulas already obtained with ON bases.

1. If $\{\bar{x}_1,\bar{x}_2,\ldots,\bar{x}_n\}$ is any ON basis in a vector space (real or complex) with inner product, then the components of any vector $\bar{x}$ in that basis are

$$\boxed{\bar{x} = \sum_{i=1}^{n}(\bar{x}_i,\bar{x})\bar{x}_i}.$$

This expansion is called the *Fourier expansion*, and the expansion coefficients $(\bar{x}_i,\bar{x})$ are usually called the *Fourier coefficients*. The component $(\bar{x}_i,\bar{x})\bar{x}_i$ is the projection of $\bar{x}$ along the unit vector $\bar{x}_i$.

2. The inner product of two vectors $\bar{x}$ and $\bar{y}$ in this ON basis takes the form

$$\boxed{(\bar{x},\bar{y}) = \sum_{i=1}^{n}(\bar{x},\bar{x}_i)(\bar{x}_i,\bar{y})}.$$

This expression is called *Parseval's identity* [4].

3. The norm of any vector $\bar{x}$ in this ON basis is

$$\boxed{||\bar{x}||^2 = \sum_{i=1}^{n}|(\bar{x}_i,\bar{x})|^2}$$

(If the vector space is real, then there are no absolute value bars). We see that when an ON set is a basis, then in Bessel's inequality there remains only the equality sign.

Here, we can derive two important inequalities relevant for the norm in inner-product vector spaces.

The first example is the well-known *Cauchy–Schwarz* inequality:
For any two vectors $\bar{x}$ and $\bar{y}$ in an arbitrary inner-product vector space, we have

$$\boxed{|(\bar{x},\bar{y})| \leq ||\bar{x}||\,||\bar{y}||}$$

There are different proofs, but the simplest one uses Bessel's inequality:

If $\bar{y} = \bar{0}$, then both sides are 0 and we get the trivial equality. So we assume $\bar{y} \neq \bar{0}$ and then use Bessel's inequality $\sum_{i=1}^{k} |(\bar{x}_i,\bar{x})|^2 \leq ||\bar{x}||^2$ for $k = 1$ $\bar{x}_1 = \frac{\bar{y}}{||\bar{y}||}$ to obtain $|(\frac{\bar{y}}{||\bar{y}||},\bar{x})|^2 \leq ||\bar{x}||^2$. Multiplying both sides with $||\bar{y}||^2$, we get $|(\bar{x},\bar{y})|^2 \leq ||\bar{x}||^2||\bar{y}||^2$, which implies the Cauchy–Schwarz inequality.

The second example is the *triangle inequality* in inner-product vector spaces (which is one of three basic properties of the norm besides positive definiteness $||\bar{x}|| > 0$ if $\bar{x} \neq \bar{0}$ and homogeneity [3] $||a\bar{x}|| = |a|\,||\bar{x}||$): $||\bar{x}+\bar{y}|| \leq ||\bar{x}|| + ||\bar{y}||$, for any $\bar{x}$ and $\bar{y}$.

Proof Using the above Cauchy–Schwarz inequality $|(\bar{x},\bar{y})| \leq ||\bar{x}||\,||\bar{y}||$:

$$||\bar{x}+\bar{y}||^2 = (\bar{x}+\bar{y},\bar{x}+\bar{y}) = ||\bar{x}||^2 + (\bar{x},\bar{y}) + (\bar{y},\bar{x}) + ||\bar{y}||^2 \leq ||\bar{x}||^2 + 2||\bar{x}||\,||\bar{y}|| + ||\bar{y}||^2$$
$$= (||\bar{x}|| + ||\bar{y}||)^2.$$

[In complex vector space $(\bar{x},\bar{y}) + (\bar{y},\bar{x}) = (\bar{x},\bar{y}) + (\bar{x},\bar{y})^* = 2Re(\bar{x},\bar{y}) \leq 2|(\bar{x},\bar{y})|$, and in the real one $2(\bar{x},\bar{y}) \leq |(\bar{x},\bar{y})|$. Taking the square roots, we get the afore-mentioned triangle inequality.

We shall now describe the Gram–Schmidt procedure, which enables us to start with an arbitrary basis $X = \{\bar{x}_1,\bar{x}_2,\ldots,\bar{x}_n\}$ in an (real or complex) inner-product vector space V_n and obtain the corresponding orthonormal (ON) basis $Y = \{\bar{y}_1,\bar{y}_2,\ldots,\bar{y}_n\}$ in the same space, but such that every $\bar{y}_m$, $m = 1,2,\ldots,n$, is a linear combination of the first m vectors $\{\bar{x}_1,\bar{x}_2,\ldots,\bar{x}_m\}$:

$$\boxed{\bar{y}_m \in L(\bar{x}_1,\bar{x}_2,\ldots,\bar{x}_m), \quad m = 1,2,\ldots,n}$$

Since X is a basis (linearly independent and spanning the set), it follows that every one of its members is a nonzero vector, so that it can be made a unit vector by dividing it by its norm.

So, we take as $\bar{y}_1$ the normalized $\bar{x}_1$:

$$\bar{y}_1 = \frac{\bar{x}_1}{||\bar{x}_1||}.$$

So, $\bar{y}_1$ is of unit norm and it belongs to $L(\bar{x}_1)$:

$$\bar{y}_1 = \frac{1}{||\bar{x}_1||}\bar{x}_1$$

(a linear combination of $\bar{x}_1$).

To obtain the second vector $\bar{y}_2$ [it must be orthogonal to $\bar{y}_1$ and of unit norm and also it must belong to $L(\bar{x}_1, \bar{x}_2)$], we form a linear combination of $\bar{x}_1$ and $\bar{x}_2$ to get $\bar{y}_2' = x_2 - a_1 \frac{\bar{x}_1}{\|\bar{x}_1\|}$. The unknown coefficient a_1 should be determined from the orthogonality condition $(\bar{y}_1, \bar{y}_2') = 0$. So, we multiply $\bar{y}_2'$ from the left with $\bar{y}_1$ and obtain $(\bar{y}_1, \bar{y}_2') = (\bar{y}_1, \bar{x}_2) - a_1(\bar{y}_1, \bar{y}_1) = (\bar{y}_1, \bar{x}_2) - a_1$. This expression is 0 if $a_1 = (\bar{y}_1, \bar{x}_2)$. The vector $\bar{y}_2'$ cannot be zero since it is a linear combination of two linearly independent vectors $\bar{x}_1$ and $\bar{x}_2$ where at least one coefficient (that of $\bar{x}_2$) is certainly different from zero. Therefore, $\bar{y}_2$ is the normalized $\bar{y}_2'$:

$$\bar{y}_2 = \frac{\bar{y}_2'}{\|\bar{y}_2'\|} = (\bar{x}_2 - (\bar{y}_1, \bar{x}_2)\bar{y}_1)/\|\bar{x}_2 - (\bar{y}_1, \bar{x}_2)\bar{y}_1\|.$$

This vector $\bar{y}_2$ is of unit norm, it is orthogonal to $\bar{y}_1$, and it is a linear combination of $\bar{x}_1$ and $\bar{x}_2$. The vector $\bar{y}_2'$ is the difference between the vector $\bar{x}_2$ and the projection $(\bar{y}_1, \bar{x}_2)\bar{y}_1$ of $\bar{x}_2$ onto the unit vector $\bar{y}_1$. This difference is called the *normal* from $\bar{x}_2$ onto the line determined by $\bar{y}_1$.

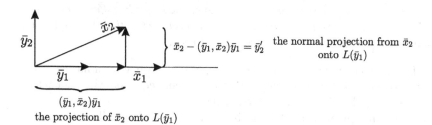

Thus, the vector $\bar{y}_2$ is the unit normal of $\bar{x}_2$ onto the subspace spanned by the previously found vector $\bar{y}_1$ of the desired ON basis. The subspace spanned by $\bar{x}_1$ and $\bar{x}_2$ is the same as that spanned by $\bar{y}_1$ and $\bar{y}_2$, but $\bar{y}_1$ and $\bar{y}_2$ are normalized (unit) vectors and are orthogonal to each other.

To determine $\bar{y}_3$ and the further vectors $\bar{y}_4, \bar{y}_5, \ldots, \bar{y}_n$, we follow the same idea: $\bar{y}_3$ is the normalized (unit) normal from $\bar{x}_3$ onto the subspace spanned by $\bar{y}_1$ and $\bar{y}_2$ (the previously found vectors from the ON basis).

The normal, nonzero vector

$$\bar{y}_3' = \bar{x}_3 - \underbrace{[(\bar{y}_1, \bar{x}_3)\bar{y}_1 + (\bar{y}_2, \bar{x}_3)\bar{y}_2]}_{}$$

projection of $\bar{x}_3$ onto $L(\bar{y}_1, \bar{y}_2) = L(\bar{x}_1, \bar{x}_2)$

and (the normalized normal)

$$\bar{y}_3 = \frac{\bar{y}_3'}{\|\bar{y}_3'\|}.$$

Thus, $\bar{y}_3$ is of unit norm, it is obviously orthogonal to both $\bar{y}_1$ and $\bar{y}_2$ [$(\bar{y}_1, \bar{y}_3') = (\bar{y}_1, \bar{x}_3) - (\bar{y}_1, \bar{x}_3) = 0$ and similarly for $\bar{y}_2$] and it belongs to $L(\bar{x}_1, \bar{x}_2, \bar{x}_3)$.

For further vectors of the ON basis, we do the same construction:

$$\bar{y}_i' = \bar{x}_i - [(\bar{y}_1, \bar{x}_i)\bar{y}_1 + (\bar{y}_2, \bar{x}_i)\bar{y}_2 + \ldots + (\bar{y}_{i-1}, x_i)\bar{y}_{i-1}], \quad i = 4, 5, \ldots, n,$$

the nonzero normal from $\bar{x}_i$ onto $L(\bar{y}_1, \bar{y}_2, \ldots, \bar{y}_{i-1})$ and

$$\bar{y}_i = \frac{\bar{y}_i'}{\|\bar{y}_i'\|}.$$

So, $\bar{y}_i$ is the normalized unit, normal from the corresponding $\bar{x}_i$ onto the subspace $L(\bar{y}_1, \bar{y}_2, \ldots, \bar{y}_{i-1})$ spanned by the previously found vectors.

The Gramm–Schmidt procedure for orthonormalization enables us to use only orthonormal bases in every (real or complex) inner-product vector space.

Example Normalized Legendre polynomials.

The linear differential equation of the second order $(1 - t^2)y''(t) - 2ty'(t) + v(v + 1)y(t) = 0$, where v is a real number, is called the *Legendre differential equation*, and every solution of this equation is called a *Legendre function*. This equation is found in many problems in physics and technology.

The Legendre equation has a polynomial as its solution if v is an integer. The Legendre polynomial $P_n(t)$ is a solution of the Legendre equation with parameter $v = n \in \{0, 1, 2, \ldots\}$ and with the property $P_n(1) = 1$.

The Legendre polynomial $P_n(t)$ can be obtained by the so-called *Rodrigues formula*

$$\boxed{P_n(t) = \frac{1}{2^n n!} \frac{d^n}{dt^n}[(t^2 - 1)^n], \quad n = 0, 1, 2, \ldots}$$

This formula gives immediately $P_0(t) = 1$ and $P_1(t) = \frac{1}{1!2} \frac{d}{dt}(t^2 - 1) = t$. The rest of the polynomials can be more easily calculated by the *recurrence formula*

$$\boxed{P_{n+1}(t) = \frac{2n+1}{n+1} t P_n(t) - \frac{n}{n+1} P_{n-1}(t).}$$

So for
$$P_2(t) = \frac{3}{2} t P_1(t) - \frac{1}{2} P_0(t) = \frac{3}{2} t^2 - \frac{1}{2} = \frac{1}{2}(3t^2 - 1),$$

and for
$$P_3(t) = \frac{5}{3} t P_2(t) - \frac{2}{3} P_1(t) = \frac{5}{3} t \frac{1}{2}(3t^2 - 1) - \frac{2}{3} t =$$
$$= \frac{5}{2} t^3 - \frac{5}{6} t - \frac{4}{6} t = \frac{5}{2} t^3 - \frac{3}{2} t = \frac{1}{2}(5t^3 - 3t)$$

If we define an inner product in the vector space $P(\mathbb{R})$ of real polynomials of real variable t on the interval $(-1, +1)$ as

$$(p(t), q(t)) = \int_{-1}^{+1} p(t)q(t)dt,$$

then we can easily deduce that Legendre polynomials are orthogonal

$$\int_{-1}^{+1} P_n(t)P_m(t)dt = 0 \quad (m \neq n)$$

[assume $m > n$, use Rodrigues formula for $P_n(t)$ and $P_m(t)$, and integrate n times by parts]. The square of the norm of $P_n(t)$ is

$$\int_{-1}^{+1} P_n^2(t) = \frac{2}{2n+1}, \quad n = 0,1,2,\ldots.$$

The normalized Legendre polynomials

$$\boxed{y_n(t) = \sqrt{\frac{2n+1}{2}}P_n(t)}, \quad n = 0,1,2,\ldots$$

form an orthonormal basis in $P(\mathbb{R})$, as well as in the Hilbert space $L_2(-1,+1)$ of square integrable real functions on the interval $(-1,+1)$.

Now, we shall show that the normalized Legendre polynomials $y_n(t) = \sqrt{\frac{2n+1}{2}} P_n(t)$ can be obtained by the Gram–Schmidt procedure for orthonormalization from the standard basis $\{1,t,t^2,t^3,\ldots\}$ in $P(\mathbb{R})$ with the above inner product. For the sake of economy, we shall derive only the first four of them:

$$y_0(t) = \sqrt{\frac{1}{2}}, \ y_1 = \sqrt{\frac{3}{2}}t, \ y_2 = \sqrt{\frac{5}{2}} \cdot \frac{1}{2}(3t^2 - 1), \ y_3 = \sqrt{\frac{7}{2}} \cdot \frac{1}{2}(5t^3 - 3t).$$

Let us start with the first four vectors of the standard basis $X = \{x_0(t), x_1(t), x_2(t), x_3(t)\} = \{1,t,t^2,t^3\}$, and apply the Gram–Schmidt procedure.

The first vector from the corresponding orthonormal basis is obviously

$$y_0(t) = \frac{x_0(t)}{||x_0(t)||} = 1/(\int_{-1}^{+1} dt)^{1/2} = 1\sqrt{2} = \sqrt{\frac{1}{2}}.$$

For the second vector from the ON basis, we calculate the first normal

$$y_1'(t) = x_1(t) - (y_0(t), x_1(t)) = t - \frac{1}{2}\int_{-1}^{+1} t\,dt = t - \frac{1}{2}\left[\frac{t^2}{2}\right]_{-1}^{+1} = t - 0 = t,$$

and then the normalized normal is

$$y_1(t) = \frac{y_1'(t)}{||y_1'(t)||} = \frac{t}{(\int_{-1}^{+1} t^2\,dt)^{1/2}} = \frac{t}{\left\{\left[\frac{t^3}{3}\right]_{-1}^{+1}\right\}^{1/2}} = \frac{1}{\sqrt{\frac{2}{3}}} = \sqrt{\frac{3}{2}}t.$$

For the third vector from the corresponding ON basis, we calculate the normal from $x_2(t)$ onto the subspace spanned by the previously found ON vectors $y_0(t)$ and $y_1(t)$:

$$y_2'(t) = x_2(t) - [(y_0(t), x_2(t))y_0(t) + (y_1(t), x_2(t))y_1(t)] =$$

$$= t^2 - [\frac{1}{2}\int_{-1}^{+1} t^2 dt + \frac{3}{2}t \int_{-1}^{+1} t^3 dt] = t^2 - \frac{1}{2}\left[\frac{t^3}{3}\right]_{-1}^{+1} - \frac{3}{2}t \left[\frac{t^4}{4}\right]_{-1}^{+1} =$$

$$= t^2 - \frac{1}{3} = \frac{1}{3}(3t^2 - 1).$$

[Notice that $P_2(t) = \frac{1}{2}(3t^2 - 1)$ and $y_2'(t) = \frac{1}{3}(3t^2 - 1)$ differ by a factor, $P_2(t) = \frac{3}{2}y_2'(t)$, (colinear vectors), but their normalized vector must be the same]. The normalized normal is

$$y_2(t) = \frac{y_2'(t)}{||y_2'(t)||},$$

but it is more practical to calculate the square of the norm

$$||y_2'(t)||^2 = \int_{-1}^{+1}(t^2 - \frac{1}{3})^2 dt = \int_{-1}^{+1} t^4 dt - \int_{-1}^{+1} \frac{2}{3}t^2 dt + \int_{-1}^{+1}\frac{1}{9} dt =$$

$$= \left[\frac{t^5}{5}\right]_{-1}^{+1} - \frac{2}{3}\left[\frac{t^3}{3}\right]_{-1}^{+1} + \frac{1}{9}[t]_{-1}^{+1} = \frac{2}{5} - \frac{2}{3}\cdot\frac{2}{3} + \frac{2}{9} = \frac{2}{5} - \frac{2}{9} = \frac{8}{45}.$$

So $||y_2'(t)|| = \sqrt{\frac{8}{45}} = \frac{2}{3}\sqrt{\frac{2}{5}}$, and $y_2(t) = \sqrt{\frac{5}{2}}\frac{3}{2}y'(t) = \sqrt{\frac{5}{2}}P_2(t) = \sqrt{\frac{5}{2}}\cdot\frac{1}{2}(3t^2 - 1)$.
For the fourth vector from the corresponding orthonormal basis, we first calculate the normal from $x_3(t)$ onto the subspace $L(y_0(t), y_1(t), y_2(t))$ spanned by the previously found ON vectors $y_0(t), y_1(t), y_2(t)$

$$y_3'(t) = x_3(t) - [(y_0(t), x_3(t))y_0(t) + (y_1(t), x_3(t))y_1(t) + (y_2(t), x_3(t))y_2(t)] =$$

$$= t^3 - \left[\frac{1}{2}\int_{-1}^{+1} t^3 dt + \frac{3}{2}t \int_{-1}^{+1} t^4 dt + \frac{5}{8}(3t^2 - 1)\int_{-1}^{+1}(3t^2 - 1)t^3 dt\right] =$$

$$= t^3 - \frac{3}{2}t\cdot\frac{2}{5} = t^3 - \frac{3}{5}t = \frac{1}{5}(5t^3 - 3t).$$

[Again $P_3(t) = \frac{1}{2}(5t^3 - 3t)$ and $y_3'(t) = \frac{1}{5}(5t^3 - 3t)$ are colinear vectors $P_3(t) = \frac{5}{2}y_3'(t)$]. The square of the norm of $y_3'(t)$ is

$$||y_3'(t)||^2 = \int_{-1}^{+1}\frac{1}{25}(5t^3 - 3t)^2 dt = \frac{1}{25}\int_{-1}^{+1}(25t^6 - 30t^4 + 9t^2) dt =$$

$$= \frac{1}{25}(25\cdot\frac{2}{7} - 30\cdot\frac{2}{5} + 9\cdot\frac{2}{3}) = \frac{1}{25}(\frac{50}{7} - 12 + 6) = \frac{1}{25}(\frac{50}{7} - \frac{42}{7}) =$$

$$= \frac{1}{25}\cdot\frac{8}{7} = \frac{4}{25}\cdot\frac{2}{7}.$$

The normalized normal is

$$y_3(t) = \sqrt{\frac{7}{2}} \cdot \frac{5}{2} \cdot \frac{1}{5}(5t^3 - 3t) = \sqrt{\frac{7}{2}} \cdot \frac{1}{2}(5t^3 - 3t) = \sqrt{\frac{7}{2}} P_3(t),$$

and it is exactly the normalized Legendre polynomial $\sqrt{\dfrac{7}{2}} P_3(t)$.

3.4 Direct and Orthogonal Sums of Subspaces and the Orthogonal Complement of a Subspace

3.4.1 Direct and Orthogonal Sums of Subspaces

Consider two subspaces V' and V'' of the vector space V which are such that they have in common only the zero vector $\bar{0}$:

$$V' \cap V'' = \{\bar{0}\}.$$

The set of all vectors $\bar{x}$ from V that can be expressed in the form $\bar{x} = \bar{x}_1 + \bar{x}_2$, $\bar{x}_1 \in V'$, $\bar{x}_2 \in V''$ is called the *direct sum* of V' and V'' and is denoted as $V' + V''$:

$$V' + V'' = \{\bar{x}_1 + \bar{x}_2 \mid \bar{x}_1 \in V' \text{ and } \bar{x}_2 \in V'' \text{ with } V' \cap V'' = \{\bar{0}\}\}.$$

Obviously, the set $V' + V''$ is a subspace itself, since it is closed to the addition of its vectors, as well as to the multiplication of its vectors with scalars: take $\bar{x}_1 + \bar{x}_2 \in V' + V''$ and $\bar{y}_1 + \bar{y}_2 \in V' + V''$, $a \in \mathbb{R}$ or $\mathbb{C}$, then $(\bar{x}_1 + \bar{x}_2) + (\bar{y}_1 + \bar{y}_2) = (\bar{x}_1 + \bar{y}_1) + (\bar{x}_2 + \bar{y}_2) \in V' + V''$ and $a(\bar{x}_1 + \bar{x}_2) = a\bar{x}_1 + a\bar{x}_2 \in V' + V''$.

In the subspace $V' + V''$, every vector $\bar{x}$ is uniquely decomposed into the sum $\bar{x}_1 + \bar{x}_2$, $\bar{x}_1 \in V'$, $\bar{x}_2 \in V''$, due to $V' \cap V'' = \{\bar{0}\}$. To see this, let us assume $V' \cap V'' \neq \{\bar{0}\}$ and let $\bar{y}$ be any vector from this intersection. Then, the decomposition $\bar{x} = \bar{x}_1 + \bar{x}_2$ is not unique since $\bar{x} = (\bar{x}_1 + \bar{y}) + (\bar{x}_2 - \bar{y})$. Thus, this decomposition is unique only if $V'' \cap V'' = \{\bar{0}\}$.

Two examples

1. In the space $\mathbb{R}^3$ consider a plane $(\mathbb{R}^2)$ through the origin and a line $(\mathbb{R}^1)$ also through the origin, but not lying in the plane. Obviously,

$$\mathbb{R}^3 = \mathbb{R}^1 + \mathbb{R}^2.$$

2. The space $\mathbb{R}_{n \times n}$ of all square $n \times n$ real matrices is the direct sum of the subspaces of all symmetric $(A^T = A)$ and of all skew-symmetric $(A^T = -A)$ matrices.

The set of all symmetric matrices $(A^T = A)$ is indeed a subspace since it is closed under the addition of matrices $[(A + B)^T = A^T + B^T = A + B]$, as well as under the

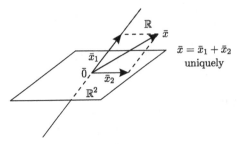

$$\bar{x} = \bar{x}_1 + \bar{x}_2$$
uniquely

multiplication of matrices with numbers $[(aA)^T = aA^T = aA]$. And similarly for the set of skew-symmetric matrices $(A^T = -A)$.

Each matrix $A \in \mathbb{R}_{n \times n}$ can be uniquely written as the sum of a symmetric and a skew-symmetric matrix:

$$A = \frac{A + A^T}{2} + \frac{A - A^T}{2},$$

so that $\left(\dfrac{A + A^T}{2}\right)^T = \dfrac{A^T + A}{2}$ and $\left(\dfrac{A - A^T}{2}\right)^T = \dfrac{A^T - A}{2} = -\dfrac{A - A^T}{2}.$

It is also obvious that only the zero matrix can be symmetric and skew-symmetric at the same time.

The dimension of the space $\mathbb{R}_{n \times n}$ is n^2 since its standard basis has n^2 matrices (the $n \times n$ matrices which have only one element $= 1$, all others being zeros). The standard basis in the subspace of symmetric matrices consists of n matrices that have one element $= 1$ on the main diagonal (the rest of the elements are zeros) and $(n-1) + (n-2) + \ldots + 2 + 1 = \frac{(n-1)n}{2}$ matrices that have two elements equal 1, the elements that are placed symmetrically with respect to the main diagonal (remember that symmetric matrices have $a_{ij} = a_{ji}$, $i \neq j$). Similarly, the standard basis for the subspace of skew-symmetric matrices consists of $\frac{(n-1)n}{2}$ matrices that have 1 and (-1) placed symmetrically to the main diagonal (remember that $a_{ij} = -a_{ji}$, $i \neq j$, and $a_{ii} = 0$ are the characteristic properties of skew-summetric matrices).

So, the sum of dimensions of the two subspaces

$$\left[n + \frac{n(n-1)}{2}\right] + \frac{n(n-1)}{2} = n + n(n-1) = n^2$$

is equal to the dimension of the space $\mathbb{R}_{n \times n}$ which is the direct sum of these subspaces. $\triangle$

This statement is generally valid:
The dimension of the subspace $V' + V''$ is equal to the sum of the dimensions of V' and V''

$$\dim(V' + V'') = \dim V' + \dim V'',$$

again due to $V' \cap V'' = \{\bar{0}\}$.

To prove this almost obvious statement, we choose a basis $\{\bar{f}_1, \bar{f}_2, \ldots, \bar{f}_m\}$ in V' ($\dim V' = m$) and a basis $\{\bar{g}_1, \bar{g}_2, \ldots, \bar{g}_n\}$ in V'' ($\dim V'' = n$) and show that the set $B = \{\bar{f}_1, \bar{f}_2, \ldots, \bar{f}_m, \bar{g}_1, \bar{g}_2, \ldots, \bar{g}_n\}$ is a basis in $V' + V''$—a linearly independent and generating set, so that $\dim(V' + V'') = m + n$. To prove that they are LIND, we make the LIND test

$$a_1 \bar{f}_1 + a_2 \bar{f}_2 + \ldots + a_m \bar{f}_m + b_1 \bar{g}_1 + b_2 \bar{g}_2 + \ldots + b_n \bar{g}_n = \bar{0}.$$

Transfering the linear combination with bs on to the other side, we conclude that one vector from V' is equal to another vector from V'', so that they both must be $\bar{0}$, since $V' \cap V'' = \{\bar{0}\}$. This immediately implies that all as and all bs are zeros, since the $\bar{f}$s and $\bar{g}$s are bases. Furthermore, the set B is a generating set by definition of $V' + V''$, since every $\bar{x} \in V' + V''$, i.e., $\bar{x} = \bar{x}_1 + \bar{x}_2$, $\bar{x}_1 \in V'$, $\bar{x}_2 \in V''$, is their linear combination. $\quad\quad \Delta$

Two subspaces V' and V'' in an inner-product vector space V (real, i.e., Euclidean or complex, i.e., unitary) are orthogonal, $V' \perp V''$, if $(\bar{x}, \bar{y}) = 0$ for every $\bar{x} \in V'$ and every $\bar{y} \in V''$. They have no common vectors, i.e., $V' \cap V'' = \{\bar{0}\}$, since only the zero vector $\bar{0}$ is orthogonal to itself. Thus, we have that the set of vectors $\bar{x} \in V$ of the form $\{\bar{x}_1 + \bar{x}_2, \ \bar{x}_1 \in V', \bar{x}_2 \in V'', \ (\bar{x}_1, \bar{x}_2) = 0\}$ is the direct sum $V' + V''$. But, since $V' \perp V''$, we call such a sum the *orthogonal sum* of V' and V'' and denote it as $V' \oplus V''$. All that we said about the direct sum $V' + V''$ is naturally valid now, but with one stronger condition $(\bar{x}_1, \bar{x}_2) = 0$. So, $\bar{x} = \bar{x}_1 + \bar{x}_2$ is the unique decomposition and $\dim(V' \oplus V'') = \dim V' + \dim V''$.

Both concepts of direct and orthogonal sums of two subspaces in V can be extended to any collection of subspaces. If $V', V'', \ldots, V^{(k)}$ is a set of subspaces in V which are all disjoint to each other $V^{(i)} \cap V^{(j)} = \{\bar{0}\}$, $i \neq j = 1, 2, \ldots, k$, then the set of vectors from V that can be expressed as a sum of k vectors, each of them from one of the subspaces, is itself a subspace of V denoted as $V' + V'' + \ldots + V^{(k)}$ and called their direct sum. Each vector from that direct sum has unique components in these subspaces.

If V is an inner-product vector space and if all subspaces are orthogonal to each other, $V^{(i)} \cap V^{(j)}$, $i \neq j = 1, 2, \ldots, k$, then the above direct sum becomes the orthogonal sum $V' \oplus V'' \oplus \cdots \oplus V^{(k)}$.

One of the main tasks in the theory of unitary spaces with regard to applications in Quantum Mechanics is to decompose the whole state space into a certain orthogonal sum of its subspaces.

3.4.2 The Orthogonal Complement of a Subspace

Definition Consider a (real or complex) inner-product vector space V and one of its subspaces W. The set of all vectors from V which are orthogonal to every vector from W is called the *orthocomplement* of W, and it is denoted by $W^\perp$.

It is easy to show that the orthocomplement $W^\perp$ is also a subspace of V. The proof follows from the fact that the inner product in V is linear in the second factors regardless if V is a real or complex vector space. Take any two vectors $\bar{x}$ and $\bar{y}$ from

$W^\perp$. To show that $W^\perp$ is a subspace in V, we have to prove that an arbitrary linear combination $a\bar{x} + b\bar{y}$ of $\bar{x}$ and $\bar{y}$ (a and b are any scalars from $\mathbb{R}$ or $\mathbb{C}$) also belongs to $W^\perp$. Indeed, the fact that $\bar{x}$ and $\bar{y}$ are from $W^\perp$ is equivalent to $(\bar{z}, \bar{x}) = 0$ and $(\bar{z}, \bar{y}) = 0$, for any $\bar{z} \in W$, so that $(\bar{z}, a\bar{x}, b\bar{y}) = a(\bar{z}, \bar{x}) + b(\bar{z}, \bar{y}) = 0$ or $a\bar{x} + b\bar{y} \in W^\perp$.

Obviously, the orthocomplement of $W^\perp$ is W itself:

$$\boxed{(W^\perp)^\perp = W}.$$

To prove this, let us start with the statement that W is a subspace of $(W^\perp)^\perp$ since every vector in W is orthogonal to all vectors in $(W^\perp)^\perp$. To demonstrate that they are equal, we shall compare their dimensions. If the dimensions are equal, this would imply that the subspace W must be equal to $(W^\perp)^\perp$. We shall show below that the dimensions of a subspace and its orthocomplement add up to the dimension of the whole space V. So, $\dim(W) + \dim(W^\perp) = \dim(V)$, as well as $\dim(W^\perp) + \dim((W^\perp)^\perp) = \dim(V)$. Consequently, $\dim((W^\perp)^\perp) = \dim(V) - \dim(W^\perp) = \dim(V) - [\dim(V) - \dim(W)] = \dim(W)$. Δ

Theorem Every vector $\bar{v} \in V$ can be written in one and only one way as the sum

$$\bar{v} = \bar{w} + \bar{w}', \quad \bar{w} \in W \text{ and } \bar{w}' \in W^\perp, \text{ [of course } (\bar{w}, \bar{w}') = 0].$$

In other words, we say that W and $W^\perp$ are orthogonally added to form V:

$$\boxed{W \oplus W^\perp = V}.$$

As a consequence, $\dim(W) + \dim(W^\perp) = \dim(V)$.

Proof Take any basis in W, and using the Gramm–Schmidt procedure for orthonormalization find out the equivalent orthonormal (ON) basis $\{\bar{x}_1, \bar{x}_2, \ldots, \bar{x}_n\}$ (n is the dimension of W). We show now that for every $\bar{v} \in V$ we can uniquely determine vectors

$$\bar{w} = \sum_{k=1}^{n} (\bar{x}_k, \bar{v})\bar{x}_k \text{ and } \bar{w}' = \bar{v} - \bar{w} = \bar{v} - \sum_{k=1}^{n} (\bar{x}_k, \bar{v})\bar{x}_k.$$

($\bar{w}$ is the sum of projections of vector $\bar{v}$ along n unit orthogonal vectors which make the ON basis in W).

Obviously, $\bar{w}$ belongs to W since it is a linear combination of basis vectors in W. Furthermore, $\bar{w}'$ belongs to $W^\perp$, since it is orthogonal to all basis vectors in W and consequently to all vectors in W:

$$(\bar{x}_j, \bar{w}') = (\bar{x}_j, \bar{v}) - \sum_{k=1}^{n} (\bar{x}_k, \bar{v})\underbrace{(\bar{x}_j, \bar{x}_k)}_{\delta_{jk}} = (\bar{x}_j, \bar{v}) - (\bar{x}_j, \bar{v}) = 0, \quad j = 1, 2, \ldots, n.$$

The components $\bar{w}$ and $\bar{w}'$ of $\bar{v}$ are unique, since it can be shown that they do not depend on a particular choice of ON basis $\{\bar{x}_1, \bar{x}_2, \ldots, \bar{x}_n\}$.

To show this uniqueness, consider another ON basis in W

$$m = 1, 2, \ldots, n,$$

$$\bar{y}_m = \sum_{k=1}^{n} r_{mk} \bar{x}_k, \quad \mathscr{R} = [r_{mk}]_{n \times n}.$$

The condition (see later) that $\{\bar{y}_1, \bar{y}_2, \ldots, \bar{y}_n\}$ is an ON basis is $\mathscr{R}^{\dagger} \mathscr{R} = I_n$ for the complex inner-product vector space, or $\mathscr{R}^T \mathscr{R} = I_n$ for the real space. Let us calculate in the complex case the projection $\bar{w}_{new}$ of $\bar{v}$ step into W using this new basis:

$$\bar{w}_{new} = \sum_{m=1}^{n} (\bar{y}_m, \bar{v}) \bar{y}_m = \sum_{m=1}^{n} \left[\left(\sum_{j=1}^{n} r_{mj} \bar{x}_j, \bar{v} \right) \sum_{k=1}^{n} r_{mk} \bar{x}_k \right] =$$

$$= \sum_{j,k=1}^{n} \sum_{m=1}^{n} r_{mj}^{*} r_{mk} (\bar{x}_j, \bar{v}) \bar{x}_k = \sum_{j,k=1}^{n} \underbrace{\sum_{m=1}^{n} \{\mathscr{R}^{\dagger}\}_{jm} \{\mathscr{R}_{mk}\}}_{\delta_{jk}} (\bar{x}_j, \bar{v}) \bar{x}_k =$$

$$= \sum_{k=1}^{n} (\bar{x}_k, \bar{v}) \bar{x}_k = \bar{w}. \quad \Delta$$

We have already called the component $\bar{w}$ of $\bar{v}$ the *projection* of $\bar{v}$ into the subspace W, and, as just shown, it is obtained as the sum of projections of $\bar{v}$ along n unit vectors of any ON basis in W.

The other component $\bar{w}' = \bar{v} - \bar{w} \in W^{\perp}$ is called the *normal* of the vector $\bar{v}$ onto the subspace W, since it is orthogonal to all vectors in W.

The length of the normal $\bar{w}'$ represents the shortest distance of vector $\bar{v}$ to the subspace W: precisely, if $\bar{z}$ is any vector in W distinct from $\bar{w}$, then $||\bar{w}'|| = ||\bar{v} - \bar{w}|| < ||\bar{v} - \bar{z}||$, $\bar{z} \in W$, $\bar{z} \neq \bar{w}$. We say that the projection $\bar{w}$ is the best approximation of the vector $\bar{v}$ by any vector from W.

Proof Notice that the vector $\bar{w} - \bar{z}$ is also from W, and as such it is orthogonal to the normal $\bar{w}' = \bar{v} - \bar{w}$. But for orthogonal vectors the Pythagorean theorem $||\bar{x}||^2 + ||\bar{y}||^2 = ||\bar{x} + \bar{y}||^2$ is valid:

$$||\bar{v} - \bar{w}||^2 + ||\bar{w} - \bar{z}||^2 = ||\bar{v} - \bar{w} + \bar{w} - \bar{z}||^2 = ||\bar{v} - \bar{z}||^2.$$

If we take away the always positve $||\bar{w}^2 - \bar{z}||^2$, we get

$$||\bar{v} - \bar{w}||^2 < ||\bar{v} - \bar{z}||^2 \Rightarrow ||\bar{v} - \bar{w}|| < ||\bar{v} - \bar{z}||. \quad \Delta$$

As an alternative for finding $W^{\perp}$ in the case of $\boxed{V = \mathbb{R}^n}$ (the space of real matrix-columns of length n), consider a subspace W in $\mathbb{R}^n$, $\dim W = m < n$, and take any basis $\{\bar{y}_1, \bar{y}_2, \ldots, \bar{y}_m\}$ in W. For the purpose of calculating $W^{\perp}$, we shall write a matrix A of the type $m \times n$ which has as its m rows the transposed $\{\bar{y}_1^T, \bar{y}_2^T, \ldots, \bar{y}_m^T\}$

$$A = \begin{bmatrix} y_{11} & y_{12} & \cdots & y_{1n} \\ y_{21} & y_{22} & \cdots & y_{2n} \\ \vdots & \vdots & \ddots & \vdots \\ y_{m1} & y_{m2} & \cdots & y_{mn} \end{bmatrix}.$$

We are looking for the subspace $W^{\perp}$ in $\mathbb{R}^n$ which consists of all vectors from $\mathbb{R}^n$ orthogonal to all vectors of the above basis. (If a vector is orthogonal to all vectors of a basis in W, then it is orthogonal to all vectors in the subspace W.)

Consider now the *kernel* of A, ker A, which is made up of all vectors $\bar{x}$ from $\mathbb{R}^n$ that are mapped by A onto the zero-vector $\bar{0}_m$ in $\mathbb{R}^m$ (the $m \times n$ matrix A is a linmap $A : \mathbb{R}^n \to \mathbb{R}^m$):

$$\ker A = \{\bar{x} \mid \bar{x} \in \mathbb{R}^n \text{ and } A\bar{x} = \bar{0}_m\}.$$

The matrix equation $A\bar{x} = \bar{0}_m$ is a concise form of a homogeneous linear system of m equations with n unknowns. The solution space of this system is a subspace of $\mathbb{R}^n$ of dimension $m - n$ (the total number of columns in A minus the number of linearly independent columns which is equal to the number of LIND rows—a basis in W). Obviously, every vector $\bar{x}$ from the solution space, i.e., ker A, is orthogonal to all vectors from the basis $\{\bar{y}_1, \bar{y}_2, \ldots, \bar{y}_m\}$:

$$(\bar{y}_i, \bar{x}) = \bar{y}_i^T \bar{x} = y_{i1}x_1 + y_{i2}x_2 + \cdots + y_{in}x_n = 0, \quad i = 1, 2, \ldots, m,$$

so it is a vector from $\mathbb{R}^n$ orthogonal to the whole W. Therefore, $\bar{x}$ belongs to $W^{\perp}$. There are no vectors in $W^{\perp}$ that are not from ker A, since the matrix equation $A\bar{x} = \bar{0}_m$ is in fact a search for all $\bar{x} \in \mathbb{R}^n$ that are orthogonal to all vectors from the basis $\{\bar{y}_1, \bar{y}_2, \ldots, \bar{y}_m\}$. Thus,

$$\boxed{\ker A = W^{\perp}}.$$

The dimension m of W and the dimension $(n - m)$ of $W^{\perp}$ add up to the dimension n of the whole space $\mathbb{R}^n$, since they have no common vectors except $\bar{0}_n$, because $\bar{0}_n$ is the only vector in $\mathbb{R}^n$ orthogonal to itself. So, $\mathbb{R}^n$ is the orthogonal sum of W and $W^{\perp}$:

$$\boxed{\mathbb{R}^n = W \oplus W^{\perp}}. \ \Delta$$

Chapter 4
Dual Spaces and the Change of Basis

From now on we shall call every linmap $U_n \rightarrow U_n$ a *linear operator* or simply an *operator*.

Linear operators in unitary spaces like projection operators (projectors), positive, Hermitian, and unitary operators are of fundamental importance for applications of linear algebra in Quantum Mechanics and Quantum statistical physics. The basic problem for these operators is to investigate the algebraic operations that can be performed among them and consequently to define the algebraic structures that they form. The second task is to find ON bases in which their representing matrices take the simplest (*canonical*) form. To say this in a more precise way, we shall try to break up the unitary space into an orthogonal sum of subspaces in each of which the operator under investigation acts simply as a multiplicative constant (the so-called *Eigen problem*). These constants form the spectrum of the operator which gives the quantum characteristics of the physical systems which are being studied.

We shall first study linear operators in unitary spaces, which is an easier task than to study linear operators in Euclidean spaces. The reason for such an approach is that the field $\mathbb{C}$ of complex numbers is algebraically closed (every algebraic equation with complex numbers has all its solutions in the same field). Linear operators in Euclidean spaces are important in Classical physics, and we shall consider them after linear operators in unitary spaces to benefit from the solutions obtained previously.

4.1 The Dual Space U_n^* of a Unitary Space U_n

The set of all linear operators in a finite-dimensional unitary space U_n is denoted by $\hat{L}(U_n, U_n)$. We know that in this set one can perform operations of addition and multiplication (*composition*) of operators:

$$(A+B)\bar{x} = A\bar{x} + B\bar{x}, \ \bar{x} \in U_n \ \text{and} \ (AB)\bar{x} = A(B\bar{x}), \ \bar{x} \in U_n,$$

as well as multiplication with scalars

$$(aA)\bar{x} = a(A\bar{x}), \quad \bar{x} \in U_n, \quad a \in \mathbb{C}.$$

Since the identity operator $I_n\bar{x} = \bar{x}$, $\bar{x} \in U_n$, belongs to this set, the usual term for the algebraic structure that they form is the *algebra with unity*.

Remark In representation theory, we discussed the set $\hat{L}(V_n(F), W_m(F))$ of linmaps (Sect. 2.6.1) and proved that this set is a vector space over the field F.

But, in the space $\hat{L}(V_n(F), V_n(F))$ of linear operators acting in $V_n(F)$, we have also multiplication (*composition*) of operators

$$(AB)\bar{x} = A(B\bar{x}), \quad \forall \bar{x} \in V_n(F).$$

This is obviously a *linear operator*

$$(AB)(a\bar{x}_1 + b\bar{x}_2) = A[B(a\bar{x}_1 + b\bar{x}_2)] = A(aB\bar{x}_1 + bB\bar{x}_2) =$$
$$= aA(B\bar{x}_1) + bA(B\bar{x}_2) = a(AB)\bar{x}_1 + b(AB)\bar{x}_2.$$

This multiplication is *associative*

$$[(AB)C]\bar{x} = (AB)(C\bar{x}) = A(BC\bar{x}) = [A(BC)]\bar{x},$$

and *distributive* with respect to the addition of operators $(A + B)C = AC + BC$ and $A(B + C) = AB + AC$:

$$[(A + B)C]\bar{x} = (A + B)(C\bar{x}) = A(C\bar{x}) + B(C\bar{x}) = (AC)\bar{x} + (BC)\bar{x} = (AC + BC)\bar{x}$$

and similarly for the other distributive law. (Since multiplication of operators is not, in general, commutative $AB \neq BA$, we need to prove both distributive laws). As far as multiplication with scalars is concerned, the new operation of multiplication of operators is related to the previous one as follows: $a(AB) = (aA)B = A(aB)$:

$$a(AB)\bar{x} = a[A(B\bar{x})] = (aA)(B\bar{x}) = [(aA)B]\bar{x}$$
$$a(AB)\bar{x} = a[A(B\bar{x})] = A[a(B\bar{x})] = A(aB\bar{x}) = [A(aB)]\bar{x}.$$

Definition (of the algebra with unity over the field F)

The vector space $\hat{L}(V_n(F), V_n(F))$ of linear operators in $V_n(F)$ becomes an *algebra with unity* over F since (1) in the vector space $\hat{L}(V_n(F), V_n(F))$ is defined another closed binary operation $(AB)\bar{x} = A(B\bar{x})$, $\forall \bar{x} \in V_n(F)$, which might be called the multiplication (or composition) of operators, such that it is associative, distributive with respect to the addition of operators (both ways) and with the property $a(AB) = (aA)B = A(aB)$ with respect to multiplication of operators with scalars from F; and (2) $\neq F$ since there exists an operator I_v, which we call the multiplicative unity, such that $I_vA = AI_v = A$ for every operator A.

A particular example All square $n \times n$ matrices with elements from the field F make a set $F_{n \times n} = \hat{L}(F^n, F^n)$, which is an algebra with unity, where the basic binary operations are the addition and multiplication of matrices, and the unity is the unit matrix I_n. But, when we consider $\hat{L}(U_n, U_n)$ as a vector space (forgetting about multiplication of operators), we shall use the term *superspace*. There is a possibility to define an inner product in the superspace $\hat{L}(U_n, U_n)$ to become a *unitary space*. But, for this purpose, we need a unitary operation $A \rightarrow A^\dagger$ in $\hat{L}(U_n, U_n)$ called adjoining: for any $A, B \in \hat{L}(U_n, U_n)$ then

$$(A, B) = \text{tr}(A^\dagger B) \quad \text{is an inner product.}$$

The adjoining is an essential bijective map in $\hat{L}(U_n, U_n)$, since all important operators in $\hat{L}(U_n, U_n)$ are defined by means of the adjoint operator (Sect. 4.2)

The most natural way to define this map is through the concept of the *dual space* U_n^* of the given unitary space U_n.

Definition A linear functional f on a unitary space U_n is a scalar-valued function $f : U_n \rightarrow \mathbb{C}$, which is linear:

$$f(a\bar{x} + b\bar{y}) = af(\bar{x}) + bf(\bar{y}), \quad \forall a, b \in \mathbb{C}, \quad \forall \bar{x}, \bar{y} \in U_n.$$

But, $\mathbb{C}$ is itself a one-dimensional unitary space $\mathbb{C}^1$ [$\mathbb{C}$ is a field, so we have addition of vectors (complex numbers) and multiplication of vectors (complex numbers) with scalars (again complex numbers) and also the inner product is $(a, b) = a^* b$, with the standard basis, which is the number 1]. Therefore, f is an element in the space of linmaps $\hat{L}(U_n, \mathbb{C}^1)$. This space we denote as U_n^* and call it the dual space of U_n : $U_n^* = \hat{L}(U_n, \mathbb{C}^1)$. The dimension of the dual space is equal to the product of the dimensions of U_n and $\mathbb{C}^1$, so it is equal to $n \cdot 1 = n$. We shall show that U_n^* is also a unitary space. So, we have two unitary spaces U_n and U_n^* of the same dimension and we expect to be able to find a bijection between them that does not depend on the choice of bases in them (there are infinitely many bijections that are basis dependent).

To find this desired basis-independent bijection, we shall first search for a matrix that represents f in a chosen ON basis $\{\bar{u}_1, \bar{u}_2, \ldots, \bar{u}_n\}$ in U_n and the standard basis (the number 1) in $\mathbb{C}^1$.

The general method for finding the representative matrix φ for f is to apply f to the chosen ON basis in U_n and then to expand the images so obtained in $\mathbb{C}^1$ in the standard basis in $\mathbb{C}^1$. We obtain the representative matrix φ by transposing the matrix of expansion coefficients:

$$\left.\begin{array}{l} f(\bar{u}_1) = f(\bar{u}_1) \cdot 1 \\ f(\bar{u}_2) = f(\bar{u}_2) \cdot 1 \\ \cdots \\ f(\bar{u}_n) = f(\bar{u}_n) \cdot 1 \end{array}\right\} \Rightarrow \varphi = [f(\bar{u}_1)\, f(\bar{u}_2)\, \cdots\, f(\bar{u}_n)].$$

Thus, φ is a $1 \times n$ matrix (*a row matrix*) whose elements are the images by f of the chosen ON basis in U_n.

To verify this result, we have to demonstrate that φ maps the representative column $\eta = [(\bar{u}_1, \bar{y})\, (\bar{u}_2, \bar{y}) \ldots (\bar{u}_n, \bar{y})]^T$ of an arbitrary vector $\bar{y} \in U_n$ in the ON basis $\{\bar{u}_1, \bar{u}_2, \ldots, \bar{u}_n\}$ onto the representative of $f(\bar{y})$ in the standard basis $\{1\}$ in $\mathbb{C}^1$: $f(\bar{y}) \cdot 1 = f(\bar{y})$. Indeed, $\varphi_\eta = \sum_{i=1}^n f(\bar{u}_i)(\bar{u}_i \bar{y})$, while $f(\bar{y}) \cdot 1 = f(\bar{y}) = f(\sum_{i=1}^n (\bar{u}_i, \bar{y})\bar{u}_i) = \sum_{i=1}^n (\bar{u}_i, \bar{y}) f(\bar{u}_i)$, so

$$\varphi_\eta = f(\bar{y}) \cdot 1.$$

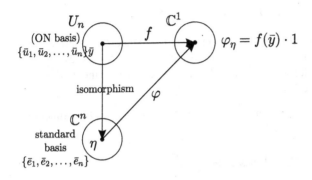

We see that φ represents $f(\bar{y})$ in the chosen ON basis in U_n and $\mathbb{C}^1$ and, at the same time, its elements determine $f(\bar{y})$ which is usually given by the images $\{f(\bar{u}_1, f(\bar{u}_2), \ldots, f(\bar{u}_m))\}$ of a basis in the domain.

Every linear functional $f \in U_n^*$ can be connected with one and only one vector $\bar{x} \in U_n$, i.e., we shall now establish a bijection between U_n and U_n^*, the so-called *dualism D*.

First, every $\bar{x} \in U_n$ defines one linear functional $f(\bar{y})$ in the following manner:

$$f(\bar{y}) = (\bar{x}, \bar{y}), \ \forall \bar{y} \in U_n.$$

This is obviously a map $U_n \to \mathbb{C}^1$ determined by $\bar{x}$ which maps the vector $\bar{y} \in U_n$ onto the scalar $(\bar{x}, \bar{y}) \in \mathbb{C}^1$, and due to the linearity of the inner product with respect to the second factor it is a *linear functional*.

Let $\bar{y}$ take on the values of the ON basis $\{\bar{u}_1, \bar{u}_2, \ldots, \bar{u}_n\}$, and we get $f(\bar{u}_1) = (\bar{x}, \bar{u}_1), \ldots, f(\bar{u}_n) = (\bar{x}, \bar{u}_n)$. So, the representing row of f is

$$\varphi = [(\bar{x}, \bar{u}_1)(\bar{x}, \bar{u}_2) \ldots (\bar{x}, \bar{u}_n)] = [(\bar{u}_1, \bar{x})^* (\bar{u}_2, \bar{x})^* \ldots (\bar{u}_n, \bar{x})^*]$$

However, we can make a reverse statement. Let $f(\bar{y})$ be a linear functional defined in U_n. Then there exists a unique vector $\bar{x} \in U_n$ such that it reproduces this functional via the inner product, i.e., by *the fundamental formula of dualism (FFD)* $f(\bar{y}) = (\bar{x}, \bar{y}), \ \forall \bar{y} \in U_n$.

To show this, let us first choose an ON basis $\{\bar{u}_1, \bar{u}_2, \ldots, \bar{u}_n\}$ in U_n. We now search for a vector $\bar{x} \in U_n$ which determines $f(\bar{y})$ via the inner product (FFD) $f(\bar{y}) = (\bar{x}, \bar{y}), \forall \bar{y} \in U_n$. We apply the functional f onto the chosen ON basis $f(\bar{u}_i) \overset{(FFD)}{=} (\bar{x}, \bar{u}_i)$ or $f^*(\bar{u}_i) = (\bar{u}_i, \bar{x})$. The Fourier expansion of $\bar{x}$ is $\bar{x} = \sum_{i=1}^n (\bar{u}_i, \bar{x})\bar{u}_i$, so that we finally have

$$\bar{x} = \sum_{i=1}^{n} f^*(\bar{u}_i)\bar{u}_i \;.$$

Thus, the vector $\bar{x}$, for which we are searching, is represented in the chosen ON basis by the *column-matrix*

$$\xi = \varphi^\dagger = [f^*(\bar{u}_1)\, f^*(\bar{u}_2) \ldots f^*(\bar{u}_n)]^T$$

which is the *adjoint* (transposed and complex conjugated) of the row-matrix $\varphi = [f(\bar{u}_1)\, f(\bar{u}_2) \ldots f(\bar{u}_n)]$ which represents $f(\bar{y})$ in the same ON basis in U_n.

Therefore, $\bar{x}$ is uniquely determined by $f(\bar{y})$ and vice versa. By the fundamental formula of dualism

$$f(\bar{y}) = [D\bar{x}](\bar{y}) = (\bar{x}, \bar{y}), \; \bar{y} \in U_n.$$

We now have to show that this bijection D between U_n and U_n^* (defined by saying that the dual vectors $\bar{x} \in U_n$ and $D\bar{x} = f \in U_n^*$ are connected so that their representing matrices are the adjoints of each other in a given ON basis in U_n) does not depend on the chosen basis (it is a basis invariant map). It is almost obvious because the ON basis was arbitrary, but to prove it exactly, we have to choose another ON basis $\{\bar{u}'_1, \bar{u}'_2, \ldots, \bar{u}'_n\}$ connected with the first one by the unitary replacement matrix $\mathcal{R}$

$$\bar{u}'_i = \sum_{j=1}^{n} r_{ij}\bar{u}_j, \; \mathcal{R} = [r_{ij}]_{n \times n}, \; \mathcal{R}^\dagger = \mathcal{R}^{-1}.$$

If ξ is the representing column of $\bar{x}$ in the first ON basis, then this column will change into $\xi' = \mathcal{R}^*\xi$ in the second one (contravariant unitary vector) (see the change of basis, Sect. 4.4). The representing row of $D\bar{x} = f$ is $\varphi = \xi^\dagger$ in the first basis, and this will change into $\varphi' = \xi^\dagger \mathcal{R}^T$ in the second basis. Verify:

$$(f(\bar{u}'_i) = f(\sum_{j=1}^{n} r_{ij}\bar{u}_j) = \sum_{j=1}^{n} r_{ij} f(\bar{u}_j) = \sum_{j=1}^{n} f(\bar{u}_j)\{\mathcal{R}^T\}_{ji}, i = 1, 2, \ldots, n,$$

$$\Rightarrow \varphi' = \varphi \mathcal{R}^T = \xi^\dagger \mathcal{R}^T)$$

The representing row of f changes like the ON basis, and for this reason it is called a *covariant unitary vector* (Sect. 4.5). This row can be written as $(\mathcal{R}^*\xi)^\dagger$; thus, the relation between the representing matrices of $\bar{x}$ and $D\bar{x} = f$ in the second ON basis is again *just adjoining*.

But, this bijection D between U_n and U_n^* is an antilinear map, since the inner product is antilinear in the first factor: Let $[D\bar{x}_1](\bar{y}) = f_1(\bar{y})$ and $[D\bar{x}_2](\bar{y}) = f_2(\bar{y})$, $\forall \bar{y} \in U_n$, then for $\forall a, b \in \mathbb{C}$ we have

$$[D(a\bar{x}_1 + b\bar{x}_2)](\bar{y}) \overset{(FFD)}{=} a(\bar{x}_1 + b\bar{x}_2, \bar{y}) = a^*(\bar{x}_1, \bar{y}) + b^*(\bar{x}_2, \bar{y})$$
$$= a^*[D\bar{x}_1](\bar{y}) + b^*[D\bar{x}_2](\bar{y}),$$

or forgetting about the argument $\bar{y}$

$$D(a\bar{x}_1 + b\bar{x}_2) = a^* D\bar{x}_1 + b^* D\bar{x}_2.$$

For an antilinear bijection, we use the term dualism between unitary spaces (some authors prefer the term conjugate isomorphism [4]), while isomorphism is always a linear bijection.

From the definition of D it follows that we have an inner product in U_n^* defined by that in U_n: the inner product of two dual vectors $D\bar{x}_1$ and $D\bar{x}_2$ is the complex conjugate of the inner product of their two preimages $\bar{x}_1$ and $\bar{x}_2$

$$(\bar{x}_1, \bar{x}_2) = \xi_1^\dagger \xi_2 = (\xi_1^T \xi_2^*)^* = (\varphi_1^* \varphi_2^T)^* = (D\bar{x}_1, D\bar{x}_2)^*,$$

where $\bar{x}_1$ and $\bar{x}_2$ are represented by the columns ξ_1 and ξ_2 in the chosen ON basis in U_n, and $D\bar{x}_1$ and $D\bar{x}_2$ are represented by the rows $\varphi_1 = \xi_1^\dagger$ and $\varphi_2 = \xi_2^\dagger$ in the same basis. (Observe that the standard inner product in the column space is $\xi_1^\dagger \xi_2$, while in the row space it is $\varphi_1^* \varphi_2^T$.)

But, due to the positive definiteness of the inner product, the bijection D preserves the norm of dual vectors $(\bar{x}, \bar{x}) = (D\bar{x}, D\bar{x}) \Rightarrow ||\bar{x}|| = ||D\bar{x}||$. This means that D is an isometry (a norm preserving map).

Every ON basis in U_n defines its dual basis in U_n^* which is also orthonormal

$$\{\bar{u}_1, \bar{u}_2, \ldots, \bar{u}_n\} \overset{D}{\to} \{D\bar{u}_1, D\bar{u}_2, \ldots, D\bar{u}_n\} \text{ and } (D\bar{u}_i, D\bar{u}_j) = (\bar{u}_i, \bar{u}_j)^* = \delta_{ij}.$$

Dual vectors $\bar{x}$ and $f = D\bar{x}$ are represented in any pair of dual ON bases by matrix-columns that are the complex conjugate of each other due to the antilinear property of D:

$$\bar{x} = \sum_{i=1}^n (\bar{u}_i, \bar{x})\bar{u}_i \text{ and } f = D\bar{x} = D\sum_{i=1}^n (\bar{u}_i, \bar{x})\bar{u}_i = \sum_{i=1}^n (\bar{u}_i, \bar{x})^* D\bar{u}_i.$$

In another ON basis in U_n obtained by a unitary replacement matrix $\mathcal{R}$, the vector $\bar{x} \in U_n$ is represented by the column $\mathcal{R}^* \xi$, where $\xi = [(\bar{u}_1, \bar{x})(\bar{u}_2, \bar{x}) \ldots (\bar{u}_n, \bar{x})]^T$, while he dual vector $D\bar{x}$ is represented by the column $\mathcal{R}\xi^* = (\xi^\dagger \mathcal{R}^T)^T$ (see Sect. 4.5). But since $\mathcal{R}\xi^* = (\mathcal{R}^* \xi)^*$, we see that the dualism D is again represented by the complex conjugation K of the representing columns.

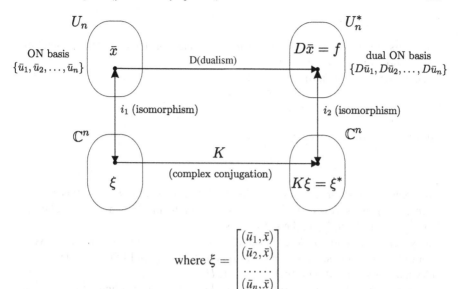

$$\text{where } \xi = \begin{bmatrix} (\bar{u}_1, \bar{x}) \\ (\bar{u}_2, \bar{x}) \\ \ldots\ldots \\ (\bar{u}_n, \bar{x}) \end{bmatrix}$$

The two isomorphisms i_1 and i_2 are induced by the choice of dual bases in U_n and U_n^*.

We shall now summarize all the relevant statements about the dualism D that we have proved so far.

Theorem (about the dualism D between U_n and U_n^*) A linear functional f in a finite-dimensional unitary space U_n is a linear map $f : U_n \to \mathbb{C}^1$. The set $\hat{L}(U_n, \mathbb{C}^1)$ of all such linear functionals is a vector space of dimension n, denoted by U_n^* and called the *dual* of U_n. We can establish a bijective map between U_n and U_n^* using the inner product in U_n:

$$\bar{x} \in U_n, \ f \in U_n^* \text{ and } f(\bar{y}) = (\bar{x}, \bar{y}), \ \forall \bar{y} \in U_n$$

(the fundamental formula for dualism—FFD). This bijective map is called the dualism between U_n and U_n^*, and it is denoted by D:

$$D\bar{x} = f, \ f = D^{-1}\bar{x}.$$

The inner product in U_n^* is given by $(f_1, f_2) = (D\bar{x}_1, D\bar{x}_2) = (\bar{x}_1, \bar{x}_2)^*$, so U_n^* is also a unitary space.

In every ON basis $\{\bar{u}_1, \bar{u}_2, \ldots, \bar{u}_n\}$ in U_n, the vector $\bar{x} \in U_n$ and its dual partner $f = D\bar{x} \in \hat{L}(U_n, \mathbb{C}^1)$ are represented by adjoint matrices:

$$\bar{x} \to \xi = [(\bar{u}_1, \bar{x}) (\bar{u}_2, \bar{x}) \ldots (\bar{u}_n, \bar{x})]^T = [f^*(\bar{u}_1) f^*(\bar{u}_2) \ldots f^*(\bar{u}_n)]^T \text{ and}$$
$$f \to \varphi = \xi^\dagger = [(\bar{u}_1, \bar{x})^* (\bar{u}_2, \bar{x})^* \ldots (\bar{u}_n, \bar{x})^*] = [f(\bar{u}_1) f(\bar{u}_2) \ldots f(\bar{u}_n)].$$

In any pair of dual ON bases $\{\bar{u}_1, \bar{u}_2, \ldots, \bar{u}_n\}$ and $\{D\bar{u}_1, D\bar{u}_2, \ldots, D\bar{u}_n\}$, the vector $\bar{x} \in U_n$ and its dual partner $f = D\bar{x} \in U_n$ are represented by the complex conjugate columns

$$\bar{x} \to \xi \text{ and } f \to \xi^*.$$

So, D itself is represented on each pair of dual bases as the complex conjugation K. Therefore, it is a basis invariant map.

The dualism D is an antilinear bijection

$$D(a\bar{x}_1 + b\bar{x}_2) = a^*D\bar{x}_1 + b^*D\bar{x}_2.$$

It is also a norm-preserving map (an isometry):

$$||\bar{x}|| = ||D\bar{x}||. \quad \Delta$$

We treat f as a linear functional in U_n and also as a vector in U_n^*. For this reason, we represent f as the row φ in an ON basis in U_n in the first case and as the column φ^T in the dual basis in U_n^* in the second case.

However, the relation between U_n and U_n* is a more symmetric one, since every vector in U_n can be considered as a linear functional in U_n^*. In FFD $f(\bar{y}) = (\bar{x}, \bar{y})$, we can fix that $\bar{y}$ and let f run through the whole U_n^*.

In this way $\bar{x} = D^{-1}f$ runs through the whole U_n. So, we get a linear functional in U_n^*: $f(\bar{y})$ for fixed $\bar{y}$ and running f associate to every f a complex number $(D^{-1}f, \bar{y})$, which is a linear map because to the linear combination $(af_1 + bf_2)$ it associates the same linear combination of the corresponding numbers

$$(D^{-1}(af_1 + bf_2), \bar{y}) = a(D^{-1}f_1, \bar{y}) + b(D^{-1}f_2, \bar{y})$$

(note that the inner product is antilinear in the first factor, but this is compensated by the antilinear property of the map $D^{-1}: U_n^* \to U_n$).

Thus, every $\bar{y} \in U_n$ defines in this way one linear functional y' in the space U_n^*: $y'(f) = (D^{-1}f, \bar{y})$.

However, for linear functionals in U_n^*, i.e., the vectors from "the second dual" U_n^{**}, there is also the FFD for this case $y'(f) = (g, f,)$ (the inner product in U_n^*), where $g \in U_n^*$, and y' is its image with respect to the dualism D' between U_n^* and U_n^{**}: $y = D'g \in U_n^{**}$.

Equating the right hand sides of the above expressions for $y'(f)$, we get

$$(D^{-1}f, \bar{y}) = (g, f) \text{ or, further,}$$
$$(\bar{x}, \bar{y}) = (D\bar{x}, g)^* \text{ or } (\bar{x}, \bar{y}) = (D^{-1}D\bar{x}, D^{-1}g) = (\bar{x}, D^{-1}g).$$

Since $\bar{x}$ runs through the whole U_n, we can conclude $\bar{y} = D^{-1}g$ or $g = D\bar{y}$. We have above $g = (D')^{-1}y'$, which together gives

$$D\bar{y} = (D')^{-1}y' \text{ or}$$
$$\boxed{y' = D'D\bar{y}}.$$

Here, $D'D$ is the composition of two invariant dualisms (invariant because they do not depend on a choice of bases, but follow from the structure of unitary spaces that are involved, more precisely from the inner products in them). This composition is an invariant isomorphism between U_n and U_n^{**}, which enables us to identify them. After this identification, we can consider vectors in U_n as linear functionals in U_n^* (see the general case of $V_n(F)$, Sect. 4.6).

4.2 The Adjoint Operator

Theorem (about adjoint operators) Let A be a linear operator in a finite-dimensional unitary space U_n, i.e., $A \in \hat{L}(U_n, U_n)$. Then there always exists a unique linear operator $A^\dagger$ in the same space U_n such that

$$\boxed{(\bar{x}, A\bar{y}) = (A^\dagger \bar{x}, \bar{y})} \text{ for all } \bar{x}, \bar{y} \in U_n.$$

We call $A^\dagger$ the *(Hermitian) adjoint* [4] of A.

The representing matrices of A and $A^\dagger$ in some ON basis in U_n are $\mathscr{A}$ and $(\mathscr{A}^*)^T = \mathscr{A}^\dagger$, where the matrix $\mathscr{A}^\dagger$ is also called the *adjoint* of $\mathscr{A}$. (An alternative name for $\mathscr{A}^\dagger$ is the *Hermitian conjugate of $\mathscr{A}$*) [5].

As far as some basic operations with adjoint operators are concerned, we have

- $(aA + bB)^\dagger = a^*A^\dagger + b^*B^\dagger$, $a, b \in \mathbb{C}$, $A, B \in \hat{L}(U_n, U_n)$,
- $(AB)^\dagger = B^\dagger A^\dagger$,
- $(A^\dagger)^\dagger = A$, *and also*
- $ranA^\dagger = (kerA)^\perp$.

Proof Let $\bar{x}$ be an arbitrary, but fixed, vector from U_n. We note that there is a map from U_n to $\mathbb{C}^1$ defined by A and $\bar{x}$ in the form

$$(*) \ (\bar{x}, A\bar{y}) = f_{A,\bar{x}}(\bar{y}), \ \forall \bar{y} \in U_n.$$

Since A is a linear map and the second factor in the inner product is in a linear position, this map adjoins a complex number $(\bar{x}, A\bar{y})$ to every vector $\bar{y} \in U_n$ in a *linear* fashion:

$$f_{A,\bar{x}}(a\bar{y}_1 + b\bar{y}_2) = (\bar{x}, A(a\bar{y}_1 + b\bar{y}_2)) = (\bar{x}, aA\bar{y}_1 + bA\bar{y}_2) =$$
$$= a(\bar{x}, A\bar{y}_1) + b(\bar{x}, A\bar{y}_2) = a f_{A,\bar{x}}(\bar{y}_1) + b f_{A,\bar{x}}(\bar{y}_2).$$

So, $f_{A,\bar{x}}(\bar{y})$ is a linear functional in U_n.

We have already established that for the linear functional $f_{A,\bar{x}}$ there exists a unique vector $\bar{x}' \in U_n$ such that

$$(**) \ f_{A,\bar{x}}(\bar{y}) = (\bar{x}', \bar{y}), \ \forall \bar{y} \in U_n(\text{FFD}), \text{ or}$$
$$\bar{x}' = D^{-1} f_{A,\bar{x}}.$$

In this way, we have a unique map $\bar{x} \to \bar{x}'$ on the whole U_n and this map depends on the operator A. We denote this map by $A^\dagger$, i.e.,

$$(***)\, A^\dagger \bar{x} = \bar{x}, \quad \forall \bar{x} \in U_n.$$

Comparing the two expressions $(*)$ and $(**)$

$$(\bar{x}, A\bar{y}) = f_{A,\bar{x}}(\bar{y}) \text{ and } f_{A,\bar{x}}(\bar{y}) = (\bar{x}', \bar{y}) = (A^\dagger \bar{x}, \bar{y}),$$

we finally have using $(***)$

$$\boxed{(\bar{x}, A\bar{y}) = (A^\dagger \bar{x}, \bar{y}), \quad \forall \bar{x}, \bar{y} \in U_n}.$$

It is easy to show that the map $A^\dagger$ is a linear one. That is to say, for $a, b \in \mathbb{C}$, $\bar{x}, \bar{y}, \bar{z} \in U_n$ we have

$$\begin{aligned}(A^\dagger(a\bar{x}+b\bar{y}), \bar{z}) &= (a\bar{x}+b\bar{y}, A\bar{z}) = a^*(\bar{x}, A\bar{z}) + b^*(\bar{y}, A\bar{z}) = \\ &= a^*(A^\dagger \bar{x}, \bar{z}) + b^*(A^\dagger \bar{y}, \bar{z}) = (aA^\dagger \bar{x} + bA^\dagger \bar{y}, \bar{z}).\end{aligned}$$

Furthermore, we have $A^\dagger(a\bar{x}+b\bar{y}) - (aA^\dagger \bar{x} + bA^\dagger \bar{y}, \bar{z})) = 0$, which must be valid for any $\bar{z} \in U_n$.

But, only the zero vector is orthogonal to all vectors in U_n, so that

$$A^\dagger(a\bar{x}+b\bar{y}) - (aA^\dagger \bar{x} + bA^\dagger \bar{y}) = \bar{0}, \text{ or}$$
$$A^\dagger(a\bar{x}+b\bar{y}) = aA^\dagger \bar{x} + bA^\dagger \bar{y},$$

which proves the linearity of the map $A^\dagger$.

Before we investigate the relation between representing matrices of A and $A^\dagger$ in any ON basis in U_n, we first have to derive the general formula for the representing matrix of any operator in an ON basis in U_n.

The basic formula for the matrix that represents an operator $A \in \hat{L}(V_n, V_n)$—where V_n is not necessarily an inner-product vector space—in the chosen basis $\{\bar{v}_1, \bar{v}_2, \ldots, \bar{v}_n\}$ in V_n is

$$A \to \mathscr{A} = [a_{ij}]_{n \times n} \text{ and } A\bar{v}_j = \sum_{k=1}^{n} a_{kj} \bar{v}_k, \quad j = 1, 2, \ldots, n.$$

When V_n is an inner-product vector space (real or complex) and the chosen basis is orthonormal (ON) $\{\bar{u}_1, \bar{u}_2, \ldots, \bar{u}_n\}$, then we can calculate the elements of $\mathscr{A}$ quite easily. We multiply the basic formula $A\bar{u}_j = \sum_{k=1}^{n} a_{kj} \bar{u}_k$ from the left by $\bar{u}_i$, $i = 1, 2, \ldots, n$, and obtain an important result

$$(\bar{u}_i, A\bar{u}_j) = \sum_{k=1}^{n} a_{kj}(\bar{u}_i, \bar{u}_k) = \sum_{k=1}^{n} a_{kj}\delta_{ik} = a_{ij}.$$

So, the matrix $\mathscr{A}$ which represents the operator A in the ON basis $\{\bar{u}_1, \bar{u}_2, \ldots, \bar{u}_n\}$ in U_n has elements $a_{ij} = (\bar{u}_i, A\bar{u}_j) = (A^\dagger \bar{u}_i, \bar{u}_j) = (\bar{u}_j, A^\dagger \bar{u}_i)^* = b_{ij}^*$, where the matrix $\mathscr{B} = [b_{ij}]_{n \times n}$ represents the operator $A^\dagger$: $b_{ij} = (\bar{u}_i, A^\dagger \bar{u}_j)$.

Thus, we have $a_{ij} = b_{ij}^*$, which means

$$\mathscr{A} = (\mathscr{B}^T)^* \text{ or } \mathscr{B} = (\mathscr{A}^*)^T = \mathscr{A}^\dagger.$$

We shall now prove the following statements:

(i) $\text{ran } A^\dagger = (\ker A)^\perp$
(ii) $(aA + bB)^\dagger = a^* A^\dagger + b^* B^\dagger$, $a, b \in \mathbb{C}$; $A, B \in \hat{L}(U_n, U_n)$
  or $(A + B)^\dagger = A^\dagger + B^\dagger$ and $(aA)^\dagger = a^* A^\dagger$,
(iii) $(AB)^\dagger = B^\dagger A^\dagger$,
(iv) $(A^\dagger)^\dagger = A$.

Proofs:

(i) Let $\bar{x} \in \text{ran } A^\dagger$, which means that there exists at least one $\bar{y} \in U_n$ such that $\bar{x} = A^\dagger \bar{y}$. Then for arbitrary $\bar{u} \in \ker A$—$A\bar{u} = \bar{0}$—it follows that $(\bar{x}, \bar{u}) = (A^\dagger \bar{y}, \bar{u}) = (\bar{y}, A\bar{u}) = (\bar{y}, \bar{0}) = 0$, which implies that $\bar{x}$ is orthogonal to the whole $\ker A$, i.e., $\bar{x} \in \ker A^\perp$. Since any arbitrary vector $\bar{x}$ from $\text{ran } A^\dagger$ belongs to $\ker A^\dagger$, we have $\text{ran } A^\dagger \subseteq \ker A^\perp$. If, on the other hand, $\bar{x} \in \text{ran } A^{\dagger \perp}$, then for any $\bar{z} \in U_n$, it follows that $(A^\dagger \bar{z}, \bar{x}) = 0$ or $(\bar{z}, A\bar{x}) = 0$, which implies $A\bar{x} = \bar{0}$ since only the zero vector $\bar{0}$ is orthogonal to all vectors in U_n. It follows that $\bar{x} \in \ker A$ or $\text{ran } A^{\dagger \perp} \subseteq \ker A$.

Generally speaking, if we have an inclusion relation between two subspaces U' and U'' of U_n, i.e., $U' \subseteq U''$, then this relation obviously reverses for their orthocomplements, i.e., $(U'')^\perp \subseteq (U')^\perp$. Thus, $\ker A^\perp \subseteq \text{ran } A^\dagger$, and finally $\text{ran } A^\dagger = \ker A^\perp$.

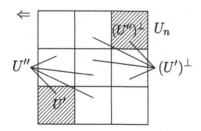

(ii) and (iii) By choosing an ON basis $\{\bar{u}_1, \bar{u}_2, \ldots, \bar{u}_n\}$ in U_n, we achieve an isomorphism i between $\hat{L}(U_n, U_n)$, here considered to be the algebra (with unity) of all linear operators acting in U_n—and $\hat{L}(\mathbb{C}, \mathbb{C})$—the algebra (with unity) of all linear operators in $\mathbb{C}$ (the algebra of complex $n \times n$ matrices). To every $A \in \hat{L}(U_n, U_n)$, there corresponds a complex matrix $\mathscr{A} = [a_{ij}]_{n \times n}$, where

$$\boxed{a_{ij} = (\bar{u}_i, A\bar{u}_j)} : A \overset{i}{\leftrightarrow} \mathscr{A}.$$

To the unique adjoint operator $A^\dagger$, there corresponds the transposed and complex conjugate matrix $(\mathscr{A}^*)^T = \mathscr{A}^\dagger$. To the sum $A + B$ of two operators, there corresponds (due to the isomorphism i of the algebras) the sum $\mathscr{A} + \mathscr{B}$ of the representing matrices.

Similarly, the product AB is represented by $\mathscr{A}\mathscr{B}$, and the product of a scalar a with an operator A, aA, is represented by $a\mathscr{A}$. The sum of $A^\dagger + B^\dagger$ is represented by $\mathscr{A}^\dagger + \mathscr{B}^\dagger$ which is equal to $(\mathscr{A} + \mathscr{B})^\dagger$, and this matrix represents $(A+B)^\dagger$. In conclusion, $(A+B)^\dagger = A^\dagger + B^\dagger$, due to the isomorphism i of the algebras. Similarly, the products $(AB)^\dagger$ and $(aA)^\dagger$ are represented by $(\mathscr{A}\mathscr{B})^\dagger$ (which is equal to $\mathscr{B}^\dagger \mathscr{A}^\dagger$), and by $(a\mathscr{A})^\dagger$ (which is equal to $a^* \mathscr{A}^\dagger$), respectively. By the isomorphism i it follows that $(AB)^\dagger = B^\dagger A^\dagger$ and $(aA)^\dagger = a^* A^\dagger$.

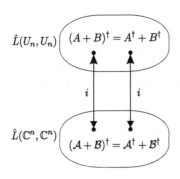

(iv) The proof that $(A^\dagger)^\dagger = A$ [the adjoint of the adjoint of A is A again—an involutive map in $\hat{L}(U_n, U_n)$] follows immediately from the definition of this map: for $\forall \bar{x}, \bar{y} \in U_n$ we have

$$(\bar{x}, A\bar{y}) \overset{def}{=} (A^\dagger \bar{x}, \bar{y}) = (\bar{y}, A^\dagger \bar{x})^* \overset{def}{=} ((A^\dagger)^\dagger \bar{y}, \bar{x})^* = (\bar{x}, (A^\dagger)^\dagger \bar{y}).$$

Furthermore, $(\bar{x}, A\bar{y}) - (\bar{x}, (A^\dagger)^\dagger \bar{y}) = 0$ or $(\bar{x}, [A - (A^\dagger)^\dagger]\bar{y}) = 0$ or $[A - (A^\dagger)^\dagger]\bar{y} = \bar{0}$ (since only the zero vector $\bar{0}$ is orthogonal to all vectors in U_n) or $A - (A^\dagger)^\dagger = \hat{0}$ (since only the zero operator $\hat{0}$ maps every vector $\bar{y} \in U_n$ onto $\bar{0}$ or

$$\boxed{(A^\dagger)^\dagger = A}. \ \Delta$$

Thus, we have finally proved the theorem about adjoint operators.

We shall conclude this Sect. 4.2 by pointing out that when we consider $\hat{L}(U_n, U_n)$ as a complex vector space (the so-called superspace) (i.e., forgetting about the multiplication of operators) we have in it a standard inner product

$$(A, B) = \text{tr}(A^\dagger B) = \text{tr}(\mathscr{A}^\dagger \mathscr{B}),$$

which makes $\hat{L}(U_n, U_n)$ a unitary space of n^2 dimensions. Here again we shall use a representation of $\hat{L}(U_n, U_n)$ in an ON basis in U_n. (Notice that the trace of the operator $A^\dagger B$ does not depend on the ON basis in which it has been calculated, Sect. 5.1). We have already discussed the inner product in $\hat{L}(\mathbb{C}^n, \mathbb{C}^n) = \mathbb{C}_{n \times n}$, i.e.,

$(\mathscr{A},\mathscr{B}) = \mathrm{tr}(\mathscr{A}^\dagger\mathscr{B}) = \sum_{i,j=1}^n a_{ij}^* b_{ij}$, and we found that it satisfies all the postulates of complex inner products and that it is a natural generalization of the standard inner product in $\mathbb{C}^n : (\bar{x},\bar{y}) = \sum_{i=1}^n x_i^* y_i$.

Summary To make a more sophisticated summary on adjoint operators in the superspace $\hat{L}(U_n,U_n)$, we can say that there is a bijective map $A \leftrightarrow A^\dagger$ of $\hat{L}(U_n,U_n)$ onto itself. The relation between A and $A^\dagger$ is given by $(\bar{x},A\bar{y}) = (A^\dagger\bar{x},\bar{y})$, $\forall \bar{x},\bar{y} \in U_n$, which is an involution $A = (A^\dagger)^\dagger$ (the relation between A and $A^\dagger$ is symmetric—A is the adjoint of $A^\dagger$), it does not depend on the choice of basis (an invariant bijection), and it is also antilinear $[(aA + bB)^\dagger = a^*A^\dagger + b^*B^\dagger]$. Since we call every antilinear bijection between unitary spaces a *dualism* (while a linear bijection is an isomorphism), it is natural to call the adjoining an *autodualism* in the superspace $\hat{L}(U_n,U_n)$. This autodualism is an isometric map since it preserves the norm $||A|| = (A,A)^{1/2} = \mathrm{tr}(A^\dagger A)^{1/2}$ and $||A^\dagger|| = (A^\dagger,A^\dagger)^{1/2} = \mathrm{tr}(AA^\dagger)^{1/2} = \mathrm{tr}(A^\dagger A)^{1/2}$ (since the trace does not change its value when we commute the factors under the trace):

$$||A|| = ||A^\dagger||.$$

So, the adjoining in the unitary superspace $\hat{L}(U_n,U_n)$ is an invariant, antilinear, and isometric bijection of $\hat{L}(U_n,U_n)$ onto itself *(an invariant and isometric autodualism)*.

4.3 The Change of Bases in $V_n(F)$

We shall now consider the case of a general vector space $V_n(F)$, where F can be the field $\mathbb{R}$ of real numbers or the field $\mathbb{C}$ of complex numbers. We shall not assume the existence of an inner product in $V_n(F)$.

Let there be given two bases in $V_n(F)$: the "old" one $v = \{\bar{v}_1,\bar{v}_2,\ldots,\bar{v}_n\}$ and the "new" one $v' = \{\bar{v}_1',\bar{v}_2',\ldots,\bar{v}_n'\}$. They determine the invertible replacement matrix $\mathscr{R} = [r_{ij}]_{n\times n}$, i.e., the unique matrix which consists of the expansion coefficients of the "new" basis in terms of the "old" one

$$\boxed{\bar{v}_i' = \sum_{j=1}^n r_{ij}\bar{v}_j}, \quad i = 1,2,\ldots,n. \quad (*)$$

This can also be written as $\boxed{[v'] = \mathscr{R}[v]}$, where

$$[v'] = \begin{bmatrix} \bar{v}_1' \\ \bar{v}_2' \\ \vdots \\ \bar{v}_n' \end{bmatrix} \quad \text{and} \quad [v] = \begin{bmatrix} \bar{v}_1 \\ \bar{v}_2 \\ \vdots \\ \bar{v}_n \end{bmatrix}$$

are columns of basis vectors.

4.3.1 The Change of the Matrix-Column ξ That Represents a Vector $\bar{x} \in V_n(F)$ (Contravariant Vectors)

Since vectors (real and complex) are such important objects in Classical and Quantum Physics, we have to know how their representatives change with the changes of coordinate systems in order to be able to formulate physical laws independently of the coordinate systems, i.e., both sides of a physical law must change in the same way (the so-called covariant formulation of physical laws).

Similarly, and with the same purpose, we shall investigate the changes of matrices that represent linear operators, dual vectors, etc.

Let us expand an arbitrary vector $\bar{x} \in V_n(F)$ in both bases v and v':

$$\bar{x} = \sum_{i=1}^{n} \xi_i \bar{v}_i \ (1) \quad \bar{x} = \sum_{k=1}^{n} \xi'_k \bar{v}'_k. \ (2)$$

These expansions establish two isomorphisms i_1 and i_2 between $V_n(F)$ and F^n:

$$\bar{x} \overset{i_1}{\leftrightarrow} \xi \quad \text{and} \quad \bar{x} \overset{i_2}{\leftrightarrow} \xi',$$

where ξ and ξ' are the matrix-columns consisting of these expansion coefficients $\xi = [\xi_1 \, \xi_2 \ldots \xi_n]^T$ and $\xi' = [\xi'_1 \, \xi'_2 \ldots \xi'_n]^T$. Our goal is to find the matrix $\mathscr{S} = [s_{ij}]_{n \times n}$ (as a function of the matrix $\mathscr{R}$) which connects these columns: $\xi' = \mathscr{S}\xi$.

We start with (2) and substitute $(*)$ in it

$$\bar{x} = \sum_{k=1}^{n} \xi'_k \bar{v}'_k = \sum_{k=1}^{n} \xi'_k \sum_{i=1}^{n} r_{ki} \bar{v}_i = \sum_{k=1}^{n} \left(\sum_{i=1}^{n} r_{ki} \xi'_k \right) \bar{v}_i.$$

Comparing this with (1), we see that we have two expansions of $\bar{x}$ in the "old" basis, so that the expansion coefficients in them must be the same

$$\xi_i = \sum_{k=1}^{n} r_{ki} \xi'_k = \sum_{k=1}^{n} \{\mathscr{R}^T\}_{ik} \xi'_k,$$

which implies $\xi = \mathscr{R}^T \xi'$.

Since the matrix $\mathscr{R}$ is invertible, we get further

$$\xi' = (\mathscr{R}^T)^{-1} \xi \quad \text{or}$$

$$\boxed{\mathscr{S} = (\mathscr{R}^{\mathscr{T}})^{-\infty}}.$$

As a mnemonic device, we can draw a figure (see next page)

So, we have proved:

Theorem (on the change of the representing column of a vector) The representing matrix-column ξ of the vector $\bar{x} \in V_n(F)$ changes (during the change of basis by the invertible replacement matrix $\mathscr{R} : [v'] = \mathscr{R}[v]$) by means of the invertible matrix $\mathscr{S} = (\mathscr{R}^T)^{-1}$, which is contragredient to

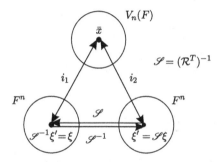

$$\mathscr{R}: \boxed{\xi' = \mathscr{S}\xi}. \ \Delta$$

In other words, every change of basis in $V_n(F)$ by the replacement matrix $\mathscr{R}$ induces an automorphism

$$\mathscr{S} = (\mathscr{R}^T)^{-1} \text{ in } F^n.$$

In terms of the expansion coefficients this reads as

$$\xi'_i = \sum_{j=1}^n s_{ij}\xi_j.$$

Thus, in comparison with $(*)$, the representing column ξ changes in the contragredient manner with respect to the change of the basis, so the representing columns of $\bar{x}$ are an example of contravariant vectors.

Remember that a contravariant vector is a map which connects a basis $v = \{\bar{v}_1, \bar{v}_2, \ldots, \bar{v}_n\}$ in $V_n(F)$ with a one-index system $\{\xi_j, \ j = 1, 2, \ldots, n\}$ of n numbers from F, such that this system goes to the new one $\xi'_i = \sum_{j=1}^n s_{ij}\xi_j$, $i = 1, 2, \ldots, n$, when the basis changes to the new one $v' = \{\bar{v}'_1, \bar{v}'_2, \ldots, \bar{v}'_n\}$ with the invertible replacement matrix $y = [r_{ij}]_{n \times n}: \bar{v}'_i = \sum_{j=1}^n r_{ij}\bar{v}_j$. The elements s_{ij}, $i, j = 1, 2, \ldots, n$ from the matrix $\zeta = [s_{ij}]_{n \times n}$ which is contragredient to $\mathscr{R}: \varphi = (\mathscr{R}^T)^{-1}$. The matrix $\mathscr{R}$ goes through the group $GL(n, F)$ of all invertible $n \times n$ matrices with elements from the field F. Thus, all bases in $V_n(F)$ are involved.

4.3.2 The Change of the n × n Matrix $\mathscr{A}$ That Represents an Operator $A \in \hat{L}(V_n(F), V_n(F))$ (Mixed Tensor of the Second Order)

We now consider an operator A which maps $V_n(F)$ into or onto itself, as well as the two bases $v = \{\bar{v}_1, \bar{v}_2, \ldots, \bar{v}_n\}$ and $v' = \{\bar{v}'_1, \bar{v}'_2, \ldots, \bar{v}'_n\}$ in $V_n(F)$ connected by the invertible replacement matrix $\mathscr{R}: [v'] = \mathscr{R}[v]$ or

$$\vec{v}'_i = \sum_{j=1}^{n} r_{ij}\vec{v}_j, \ i = 1,2,\ldots,n. (*)$$

We can represent A in both of these bases by matrices $\mathscr{A} = [a_{ij}]_{n \times n}$ and $\mathscr{A}' = [a'_{ij}]_{n \times n}$, respectively:

$$A\vec{v}_j = \sum_{k=1}^{n} a_{kj}\vec{v}_k \ \ j = 1,2,\ldots,n \ (1) \ \text{ and } \ A\vec{v}'_i = \sum_{j=1}^{n} a'_{ji}\vec{v}'_j, \ i = 1,2,\ldots,n \ (2).$$

We want to find how these matrices are related to each other in terms of the replacement matrix $\mathscr{R}$. Using $(*)$ and (2), we can obtain two different expansions of $A\vec{v}'_i$ in the basis v:

$$A\vec{v}'_i \overset{(*)}{=} A\sum_{j=1}^{n} r_{ij}\vec{v}_j = \sum_{j=1}^{n} r_{ij}A\vec{v}_j \overset{(1)}{=} \sum_{j=1}^{n} r_{ij} \sum_{k=1}^{n} a_{kj}\vec{v}_k = \sum_{k=1}^{n}\left(\sum_{k=1}^{n} r_{ij}a_{kj}\right)\vec{v}_k$$

and

$$A\vec{v}'_i \overset{(2)}{=} \sum_{j=1}^{n} a'_{ji}\vec{v}'_j \overset{(*)}{=} \sum_{j=1}^{n} a'_{ji}\sum_{k=1}^{n} r_{jk}\vec{v}_k = \sum_{k=1}^{n}\left(\sum_{j=1}^{n} a'_{jk}r_{jk}\right)\vec{v}_k, \ i = 1,2,\ldots,n.$$

Since the vector $A\vec{v}'_i$ must have a unique expansion coefficients in the two expansions in the "old" basis, we get for $i,k = 1,2,\ldots,n,$

$$\sum_{j=1}^{n} a_{kj}r_{ij} = \sum_{j=1}^{n} r_{jk}a'_{ji} \ \text{ or } \ \sum_{j=1}^{n}\{\mathscr{A}\}_{kj}\{\mathscr{R}^T\}_{ji} = \sum_{j=1}^{n}\{\mathscr{R}^T\}_{kj}\{\mathscr{A}'\}_{ji}.$$

In terms of matrices, these equalities between the (k,i) - matrix elements amount to $\mathscr{A}\mathscr{R}^T = \mathscr{R}^T\mathscr{A}'$, so that we finally have

$$\boxed{\mathscr{A}' = (\mathscr{R}^T)^{-1}\mathscr{A}\mathscr{R}^T = \mathscr{S}\mathscr{A}\mathscr{S}^{-1}},$$

where $\mathscr{S} = (\mathscr{R}^T)^{-1}$.

We have derived:

Theorem (on the change of the matrix $\mathscr{A}$ representing an operator A)

If the linear operator $A \in \hat{L}(V_n(F), V_n(F))$ is represented in the two bases $v = \{\vec{v}_1, \vec{v}_2, \ldots, \vec{v}_n\}$ and $v' = \{\vec{v}'_1, \vec{v}'_2, \ldots, \vec{v}'_n\}$ by the matrices $\mathscr{A}$ and $\mathscr{A}'$, respectively, when the two bases are connected by the invertible replacement matrix $\mathscr{R} : [v'] = \mathscr{R}[v]$, then the matrix $\mathscr{A}'$ is obtained from the matrix $\mathscr{A}$ by the similarity transformation with the matrix $\mathscr{S} = (\mathscr{R}^T)^{-1}$:

$$\boxed{\mathscr{A}' = \mathscr{S}\mathscr{A}\mathscr{S}^{-1}}. \ \Delta$$

This result can be illustrated by the following figure

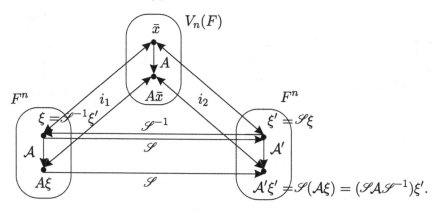

In other words, $\mathscr{A}' = \mathscr{S}\mathscr{A}\mathscr{R}^T$, or, in terms of matrix elements $(k, m = 1, 2, \ldots, n,)$

$$d'_{km} = \sum_{i,j=1}^{n} \{\mathscr{S}_{ki}\}\{\mathscr{A}_{ij}\}\{\mathscr{R}^T\}_{jm} = \sum_{i,j=1}^{n} s_{ki} r_{mj} a_{ij}.$$

So, the set of matrices that represent the operator A in all bases of $V_n(F)$ are in fact a mixed tensor of the second order (once contravariant and once covariant). Remember that a mixed tensor of the second order is a map which connects a basis $v = \{\bar{v}_1, \bar{v}_2, \ldots, \bar{v}_n\}$ in $V_n(F)$ with a two-index system $\{a_{ij}, i, j = 1, 2, \ldots, n\}$ of n^2 numbers from F, such that this system goes to the new one $a'_{km} = \sum_{i,j=1}^{n} s_{ki} r_{mj} a_{ij}$, $k, n = 1, 2, \ldots, m$, when the basis changes to the new one $v' = \{\bar{v}'_1, \bar{v}'_2, \ldots, \bar{v}'_n\}$ with the invertible replacement matrix $\mathscr{R} = [r_{ij}]_{n \times n}$: $\bar{v}'_i = \sum_{j=1}^{n} r_{ij} \bar{v}_j$. The elements $s_{ij}, i = 1, 2, \ldots, n$, form the matrix $\mathscr{S} = [s_{ij}]_{n \times n}$, which is contragredient to $\mathscr{R}$: $\mathscr{S} = (\mathscr{R}^T)^{-1}$.

To be more precise, we should say that it is a mixed tensor with respect to the general linear group $GL(n, F)$ of all invertible $n \times n$ matrices with elements from the field F, since $\mathscr{R}$ runs through that group in order to reach all the bases in $V_n(F)$ from the initial one v.

In fact, the pair $v = \{\bar{v}_1, \bar{v}_2 v, \ldots, \bar{v}_n\}$ and $\mathscr{A} = [a_{ij}]_{n \times n}$ determines the operator A by $A\bar{v}_i = \sum_{j=1}^{n} a_{ji} \bar{v}_j$ (note that an operator is given if we know the images of a basis by that operator). All other bases and the corresponding representative matrices of A follow as $\mathscr{R}[v]$ and $(\mathscr{R}^T)^{-1}\mathscr{A}\mathscr{R}^T$, $\mathscr{R} \in GL(n, F)$.

To be more sophisticated, we can say that the similarity transformation $\mathscr{S}\mathscr{A}\mathscr{S}^{-1}$, $\mathscr{S} \in GL(n, F)$ is an equivalence relation in the set F^{nn} of all $n \times n$ matrices with elements from F, since it is a reflexive, symmetric, and transitive (RST) relation due to the fact that $GL(n, F)$ is a group. Namely, it is reflexive since the unit $n \times n$ matrix I_n belongs to $GL(n, F)$: $\mathscr{A} = I_n \mathscr{A} I_n^{-1}$; it is symmetric since the group is closed to the inversion: $\mathscr{A}' = \mathscr{S}\mathscr{A}\mathscr{S}^{-1}$ and $\mathscr{A} = (\mathscr{S}^{-1})\mathscr{A}'(\mathscr{S}^{-1})^{-1}$; it is transitive since the group is closed with respect to matrix multiplication $\mathscr{A}' = \mathscr{S}\mathscr{A}\mathscr{S}^{-1}$ and $\mathscr{A}'' = \mathscr{P}\mathscr{A}'\mathscr{P}^{-1} \Rightarrow \mathscr{A}'' = (\mathscr{P}\mathscr{S})\mathscr{A}(\mathscr{P}\mathscr{S})^{-1}$.

As an equivalence relation, this similarity transformation partitions the set F^{nn} into equivalence classes. Thus, once we choose the pair $(v, \mathscr{A})$ which determines A, then the set of all matrices that represent A in all bases is the equivalence class to which $\mathscr{A}$ belongs as representative.

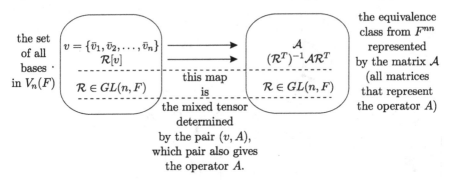

the mixed tensor
determined
by the pair (v, A),
which pair also gives
the operator A.

Remark The use of the matrices $\mathscr{R}$ and $\mathscr{S} = (\mathscr{R}^T)^{-1}$ is customary when one investigates the tensorial nature of physical objects, like vectors and operators.

Another, even more frequent, method in change-of-basis problems is to use the transition operator T which expresses the "new" basis vectors directly from those of the "old" basis $\bar{v}'_i = T\bar{v}_i$, $i = 1, 2, \ldots, n$.

To represent this operator in the "old" basis, we write $T\bar{v}_i = \bar{v}'_i \overset{(*)}{=} \sum_{j=1}^{n} r_{ij}\bar{v}_j = \sum_{j=1}^{n} t_{ji}\,\bar{v}_j \Rightarrow r_{ij} = t_{ji}$, $i = 1, 2, \ldots, n$ which gives immediately

$$\boxed{\mathscr{T} = \mathscr{R}^T}.$$

Thus, our formula for the change of the matrix-column ξ that represents the vector $\bar{x} \in V_n(F)$ is now $\xi' = (\mathscr{R}^T)^{-1}\xi = \mathscr{T}^{-1}\xi$, so $\mathscr{S} = \mathscr{T}^{-1}$.

As far as the matrix $\mathscr{A}$, that represents the operator $A \in \hat{L}(V_n(F), V_n(F))$, is concerned we have

$$\mathscr{A}' = (\mathscr{R}^T)^{-1}\mathscr{A}\mathscr{R}^T = \mathscr{T}^{-1}\mathscr{A}\mathscr{T}.$$

From this result, we can conclude that $\mathscr{T}$ represents the transition operator T also in the "new" basis:

$$\mathscr{T}' = \mathscr{T}^{-1}\mathscr{T}\mathscr{T} = \mathscr{T}.$$

4.4 The Change of Bases in Euclidean (E_n) and Unitary (U_n) Vector Spaces

The great advantage of the above formulation of the change of basis in a general vector space $V_n(F)$ is that as a consequence we get this theory in Euclidean and unitary spaces at once.

In Euclidean spaces, orthonormal (ON) bases are connected with orthogonal replacement matrices $\mathscr{R}$ which are defined by $\boxed{\mathscr{R}^{-1} = \mathscr{R}^T}$ (the proof is analogous to the case of unitary spaces below). The orthogonal $n \times n$ real matrices form the orthogonal group $O(n)$, which is the group of all automorphisms in $\mathbb{R}^n$.

Proof that orthogonal matrices form a group:

1. $I_n \in O(n)$;
2. $A \in O(n) \Rightarrow (A^{-1})^{-1} = A = (A^{-1})^T \Rightarrow A^{-1} \in O(n)$;
3. $A \in O(n)$ and $B \in O(n) \Rightarrow (AB)^{-1} = B^{-1}A^{-1} = B^T A^T = (AB)^T \Rightarrow (AB) \in O(n)$ $\triangle$)

From the definition of orthogonal matrices, it follows that $\mathscr{S} = (\mathscr{R}^T)^{-1} = \mathscr{R}$, so that the law for the change of the matrix-column ξ that represents a vector $\bar{x} \in E_n$ is $\boxed{\xi' = \mathscr{R}\xi}$, and such a law for the matrix $\mathscr{A}$ that represents an operator $A \in \hat{L}(E_n, E_n)$ is

$$\boxed{\mathscr{A}' = \mathscr{R}\mathscr{A}\mathscr{R}^{-1}}.$$

So, in Euclidean spaces, there is no difference between covariant and contravariant tensor quantities. The matrix-columns of $\bar{x} \in E_n$ change as tensors of the first order $\xi'_i = \sum_{j=1}^{n} r_{ij}\xi_j$, $i = 1, 2, \ldots, n$, and the matrices of $A \in \hat{L}(E_n, E_n)$ as tensors of the second order $a'_{km} = \sum_{i,j=1}^{n} r_{ki} r_{mj} a_{ij}$.

It is useful to call tensors by the name of the subgroup of $GL(n, F)$ through which the replacement matrix $\mathscr{R}$ goes; in this case, it is the group $O(n)$, so the above tensors are orthogonal ones. But, it is quite customary (even though it is wrong) to call them Cartesian tensors.

In unitary spaces, ON bases are connected with unitary replacement matrices $\mathscr{R}$ which are defined by $\mathscr{R}^{-1} = (\mathscr{R}^T)^* = \mathscr{R}^\dagger$.

Proof that in U_n orthonormal bases are connected by unitary $\mathscr{R}$: Since the "old" basis $u = \{\bar{u}_1, \bar{u}_2, \ldots, \bar{u}_n\}$ is ON, i.e., $(\bar{u}_i, \bar{u}_j) = \delta_{ij}$, $i, j = 1, 2, \ldots, n$, we shall find which conditions the replacement matrix $\mathscr{R} = [r_{ij}]_{n \times n}$ must satisfy, so that the "new" basis $u' = \{\bar{u}'_1, \bar{u}'_2, \ldots, \bar{u}'_n\}$ is also ON. We start with $\bar{u}'_i = \sum_{j=1}^{n} r_{ij}\bar{u}_j$, $i = 1, 2, \ldots, n$, and calculate

$$(\bar{u}'_i, \bar{u}'_k) = (\sum_{j=1}^{n} r_{ij}\bar{u}_j, \sum_{m=1}^{n} r_{km}\bar{u}_m) = \sum_{j,m=1}^{n} r^*_{ij} r_{km} \overset{\delta_{jm}}{(\bar{u}_j, \bar{u}_m)} =$$

$$= \sum_{j=1}^{n} r^*_{ij} r_{kj} = \sum_{j=1}^{n} \{\mathscr{R}^*\}_{ij} \{\mathscr{R}^T\}_{jk}.$$

To have $(\bar{u}'_2, \bar{u}'_k) = \delta_{ik}$, the matrix $\mathscr{R}$ must satisfy $\mathscr{R}^* \mathscr{R}^T = I_n$ or $\mathscr{R}\mathscr{R}^\dagger = I_n$. This means that $\mathscr{R}$ must be a unitary matrix $\mathscr{R}^{-1} = \mathscr{R}^\dagger$. $\triangle$

The unitary $n \times n$ matrices form the unitary group $U(n)$. (The proof is analogous to that for the orthogonal group $O(n)$ above). This is the group of all automorphisms in $\mathbb{C}^n$. From the definition of unitary matrices, it follows that $\mathscr{S} = (\mathscr{R}^{-1})^T = (\mathscr{R}^\dagger)^T = \mathscr{R}^*$. This means that the contragredient change here is the complex conjugate one:

$$\xi' = \mathscr{R}^* \xi \text{ and } \mathscr{A}' = \mathscr{R}^* \mathscr{A} \mathscr{R}^T.$$

In terms of matrix elements we have

$$\xi_i' = \sum_{j=1}^n r_{ij}^* \xi_j, \ i = 1,2,\ldots,n, \ \text{and} \ a_{km}' = \sum_{i,j=1}^n r_{ki}^* r_{mj} a_{ij}, \ k,m = 1,2,\ldots,n.$$

These tensors are called unitary ones, since $\mathscr{R} \in U_n$: contravariant unitary vector and mixed unitary tensor, respectively.

4.5 The Change of Biorthogonal Bases in $V_n^*(F)$ (Covariant Vectors)

Let us again consider a general vector space $V_n(F)$ (no inner product is assumed). The vector space $\hat{L}(V_n(F),F)$ of all linear functionals in $V_n(F)$ is obviously an n-dimensional space and it is denoted by $V_n^*(F)$.

To represent a linear functional $f \in V_n^*(F)$ in a basis $v = \{\bar{v}_1, \bar{v}_2, \ldots, \bar{v}_n\}$ in $V_n(F)$, we shall apply f to all basis vectors and then form the matrix-row $\mathscr{S}$ of the so-obtained n numbers (Sect. 4.1) $\mathscr{S} = [f(\bar{v}_1) f(\bar{v}_2) \ldots f(\bar{v}_n)]$.

To find how this representative row changes with the change of basis v, we shall reveal a special *bijection* that exists between the set of all bases in $V_n(F)$ and such a set in $V_n^*(F)$. This bijection is called *biorthogonality*.

To every basis v in $V_n(F)$, we can uniquely associate a set $\{f_1, f_2, \ldots, f_n\}$ of n vectors in $V_n^*(F)$ in two equivalent ways:

(1) $f_i(\bar{y}) = \eta_i \ i = 1,2,\ldots,n$, where η_i are the coordinates of an arbitrary vector $\bar{y} \in V_n(F)$ in the basis v: $\bar{y} = \sum_{i=1}^n \eta_i \bar{v}_i$.
(2) $f_i(\bar{v}_j) = \delta_{ij}, \ i,j = 1,2,\ldots,n$.

[the biorthogonal vectors (functionals) $\{f_1, f_2, \ldots, f_n\}$ are represented in the basis v by the n rows $[10\ldots0], [01\ldots0] \ldots [00\ldots1]$, respectively.]

That these two definitions are equivalent becomes obvious when we remember that $\bar{v}_j = \sum_{i=1}^n \delta_{ij} \bar{v}_i$, so (1) $\Rightarrow$ (2), and that the functional f_i is linear

$$f_i(\bar{y}) = \sum_{j=1}^n \eta_j f_i(\bar{v}_j) = \sum_{j=1}^n \eta_j \delta_{ij} = \eta_i, \ \text{so} \ (2) \Rightarrow (1).$$

Since the set $\{f_1, f_2, \ldots, f_n\}$ has n elements, which number is equal to the dimension of $V_n^*(F)$, this set will be a basis in $V_n^*(F)$ if we prove that it is a spanning set. With this purpose in mind, let us apply an arbitrary functional $f \in V_n^*(F)$ to an arbitrary vector $\bar{y} \in V_n(F)$:

$$f(\bar{y}) = f \sum_{i=1}^n \eta_i \bar{v}_i = \sum_{i=1}^n \eta_i f(\bar{v}_i) \stackrel{(1)}{=} \sum_{i=1}^n f(v_i) f_i(\bar{y}).$$

Thus, it is a spanning set $f = \sum_{i=1}^n f(\bar{v}_i)f_i$, and the representative matrix-column of f in this basis is $\mathscr{S}^T = [f(\bar{v}_1)\,f(\bar{v}_2)\dots f(\bar{v}_n)]^T$. We call the basis $\{f_1, f_2, \dots, f_n\}$ in $V_n^*(F)$ *biorthogonal* to the basis v in $V_n(F)$, because of definition (2): $f_i(\bar{v}_j) = \delta_{ij}$.

So, we can represent f by the row $\mathscr{S}$ in the basis v in $V_n(F)$ [since it is a functional in $V_n(F)$] or by the column $\mathscr{S}^T$ in the biorthogonal basis $\{f_1, f_2, \dots, f_n\}$ in $V_n^*(F)$ [since it is a vector in $V_n^*(F)$].

Remark about unitary spaces.

We have already pointed out that every ON basis in the unitary space U_n defines its dual basis in U_n^* which is also orthonormal:

$$u = \{\bar{u}_1, \bar{u}_2, \dots, \bar{u}_n\} \xrightarrow{D} \{D\bar{u}_1, D\bar{u}_2, \dots, D\bar{u}_n\}$$

and $(D\bar{u}_i, D\bar{u}_j) = (\bar{u}_i, \bar{u}_j)^* = \delta_{ij}$, $i, j = 1, 2, \dots, n$.

However, the dual basis is at the same time biorthogonal to u : the basis $\{f_1, f_2, \dots, f_n\}$ biorthogonal to u is given by formula (1)

$$\boxed{f_i(\bar{y}) = \eta_i}, \ \forall \bar{y} \in U_n, \ i = 1, 2, \dots, n,$$

but since η_i is the i-th coordinate of $\bar{y}$ in u, it is in fact the Fourier coefficient

$$\boxed{\eta_i = (\bar{u}_i, \bar{y})}$$

$$[\bar{y} = \sum_{i=1}^n \eta_i \bar{u}_i = \sum_{i=1}^n (\bar{u}_i, \bar{y})\bar{u}_i].$$

So, $f_i(\bar{y}) = (\bar{u}_i, \bar{y})$, $\forall \bar{y} \in U_n$, $i = 1, 2, \dots, n$, which is the fundamental formula for dualism (FFD), implying $f_i = D\bar{u}_i$. or, in other words, that the biorthogonal basis of u is at the same time its dual basis.

In conclusion, the set of all ON bases in U_n can be bijectively mapped onto the set of its biorthogonal bases in U_n^*, which are all ON, and dual as well. Δ

We shall now investigate how the biorthogonal basis $\{f_1, f_2, \dots, f_n\}$ changes when the basis v is changed by an invertible replacement matrix $\mathscr{R} = [r_{ij}]_{n \times n}$ into

$$v' = \{\bar{v}_1', \bar{v}_2', \dots, \bar{v}_n'\} : \bar{v}_i' = \sum_{j=1}^n r_{ij} \bar{v}_j \ (3).$$

Let us denote the corresponding replacement matrix in $V_n^*(F)$ by

$$\mathscr{S} = [s_{ij}]_{n \times n} : f_i' = \sum_{j=1}^n s_{ij} f_j \ (4).$$

We do not know the relation between $\mathscr{R}$ and $\mathscr{S}$. Let us write definition (2) $f_i'(\bar{v}_k') = \delta_{ik}$ for v' and $\{f_1', f_2', \dots, f_n'\}$, and then write $f_i'(\bar{v}_k')$ in detail:

$$f_i'(\bar{v}_k') \overset{(4)}{=} \sum_{j=1}^{n} s_{ij} f_j(\bar{v}_k') \overset{(3)}{=} \sum_{j=1}^{n} s_{ij} f_j(\sum_{m=1}^{n} r_{km} \bar{v}_m) =$$

$$= \sum_{m,j=1}^{n} s_{ij} r_{km} f_j(\bar{v}_m) \overset{(2)}{=} \sum_{m,j=1}^{n} s_{ij} r_{km} \delta_{jm} = \sum_{j=1}^{n} s_{ij} r_{kj} = \sum_{j=1}^{n} \{\mathscr{S}\}_{ij} \{\mathscr{R}^T\}_{jk} = \{\mathscr{S}\mathscr{R}^T\}_{ik}.$$

Since it is δ_{ik} by (2) above, we finally have $\mathscr{S}\mathscr{R}^T = I_n \Rightarrow \boxed{\mathscr{S} = (\mathscr{R}^T)^{-1}}$ (the contragredient matrix). Δ

Therefore, when $\mathscr{R}$ runs through the group $GL(n,F)$, so does $\mathscr{S}$ and there is a bijection $\mathscr{R} \leftrightarrow (\mathscr{R}^T)^{-1}$, so that we have a bijection between the set of all bases in $V_n(F)$ and the set of all bases in $V_n^*(F)$, called *biorthogonality*:

$$\mathscr{R} \begin{bmatrix} \bar{v}_1 \\ \bar{v}_2 \\ \vdots \\ \bar{v}_n \end{bmatrix} \leftrightarrow (\mathscr{R}^T)^{-1} \begin{bmatrix} f_1 \\ f_2 \\ \vdots \\ f_n \end{bmatrix}, \quad \mathscr{R} \in GL(n,F).$$

Since biorthogonal bases in $V_n^*(F)$ are connected by $(\mathscr{R}^T)^{-1}$ when their correspondent bases v and v' in $V_n(F)$ are related by $\mathscr{R}$, we want to know the change of representing columns $\mathscr{S}^T$ and $(\mathscr{S}')^T$ in those biorthogonal bases of a vector $f \in V_n^*(F)$. This change will be by the matrix contragredient to $(\mathscr{R}^T)^{-1}$ (cf. the theorem on the change of the representing column of a vector, Sect. 4.3.1). This is obviously the matrix $\mathscr{R}$:

$$\boxed{(\mathscr{S}')^T = \mathscr{R}\mathscr{S}^T},$$

or, in terms of coordinates, $f(\bar{v}_i') = \sum_{j=1}^{n} r_{ij} f(\bar{v}_j)$. Comparing with (3) $\bar{v}_i' = \sum_{j=1}^{n} r_{ij} \bar{v}_j$, we see that the representative column of f changes like the bases in $V_n(F)$, so f is called a covariant vector. When f is considered as a functional in $V_n(F)$, this obviously reads as

$$\boxed{\mathscr{S}' = \mathscr{S}\mathscr{R}^T}.$$

Theorem (on the change of the representing row of a functional) Representing the matrix-row $\mathscr{S} = [f(\bar{v}_1) f(\bar{v}_2) \ldots f(\bar{v}_n)]$ of a functional $f \in V_n^*(F)$ in the basis $v = \{\bar{v}_1, \bar{v}_2, \ldots, \bar{v}_n\}$ in $V_n(F)$ changes, when the basis is changed by the replacement matrix $\mathscr{R}$ into $v' = \{\bar{v}_1', \bar{v}_2', \ldots, \bar{v}_n'\}$: $\bar{v}_i' = \sum_{j=1}^{n} r_{ij} \bar{v}_j$, $i = 1,2,\ldots,n$, in the same manner as the basis, i.e., $f(\bar{v}_i') = \sum_{j=1}^{n} r_{ij} f(\bar{v}_j)$ or in the form of matrices:

$$\boxed{\mathscr{S}' = \mathscr{S}\mathscr{R}^T}.$$

For this reason, the representing rows of f are an example of covariant vectors.

Remember that a covariant vector is a map which connects a basis $v = \{\bar{v}_1, \bar{v}_2, \ldots, \bar{v}_n\}$ in $V_n(F)$ with a one-index system $\{\mathscr{S}_j, k = 1,2,\ldots,n\}$ of n numbers from F, such that this system goes to the new one $\mathscr{S}_i' = \sum_{j=1}^{n} r_{ij} \mathscr{S}_j$, $i = 1,2,\ldots,n$, when the basis changes to the new one $v' = \{\bar{v}_1, \bar{v}_2, \ldots, \bar{v}_n\}$ in the same manner with the invertible replacement matrix $\mathcal{R} = [r_{ij}]_{n \times n} \in GL(n, F)$:

$$\bar{v}_i = \sum_{j=1}^{n} r_{ij} \bar{v}_j.$$

4.6 The Relation between $V_n(F)$ and $V_n^*(F)$ is Symmetric (The Invariant Isomorphism between $V_n(F)$ and $V_n^{**}(F)$)

There cannot be established an invariant (i.e., basis independent) isomorphism between $V_n(F)$ and $V_n^*(F)$: if we choose a pair of biorthogonal bases $v = \{\bar{v}_1, \bar{v}_2, \ldots, \bar{v}_n\}$ in $V_n(F)$ and $\{f_1, f_2, \ldots, f_n\}$ in $V_n^*(F)$ by $f_i(\bar{v}_j) = \delta_{ij}$, we can define an isomorphism between $V_n(F)$ and $V_n^*(F)$ by mapping $\bar{x} = \sum_{i=1}^{n} \xi_i \bar{v}_i \in V_n(F)$ onto $f = \sum_{i=1}^{n} \xi_i f_i \in V_n^*(F)$. But, in another pair of biorthogonal bases

$$\mathcal{R} \begin{bmatrix} \bar{v}_1 \\ \bar{v}_2 \\ \vdots \\ \bar{v}_n \end{bmatrix} \quad \text{and} \quad (\mathcal{R}^T)^{-1} \begin{bmatrix} f_1 \\ f_2 \\ \vdots \\ f_n \end{bmatrix},$$

$\mathcal{R}$ is one of the matrices from $GL(n, F)$, the representing column $\xi = [\xi_1 \, \xi_2 \, \ldots \, \xi_n]^T$ of $\bar{x}$ will become $(\mathcal{R}^T)^{-1}\xi$ and that of f will be $\mathcal{R}\xi$. This means that in this new pair of biorthogonal bases the vectors $\bar{x}$ and f will no longer have the same expansions, i.e., representations, and the above isomorphism is not valid any more.

But, we can show that the dual space of $V_n^*(F)$, i.e., $V_n^{**}(F)$, the so-called second dual space, is isomorphic to $V_n(F)$, and that this isomorphism is invariant in the sense that when it is defined in one pair of biorthogonal bases it remains valid in all other pairs of biorthogonal bases. It is quite customary to identify two vector spaces that can be joined by an invariant isomorphism:

$$\boxed{V_n^{**}(F) \overset{inv}{\cong} V_n(F)}.$$

Let us study in detail this isomorphism. Consider a vector $\bar{x} \in V_n(F)$ which is represented in the basis $v = \{\bar{v}_1, \bar{v}_2, \ldots, \bar{v}_n\}$ by the column $\xi = [\xi_1 \, \xi_2 \, \ldots \, \xi_n]^T$ as a consequence of the expansion $\bar{x} = \sum_{i=1}^{n} \xi_i \bar{v}_i$. If we fix that $\bar{x}$ and let f in $f(\bar{x})$ run through the basis $\{f_1, f_2, \ldots, f_n\}$ biorthogonal to v, then by (1): $f_1(\bar{x}) = \xi_1, f_2(x) = \xi_2, \ldots, f_n(\bar{x}) = \xi_n$ and this is a linear functional in $V_n^*(F)$ denoted by $\bar{\bar{x}}(f)$, $f \in V_n^*(F)$. Let us explain this in more detail. If in $f(\bar{x})$ we fix $\bar{x}$ and let f go through the whole space $V_n^*(F)$, then we get a map $V_n^*(F) \to F$ or $f \mapsto f(\bar{x}) \in F$. This is a functional in $V_n^*(F)$ that we denote as $\bar{\bar{x}}(f)$, $f \in V_n^*(F)$. This functional is a linear one: $\bar{\bar{x}}(af_1 + bf_2) = a\bar{\bar{x}}(f_1) + b\bar{\bar{x}}(f_2)$, since

$$\bar{\bar{x}}(af_1 + bf_2) = [af_1 + bf_2](\bar{x}) = af_1(\bar{x}) + bf(\bar{x}) =$$
$$= a\bar{\bar{x}}(f_1) + b\bar{\bar{x}}(f_2)$$

(according to the rule of action of a linear combination of linear maps.) This linear functional is given by the images of the biorthogonal basis:

$$\bar{\bar{x}}(f_1) = \xi_1, \, \bar{\bar{x}}(f_2) = \xi_2, \, \ldots, \bar{\bar{x}}(f_n) = \xi_n,$$

and it is represented in that basis by the row $\xi^T = [\xi_1, \xi_2, \ldots, \xi_n]$.

Thus, the vector $\bar{x}$ in $V_n(F)$ and the functional $\bar{\bar{x}}$ in $V_n^*(F)$ determine each other (we have a bijection $\bar{x} \leftrightarrow \bar{\bar{x}}$) once the pair of biorthogonal bases $v = \{\bar{v}_1, \bar{v}_2, \ldots, \bar{v}_n\}$ and $\{f_1, f_2, \ldots, f_n\}$ is given. This bijection is obviously a homomorphism of vector spaces (it preserves linear combinations): if $\bar{x}_1 \leftrightarrow \bar{\bar{x}}_1$ and $\bar{x}_2 \leftrightarrow \bar{\bar{x}}_2$, then $a\bar{x}_1 + b\bar{x}_2 \leftrightarrow a\bar{\bar{x}}_1 + b\bar{\bar{x}}_2$. So it is an isomorphism.

When we replace v by the replacement matrix $\mathscr{R}$, then its biorthogonal basis will be replaced by $(\mathscr{R}^T)^{-1}$. The column of $\bar{x}$ in the new basis will be $(\mathscr{R}^T)^{-1}\xi$ (see the theorem in Sect. 4.3.1 on the change of the representing column of a vector) and the row of $\bar{\bar{x}}$ will be $\xi^T \mathscr{R}^{-1}$, since the representing row of this functional in $V_n^*(F)$ will change by the matrix transposed of the replacement matrix $(\mathscr{R}^T)^{-1}$ of $\{f_1, f_2, \ldots, f_n\}$, i.e., $((\mathscr{R}^{-1})^T)^T = \mathscr{R}^{-1}$ (see the theorem on the change of the representing row of a functional, Sect. 4.5).

But, $\xi^T \mathscr{R}^{-1}$ is obviously the transposition of $(\mathscr{R}^T)^{-1}\xi$: $[(\mathscr{R}^T)^{-1}\xi]^T = \xi^T \mathscr{R}^{-1}$. Therefore, in all pairs of biorthogonal bases, the vector $\bar{x}$ and the functional $\bar{\bar{x}}$ are represented by the transposed matrices (a column and a row) with the same elements. We have established the invariant isomorphism $V_n(F) \overset{inv}{\cong} V_n^{**}(F)$. We can identify $\bar{x}$ and $\bar{\bar{x}}$, and consider $\bar{x}$ as a functional in $V_n^*(F)$, since it immediately determines $\bar{\bar{x}}$ by ξ^T in $\{f_1, f_2, \ldots, f_n\}$. The relation between $V_n(F)$ and $V_n^*(F)$ becomes symmetric, since the vectors in $V_n(F)$ now play the roles of functionals in $V_n^*(F)$ (Compare, the treatment of the same problem in the special case of a unitary space, Sect. 4.1).

4.7 Isodualism—The Invariant Isomorphism between the Superspaces $\hat{L}(V_n(F), V_n(F))$ and $\hat{L}(V_n^*(F), V_n^*(F))$

There is a second invariant isomorphism related to $V_n(F)$ and $V^*(F)$, not directly in connection with these two vector spaces, but instead between their superspaces $\hat{L}(V_n(F), V_n(F))$ and $\hat{L}(V_n^*(F), V_n^*(F))$.

We would like to emphasize that $V_n(F) \overset{inv}{\cong} V_n^{**}(F)$ and $\hat{L}(V_n(F), V_n(F)) \overset{inv}{\cong} \hat{L}(V_n^*(F), V_n^*(F))$ are the only two invariant isomorphisms in the theory of dual spaces.

To define this new isomorphism, let us notice that we can uniquely attach to an arbitrary linear operator $A \in \hat{L}(V_n(F), V_n(F))$ a map A^T which acts in $V_n^*(F)$:

$$\boxed{[A^T f](\bar{y}) = f(A\bar{y}), \, \forall \bar{y} \in V_n(F)} \quad \text{or}$$

$$A^T f = f' = f \circ A,$$

i.e., A^T maps an arbitrary linear functional $f \in V_n^*(F)$ onto another functional $f' = f \circ A$, obtained by composition of the two linear mappings

$$f \in \hat{L}(V_n(F), F) \text{ and } A \in \hat{L}(V_n(F), V_n(F)).$$

Obviously, f' is a linear functional, since from the composition it follows that

$$f' = f \circ A \in \hat{L}(V_n(F), F).$$

It can be easily shown that A^T is a linear operator in $V_n^*(F)$, i.e., that it maps a linear combination of functionals onto the same linear combination of their images:

$$A^T(af_1 + bf_2) = (af_1 + bf_2) \circ A = af_1 \circ A + bf_2 \circ A = aA^T f_1 + bA^T f_2.$$

The inverse map attaches uniquely to every linear operator A^T in $V_n^*(F)$ a map A in $V_n(F)$ which maps every $\bar{y} \in V_n(F)$ (considered here as a linear functional in $V_n^*(F)$, see Sect. 4.6) onto a linear functional

$$\boxed{\bar{y}'(f) = [A\bar{y}](f) = y(A^T f), \ \forall f \in V_n^*(F)} \text{ or}$$
$$A\bar{y} = \bar{y}' = \bar{y} \circ A^T.$$

(The proofs that $\bar{y}'$ is a linear functional and that A is a linear operator are analogous to the above proofs).

We shall now demonstrate that if the linear operator A is represented by a matrix $\mathscr{A}$ in some basis $v = \{\bar{v}_1, \bar{v}_2, \ldots, \bar{v}_n\}$ in $V_n(F)$, then the linear operator A^T is represented by the transposed matrix $\mathscr{A}^T$ in the biorthogonal basis $\{f_1, f_2, \ldots, f_n\}$ $(f_i(\bar{v}_j) = \delta_{ij})$ in $V_n^*(F)$.

Let us denote by $\mathscr{A} = [a_{ij}]_{n \times n}$ and $\mathscr{C} = [c_{ij}]_{n \times n}$ the matrices that represent A and A^T in the above bases. We write the basic formula for representing A^T:

$$[A^T f_k](\bar{y}) = \sum_{i=1}^{n} c_{ik} f_i(\bar{y}), \ k = 1, 2, \ldots, n$$

and replace the l.h.s. by the definition of A^T

$$[A^T f_k](\bar{y}) = f_k(A\bar{y}).$$

We first use the expansion of $\bar{y}$ and the linearity of A, then the basic formula for its representation, followed by use of the linearity of f_k, and, finally, the two equivalent definitions of biorthogonality:

$$k = 1, 2, \ldots, n, \quad [A^T f_k](\bar{y}) = f_k(A\bar{y}) = f_k(A \sum_{i=1}^{n} \eta_i \bar{v}_i) =$$

$$= f_k(\sum_{i=1}^{n} \eta_i A \bar{v}_i) = f_k(\sum_{i=1}^{n} \eta_i \sum_{j=1}^{n} a_{ji} \bar{v}_j) = \sum_{i=1}^{n} \eta_i \sum_{j=1}^{n} a_{ji} \delta_{kj} =$$

$$= \sum_{i=1}^{n} \eta_i a_{ik} = \sum_{i=1}^{n} a_{ki} f_i(\bar{y}).$$

Comparing this result with $[A^T f_k](\bar{y}) = \sum_{i=1}^{n} c_{ik} f_i(y)$, we see immediately that $c_{ik} = a_{ik}$, $i, k = 1, 2, \ldots, n$ or

$$\boxed{\mathscr{C} = \mathscr{A}^T}. \; \Delta$$

Remember that if some functional $f \in V_n^*(F)$ is represented by the matrix-column $\mathscr{S}^T = [f(\bar{v}_1) f(\bar{v}_2) \ldots f(\bar{v}_n)]^T$ in the biorthogonal basis $\{f_1, f_2, \ldots, f_n\}$ in $V_n^*(F)$, then the functional $A^T f$ is represented by the column $\mathscr{A}^T \mathscr{S}^T$, where $\mathscr{A}^T$ represents the operator A^T in the same basis. However, in the basis v which is biorthogonal to $\{f_1, f_2, \ldots, f_n\}$, the functional f is represented by the matrix-row $\mathscr{S}$, and $A^T f$ by the row $(\mathscr{A}^T \mathscr{S}^T) = \mathscr{S} \mathscr{A}$, where $\mathscr{A}$ represents A in v. This means that A and A^T are represented in the basis v in $V_n(F)$ by matrices $\mathscr{A}_\rightarrow$ and $_\leftarrow \mathscr{A}$, respectively [the arrows indicate that the first matrix $\mathscr{A}_\rightarrow$ acts to the right on the columns that represent vectors in $V_n(F)$ in the basis v, while the second matrix $_\leftarrow \mathscr{A}$ acts to the left on the rows that represent functionals in $V_n(F)$ in the same basis v].

So, A and A^T are *represented by the same matrix* $\mathscr{A}$ in the basis v in $V_n(F)$.

This relation (the sameness of the representing matrices) is valid for all bases in $V_n(F)$, since both matrices change by the same similarity transformation $(\mathscr{R}^T)^{-1} \mathscr{A} \mathscr{R}^T$ when the column of basis vectors $[v]$ changes to $\mathscr{R}[v]$. [For the first matrix $\mathscr{A}_\rightarrow$ see the theorem on the change of the matrix $\mathscr{A}$ that represents an operator A in $V_n(F)$, and for the transposition of the second matrix, i.e., $\mathscr{A}^T$, we have $\mathscr{R} \mathscr{A}^T \mathscr{R}^{-1}$ (the biorthogonal basis $\{f_1, f_2, \ldots, f_n\}$ changes by $\mathscr{S} = (\mathscr{R}^T)^{-1}$, when $[v]$ changes into $\mathscr{R}[v]$, and the matrix $\mathscr{A}^T$ consequently by $(\mathscr{S}^T)^{-1} \mathscr{A}^T \mathscr{S}^T = \mathscr{R} \mathscr{A}^T \mathscr{R}^{-1}$) and after transposition we get the change for the second matrix $_\leftarrow \mathscr{A}$:

$$(\mathscr{R} \mathscr{A}^T \mathscr{R}^{-1})^T = (\mathscr{R}^T)^{-1} \mathscr{A} \mathscr{R}^T,$$

which is the same change as for $\mathscr{A}_\rightarrow$].

The fact that $A \in \hat{L}(V_n(F), V_n(F))$ and $A^T \in \hat{L}(V_n^*(F), V_n^*(F))$ are represented by the same matrix $\mathscr{A}$ (more exactly by $\mathscr{A}_\rightarrow$ and $_\leftarrow \mathscr{A}$, respectively) in a basis v in the space $V_n(F)$ and that this relation (the sameness) remains in all bases in $V_n(F)$, can be used as an *alternative definition of this invariant isomorphism* between the two superspaces. So, A and A^T determine each other uniquely on any pair of biorthogonal bases.

This means that for a basis $v = \{\bar{v}_1, \bar{v}_2, \ldots, \bar{v}_n\}$ in $V_n(F)$ and its biorthogonal basis $\{f_1, f_2, \ldots, f_n\}$ ($f_i(\bar{v}_i) = \delta_{ij}$, $i, j = 1, 2, \ldots, n$), the representing matrix $\mathscr{A} = [a_{ij}]_{n \times n}$ determines both A and A^T: $A\bar{v}_i = \sum_{j=1}^{n} a_{ji} \bar{v}_j$ and $A^T f_i = \sum_{j=1}^{n} a_{ij} f_j$, $i = 1, 2, \ldots, n$, since $\mathscr{A}$ represents A in v and $\mathscr{A}^T$ represents A^T in $\{f_1, f_2, \ldots, f_n\}$. Thus, A and A^T determine each other uniquely (this is a bijection).

We have only to show that this bijection preserves any linear combination, i.e., that it is a vector space homomorphism:

$$(aA_1^T + bA_2^T)f = a(A_1^T f) + b(A_2^T f) = af \circ A_1 + bf \circ A_2 = f \circ (aA_1 + bA_2) \text{ or}$$
$$aA_1 + bA_2 \rightarrow aA_1^T + bA_2^T.$$

Thus, we have established an invariant (natural) isomorphism between the two superspaces. We call this isomorphism *isodualism*, and it enables us to identify these two superspaces.

Since the vectors from $V_n^*(F)$ are represented by rows in v and the matrices that represent operators act on them from the right ($\mathscr{S}\mathscr{A}$), we use the same order in Dirac notation in Quantum Mechanics in U_n: vectors in U_n^* are denoted as $< f|$ (and called [5] *BRAs*) and the operators act on them also from the right:

$$\boxed{< f|A}.$$

The vectors in U_n are denoted as $|x >$ (and are called [5] *KETs*) and the operators act on them form the left:

$$\boxed{A|x >},$$

in analogy with $\mathscr{A}\xi$, where ξ is the column that represents $|x >$.

The identification between the two superspaces makes it possible to use the same notation for A and A^T, more precisely $A_{\rightarrow}$ for A and $_{\leftarrow}A$ for A^T as in Dirac notation. The only difference is in the direction of their action on vectors (the arrows disappear when operators act properly).

Chapter 5
The Eigen Problem or Diagonal Form of Representing Matrices

5.1 Eigenvalues, Eigenvectors, and Eigenspaces

Since many types of operators in unitary spaces (U_n), as well as in Euclidean spaces (E_n), play fundamental roles in physical applications in Quantum Mechanics and Classical Physics, respectively, it is of great importance to know when and how they act in the simplest possible way. More precisely, on which vectors do they act as multiplication with numbers, and what are these numbers? This is the shortest formulation of the *Eigen problem.*

To add more details to this problem, let us agree that the multiplication by a number (scalar) $\lambda \in F$ of all vectors in $V_n(F)$ is a map $V_n(F) \rightarrow V_n(F)$ that preserves linear combinations of vectors, and that this follows from the very structure of $V_n(F)$:

$$\lambda(a\bar{x} + b\bar{y}) = a(\lambda\bar{x}) + b(\lambda\bar{y}), \ \bar{x}, \bar{y} \in V_n(F).$$

So, it is the most simple linear operator in $V_n(F)$.

The natural question now is whether for a given linear operator A in $V_n(F)$ there exists a nonzero vector $\bar{x}$ on which A acts in this simplest way, i.e., as multiplication with a scalar λ: $A\bar{x} = \lambda\bar{x}, \bar{x} \neq \bar{0}$ or equivalently $(A - \lambda I_v)\bar{x} = \bar{0}$, where I_v is the identity operator in $V_n(F)$, [then the vector $A\bar{x}$ is in the subspace $L(\bar{x})$].

Such a vector $\bar{x}$ is called an *eigenvector* (eigen is German word meaning characteristic or proper) of the operator A, and the corresponding scalar λ is an *eigenvalue* of A.

Definition If $A \in \hat{L}(V,V)$, then a scalar λ is an eigenvalue of A if there is a nonzero vector $\bar{x}$ such that $A\bar{x} = \lambda\bar{x}$. The vector $\bar{x}$ is then an eigenvector of A corresponding to λ.

Such vectors appear in the study of vibrations, electrical systems, genetics, chemical reactions, quantum mechanics, mechanical stress, economics, and geometry.

The Eigen problem for A consists in finding all eigenvalues and all eigenvectors of that operator. (Note that some operators do not have any of these. The most famous examples are the linear operators that describe rotations in the plane counterclockwise through a positive angle $\theta < 180°$).

Before we start solving the Eigen problem, we note the following two facts:

1. If a vector $\bar{x}$ is an eigenvector of A, then all vectors of the form $a\bar{x}$, $\forall a \in F$, are also eigenvectors of A with the same eigenvalue, i.e., if $A\bar{x} = \lambda\bar{x}$, then for $\forall a \in F$ we have as well

$$A(a\bar{x}) = aA\bar{x} = a\lambda\bar{x} = \lambda(a\bar{x}).$$

In conclusion, with every eigenvector $\bar{x}$ of A the whole one-dimensional subspace $L(\bar{x}) = \{a\bar{x} | a \in F\}$ (which is spanned by $\bar{x}$) consists of eigenvectors of A.

2. The set of all eigenvectors of A which correspond to one eigenvalue λ (together with $\bar{0}$, which is never considered as an eigenvector) form a subspace E_λ, whose dimension is at least 1.

Proof

$$\text{If } A\bar{x} = \lambda\bar{x} \text{ and } A\bar{y} = \lambda\bar{y} \Leftrightarrow \bar{x}, \bar{y} \in E_\lambda, \text{ then}$$
$$A(a\bar{x} + b\bar{y}) = aA\bar{x} + bA\bar{y} = a(\lambda\bar{x}) + b(\lambda\bar{y}) = \lambda(a\bar{x} + b\bar{y}) \Leftrightarrow$$
$$a\bar{x} + b\bar{y} \in E_\lambda, \text{ i.e., } E_\lambda \text{ is a subspace. } \Delta$$

We see that the operator $(A - \lambda I_v)$ annihilates all vectors from E_λ (and only those)

$$(A - \lambda I_v)\bar{x} = \bar{0} \text{ for } \bar{x} \in E_\lambda.$$

So, E_λ is in fact the *kernel* of the operator $(A - \lambda I_v)$, i.e.,

$$\boxed{E_\lambda = \ker (A - \lambda I_v)}, \, A \in \hat{L}(V, V).$$

This means that when λ is known, $\ker (A - \lambda I_v)$ is uniquely determined. We call E_λ the *eigenspace* of A which corresponds to the eigenvalue λ.

The method for solving the Eigen problem of a linear operator $A \in \hat{L}(V_n(F), V_n(F))$ consists of two steps:

(A) First, we have to find all eigenvalues $\lambda \in F$ for which the operator $(A - \lambda I_v)$ has a non-trivial kernel: $\ker (A - \lambda I_v) \neq \{\bar{0}\}$.

The necessary and sufficient condition for this is that $\det(\mathscr{A} - \lambda I_n) = 0$, where $\mathscr{A}$ is a square $n \times n$ matrix that represents the operator A in some basis $\{\bar{v}_1, \bar{v}_2, \ldots, \bar{v}_n\}$ in $V_n(F)$, and I_n is the unit matrix in F^n.

It is very important to notice that $\det(\mathscr{A} - \lambda I_n)$ does not depend on the choice of this basis, unlike the representing matrix $\mathscr{A}$, which changes by similarity $\mathscr{A}' = \mathscr{S}\mathscr{A}\mathscr{S}^{-1}$, $\mathscr{S} = (\mathscr{R}^T)^{-1}$, when the basis changes with the replacement matrix $\mathscr{R}$ (see Sect. 4.3.2. the theorem on the change of the matrix $\mathscr{A}$ representing an operator A). It is easy to prove this statement:

$$\text{Let } \mathscr{A}' = \mathscr{S}\mathscr{A}\mathscr{S}^{-1}, \text{ then}$$
$$\det(\mathscr{A}' - \lambda I_n) = \det(\mathscr{S}\mathscr{A}\mathscr{S}^{-1} - \lambda I_n) = \det(\mathscr{S}(\mathscr{A} - \lambda I_n)\mathscr{S}^{-1}) =$$
$$= \det\mathscr{S} \cdot \det(\mathscr{A} - \lambda I_n) \cdot \det\mathscr{S}^{-1} = \det(\mathscr{A} - \lambda I_n),$$

since $\det\mathscr{S} \cdot \det\mathscr{S}^{-1} = \det(\mathscr{S}\mathscr{S}^{-1}) = \det I_n = 1. \, \Delta$

The equation $\det(\mathscr{A} - \lambda I_n) = 0$ is in fact an equation of the degree n in the unknown λ. It is called the *characteristic equation* of A. Its explicit form is

$$\det(\mathscr{A} - \lambda I_n) = (-\lambda)^n + p_1(-\lambda)^{n-1} + p_2(-\lambda)^{n-2} + \cdots + p_{n-1}(-\lambda) + p_n = 0,$$

where $p_1 = tr\mathscr{A}$ (the sum of diagonal elements of $\mathscr{A}$), and $p_n = \det\mathscr{A}$. The other coefficients $p_k, k = 2, 3, \ldots, n - 1$, in front of $(-\lambda)^{n-k}$ are the sums of the principal minors of $\mathscr{A}$ of order k. Notice that the same rule is valid for p_1 and p_n—the sum of principal minors of $\mathscr{A}$ of order 1 and the principal minor of $\mathscr{A}$ of order n. These two coefficients can be obtained immediately, the first of them from $\det(\mathscr{A} - \lambda I_n) = (a_{11} - \lambda)(a_{22} - \lambda)\cdots(a_{nn} - \lambda)+$ terms with at most $(n - 2)$ factors of the form $(a_{ii} - \lambda)$ and the second by putting $\lambda = 0$ in $\det(\mathscr{A} - \lambda I_n)$.

Remember that a minor is principal if the diagonal elements of the minor come from the diagonal of the matrix.

Example

$$\text{Let } \mathscr{A} = \begin{bmatrix} a_{11} & a_{12} & a_{13} \\ a_{21} & a_{22} & a_{23} \\ a_{31} & a_{32} & a_{33} \end{bmatrix}, \text{ then}$$

$$\det(\mathscr{A} - \lambda I_3) = -\lambda^3 + (a_{11} + a_{22} + a_{33})\lambda^2 -$$

$$- \left(\begin{vmatrix} a_{11} & a_{12} \\ a_{21} & a_{22} \end{vmatrix} + \begin{vmatrix} a_{11} & a_{13} \\ a_{31} & a_{33} \end{vmatrix} + \begin{vmatrix} a_{22} & a_{23} \\ a_{32} & a_{33} \end{vmatrix} \right) \lambda + \begin{vmatrix} a_{11} & a_{12} & a_{13} \\ a_{21} & a_{22} & a_{23} \\ a_{31} & a_{32} & a_{33} \end{vmatrix}.$$

It should be noted that all coefficients of the characteristic equation (in particular $tr\mathscr{A}$ and $\det\mathscr{A}$) are invariant under similarity transformations, so we can talk about the *trace* and *determinant* of the operator A.

All solutions of the characteristic equation are the eigenvalues of the operator A. They form the *spectrum* of A. Some of the solutions are singe-valued (we call them nondegenerate eigenvalues), while some are multivalued (i.e., degenerate eigenvalues). The multiplicity of λ as a solution of the characteristic equation is called the *algebraic multiplicity* n_λ of λ.

When F is the complex field $\mathbb{C}$, then the characteristic equation $\det(\mathscr{A} - \lambda I_n) = 0$ of degree n has exactly n solutions, if we count each solution as many times as its algebraic multiplicity.

(Remember the famous *Fundamental Theorem of Algebra* which states that every polynomial equation with coefficients in field $\mathbb{C}$ has n solutions in $\mathbb{C}$, where n is the degree of the polynomial and the solutions are counted with their algebraic multiplicities.)

(B) Second, for every eigenvalue λ, we have to find all nontrivial solutions of the homogeneous system of linear equations

$$[\mathscr{A} - \lambda I_n]\xi = \bar{0},$$

where $\xi = [\xi_1\, \xi_2\, \ldots\, \xi_n]^T$ is the matrix-column (column vector) of unknowns, and the $n \times n$ matrix $[\mathscr{A} - \lambda I_n]$ is singular (its determinant is zero), which means that it has at least one nontrivial solution.

By doing this, we are actually finding $\ker(\mathscr{A} - \lambda I_n) = E_\lambda$. This system can have at most as many linearly independent nontrivial solutions as the algebraic multiplicity n_λ of λ. The dimension of $\ker(\mathscr{A} - \lambda I_n) = E_\lambda$ is called the *geometric multiplicity* of λ, and

$$1 \leq \dim E_\lambda \leq n_\lambda.$$

Note that the operator A and all of its representing matrices have the same dimension as their kernels.

The fact that the geometric multiplicity of λ, i.e., $\dim\ker(\mathscr{A} - \lambda I_n) = \dim E_\lambda$, is smaller than or equal to its algebraic multiplicity n_λ can be proved in the following way:

Let us suppose that we have found that λ_0 is a solution of the characteristic equation $\det(\mathscr{A} - \lambda I_n) = 0$, but we have not found the algebraic multiplicity n_{λ_0} of that solution. Furthermore, suppose that we have solved the homogeneous eigensystem $\mathscr{A}\xi = \lambda_0\xi$, and have discovered $\dim\ker(\mathscr{A} - \lambda_0 I_n) = r$ linearly independent eigenvectors $\{\xi_1, \xi_2, \ldots, \xi_r\}$—the basis in E_{λ_0} in F^n. Let us now extend this system of eigenvectors to get a basis in the space F^n

$$\{\xi_1, \xi_2, \ldots, \xi_r, \xi_{r+1}, \ldots, \xi_n\}.$$

Since the initial choice of the basis $\{\bar{v}_1, \bar{v}_2, \ldots, \bar{v}_n\}$ in $V_n(F)$, which produced the representing matrix $\mathscr{A}$, meant that we established the isomorphism i between $V_n(F)$ and F^n, we can now use this isomorphism to map the basis $\{\xi_1, \xi_2, \ldots, \xi_r, \xi_{r+1}, \ldots, \xi_n\}$ in F^n onto a new basis $\{\bar{v}'_1, \bar{v}'_2, \ldots, \bar{v}'_r, \bar{v}'_{r+1}, \ldots, \bar{v}'_n\}$ in $V_n(F)$. In this new basis, the operator A is represented by a *triangular block-matrix*

$$\mathscr{A}_0 = \begin{bmatrix} \lambda_0 I_r & \mathscr{B} \\ \bar{0} & \mathscr{C} \end{bmatrix} \begin{matrix} r \\ n-r \end{matrix}$$
$$\phantom{\mathscr{A}_0 = }\begin{matrix} r & n-r \end{matrix}$$

If we now form the characteristic equation of the matrix $\mathscr{A}_0$, we will get

$$\det(\mathscr{A}_0 - \lambda I_n) = \det(\lambda_0 - \lambda)I_r \cdot \det(\mathscr{C} - \lambda I_{n-r}) = 0,$$

since the determinant of a triangular block-matrix with square matrices on its diagonal is a multiple of determinants of the blocks on the diagonal. Finally,

$$\det(\mathscr{A} - \lambda I_n) = \det(\mathscr{A}_0 - \lambda I_n) = (\lambda_0 - \lambda)^r \det(\mathscr{C} - \lambda I_{n-r}) = 0.$$

From the last result, it follows that the algebraic multiplicity of λ_0 is at least r, but it can be greater if λ_0 is also a root of $\det(\mathscr{C} - \lambda_0 I_{n-r})$. In conclusion, $n_{\lambda_0} \geq r$, or

algebraic multiplicity $\geq$ geometric multiplicity ≥ 1. Δ

Two Examples

1. Solve the Eigen problem for the following 3×3 real matrix

$$\mathscr{A} = \begin{bmatrix} 2 & 1 & 0 \\ -1 & 0 & 1 \\ 1 & 3 & 1 \end{bmatrix}.$$

The characteristic equation of $\mathscr{A}$ is

$$\det(\mathscr{A} - \lambda I_3) = \begin{bmatrix} 2-\lambda & 1 & 0 \\ -1 & -\lambda & 1 \\ 1 & 3 & 1-\lambda \end{bmatrix} =$$

$$= (2-\lambda)\begin{bmatrix} -\lambda & 1 \\ 3 & 1-\lambda \end{bmatrix} - 1 \cdot \begin{bmatrix} -1 & 1 \\ 1 & 1-\lambda \end{bmatrix} =$$

$$= (2-\lambda)(\lambda^2 - \lambda - 3) - (\lambda - 2) =$$

$$= -(\lambda - 2)(\lambda^2 - \lambda - 2) = -(\lambda - 2)^2(\lambda + 1).$$

So, the eigenvalues are $\lambda_1 = -1, \lambda_2 = \lambda_3 = 2$.
The augmented matrix of the corresponding homogeneous linear system for $\lambda_1 = -1$ is

$$\begin{bmatrix} \mathscr{A} - \lambda_1 I_3 & \begin{matrix} 0 \\ 0 \\ 0 \end{matrix} \end{bmatrix} = \begin{bmatrix} \mathscr{A} + I_3 & \begin{matrix} 0 \\ 0 \\ 0 \end{matrix} \end{bmatrix} =$$

$$= \begin{bmatrix} 3 & 1 & 0 & 0 \\ -1 & 1 & 1 & 0 \\ 1 & 3 & 2 & 0 \end{bmatrix} \sim \begin{bmatrix} 1 & 3 & 2 & 0 \\ -1 & 1 & 1 & 0 \\ 3 & 1 & 0 & 0 \end{bmatrix} \sim$$

$$\sim \begin{bmatrix} 1 & 3 & 2 & 0 \\ 0 & 4 & 3 & 0 \\ 0 & -8 & -6 & 0 \end{bmatrix} \sim \begin{bmatrix} 1 & 3 & 2 & 0 \\ 0 & 4 & 3 & 0 \\ 0 & 0 & 0 & 0 \end{bmatrix} \sim$$

$$\sim \begin{bmatrix} 1 & 3 & 2 & 0 \\ 0 & 1 & \frac{3}{4} & 0 \\ 0 & 0 & 0 & 0 \end{bmatrix} \sim \begin{bmatrix} 1 & 0 & -\frac{1}{4} & 0 \\ 0 & 1 & \frac{3}{4} & 0 \\ 0 & 0 & 0 & 0 \end{bmatrix}.$$

This reduced Gauss–Jordan form of $(\mathscr{A} + I_3)$ tells us that in this matrix the third column is linearly dependent on the first two linearly independent ones, and this is exactly

$$c_3 = -\tfrac{1}{4}c_1 + \tfrac{3}{4}c_2$$

$$\text{(verification} \begin{bmatrix} 0 \\ 1 \\ 2 \end{bmatrix} = -\tfrac{1}{4}\begin{bmatrix} 3 \\ -1 \\ 1 \end{bmatrix} + \tfrac{3}{4}\begin{bmatrix} 1 \\ 1 \\ 3 \end{bmatrix}) \text{ or}$$

$$\bar{0} = -\tfrac{1}{4}c_1 + \tfrac{3}{4}c_2 - 1 \cdot c_3 \text{ (the modified canonical expansion)}.$$

These coefficients form the basis vector in E_{-1}, i.e.,

$$E_{-1} = \{ s \begin{bmatrix} -\frac{1}{4} \\ \frac{3}{4} \\ -1 \end{bmatrix} \mid s \in \mathbb{R} \} = \ker (\mathscr{A} + I_3)$$

(see Chap. 2, 17B, about ker A).

Choosing $s = -4$, we can get a somewhat simpler basis vector in E_{-1}, i.e.,

$$\xi_1 = \begin{bmatrix} 1 \\ -3 \\ 4 \end{bmatrix}.$$

For $\lambda_2 = \lambda_3 = 2$, we obtain the augmented matrix

$$\begin{bmatrix} \mathscr{A} - 2I_3 & \begin{matrix} 0 \\ 0 \\ 0 \end{matrix} \end{bmatrix} = \begin{bmatrix} 0 & 1 & 0 & 0 \\ -1 & -2 & 1 & 0 \\ 1 & 3 & -1 & 0 \end{bmatrix} \sim$$

$$\sim \begin{bmatrix} 1 & 3 & -1 & 0 \\ 0 & 1 & 0 & 0 \\ -1 & -2 & 1 & 0 \end{bmatrix} \sim \begin{bmatrix} 1 & 3 & -1 & 0 \\ 0 & 1 & 0 & 0 \\ 0 & 1 & 0 & 0 \end{bmatrix} \sim$$

$$\sim \begin{bmatrix} 1 & 3 & -1 & 0 \\ 0 & 1 & 0 & 0 \\ 0 & 0 & 0 & 0 \end{bmatrix} \sim \begin{bmatrix} 1 & 0 & -1 & 0 \\ 0 & 1 & 0 & 0 \\ 0 & 0 & 0 & 0 \end{bmatrix} \text{ or}$$

$$c_3 = -1 \cdot c_1 + 0 \cdot c_2 \text{ or } \bar{0} = -1 \cdot c_1 + 0 \cdot c_2 - 1 \cdot c_3 \quad (*)$$

(the modified canonical expansion.)

So, there is only one basis eigenvector for $\lambda_2 = \lambda_3 = 2$ formed from the coefficients in $(*)$, it is $\xi_2 = \begin{bmatrix} -1 \\ 0 \\ -1 \end{bmatrix}$, and the kernel of $[\mathscr{A} - 2I_3]$ is one-dimensional.

The geometric multiplicity for $\lambda = 2$ is only 1, while the algebraic multiplicity of this eigenvalue is 2.

2. Solve the Eigen problem for the following 3×3 real matrix

$$\mathscr{A} = \begin{bmatrix} 0 & 0 & -2 \\ 1 & 2 & 1 \\ 1 & 0 & 3 \end{bmatrix}.$$

The characteristic equation of $\mathscr{A}$ is

$$\det(\mathscr{A} - \lambda I_3) = \begin{vmatrix} -\lambda & 0 & -2 \\ 1 & 2-\lambda & 1 \\ 1 & 0 & 3-\lambda \end{vmatrix}$$

$$= -\lambda^3 + 5\lambda^2 - \left(\begin{vmatrix} 0 & 0 \\ 1 & 2 \end{vmatrix} + \begin{vmatrix} 0 & -2 \\ 1 & 3 \end{vmatrix} + \begin{vmatrix} 2 & 1 \\ 0 & 3 \end{vmatrix} \right) \lambda +$$

$$+ \begin{vmatrix} 0 & 0 & -2 \\ 1 & 2 & 1 \\ 1 & 0 & 3 \end{vmatrix} = -\lambda^3 + 5\lambda^2 - (2+6)\lambda + 4 = 0.$$

Since $\lambda = 1$ is an obvious solution $(-1 + 5 - 8 + 4 = 0)$, we shall divide the characteristic polynomial $\lambda^3 - 5\lambda^2 + 8\lambda - 4$ by $\lambda - 1$ and obtain $\lambda^2 - 4\lambda + 4 = (\lambda - 2)^2$. So, there are three real eigenvalues $\lambda_3 = 1, \lambda_1 = \lambda_2 = 2$.

For $\lambda_1 = \lambda_2 = 2$, we get the augmented matrix of the corresponding homogeneous linear system

$$\begin{bmatrix} -2 & 0 & -2 & | & 0 \\ 1 & 0 & 1 & | & 0 \\ 1 & 0 & 1 & | & 0 \end{bmatrix} \sim \begin{bmatrix} 1 & 0 & 1 & | & 0 \\ 0 & 0 & 0 & | & 0 \\ 0 & 0 & 0 & | & 0 \end{bmatrix}$$

or $c_2 = 0 \cdot c_1 + 0 \cdot c_3 \Rightarrow \bar{0} = 0 \cdot c_1 - 1 \cdot c_2 + 0 \cdot c_3$
or $c_3 = 1 \cdot c_1 + 0 \cdot c_2 \Rightarrow \bar{0} = 1 \cdot c_1 + 0 \cdot c_2 - 1 \cdot c_3$

The two (obviously linearly independent) basis eigenvectors for $E_2 = \ker (\mathscr{A} - 2I_3)$ are made of the above coefficient of $\bar{0}$

$$\xi_1 = \begin{bmatrix} 0 \\ -1 \\ 0 \end{bmatrix} \text{ and } \xi_2 = \begin{bmatrix} 1 \\ 0 \\ -1 \end{bmatrix}. \text{ In other words,}$$

$$E_2 = \mathrm{LIN}(\xi_1, \xi_2) = \{ s \begin{bmatrix} 0 \\ -1 \\ 0 \end{bmatrix} + t \begin{bmatrix} 1 \\ 0 \\ -1 \end{bmatrix} \mid s, t \in \mathbb{R} \}.$$

The dimension of E_2 is 2; thus, the geometric multiplicity of $\lambda = 2$, so it is equal to its algebraic multiplicity.

For $\lambda_3 = 1$, we have the augmented matrix (the algebraic multiplicity is 1, so there will be only one basis vector):

$$\begin{bmatrix} -1 & 0 & -2 & | & 0 \\ 1 & 1 & 1 & | & 0 \\ 1 & 0 & 2 & | & 0 \end{bmatrix} \sim \begin{bmatrix} 1 & 0 & 2 & | & 0 \\ 1 & 1 & 1 & | & 0 \\ 0 & 0 & 0 & | & 0 \end{bmatrix} \sim \begin{bmatrix} 1 & 0 & 2 & | & 0 \\ 0 & 1 & -1 & | & 0 \\ 0 & 0 & 0 & | & 0 \end{bmatrix}$$

or $c_3 = 2c_1 - 1 \cdot c_2 \Rightarrow \bar{0} = 2c_1 - 1 \cdot c_2 - 1 \cdot c_3$,

So, the basis vector is

$$\xi_3 = \begin{bmatrix} 2 \\ -1 \\ -1 \end{bmatrix}, \text{ and}$$

$$E_1 = \mathrm{LIN}(\xi_3) = \{ r \begin{bmatrix} 2 \\ -1 \\ -1 \end{bmatrix} \mid r \in \mathbb{R} \}.$$

Therefore, the vector space $\mathbb{R}^3$ is decomposed into the direct sum of E_2 and E_1, i.e., $\mathbb{R}^3 = E_2 + E_1$, since ξ_3 is linearly independent from ξ_1 and ξ_2 (otherwise, if it were linearly dependent, it would be an eigenvector of $\lambda = 2$). Namely, the linear independency test for ξ_1, ξ_2, ξ_3 reads as

$$a\,\xi_1 + b\,\xi_2 + c\,\xi_3 = \bar{0}_3 \text{ or } a \begin{bmatrix} 0 \\ -1 \\ 0 \end{bmatrix} + b \begin{bmatrix} 1 \\ 0 \\ -1 \end{bmatrix} + c \begin{bmatrix} 2 \\ -1 \\ -1 \end{bmatrix} = \begin{bmatrix} 0 \\ 0 \\ 0 \end{bmatrix} \Rightarrow$$

$$\left. \begin{array}{l} b + 2c = 0 \\ -a - c = 0 \\ -b - c = 0 \end{array} \right\} \Rightarrow c = 0,\, a = 0,\, b = 0,$$

which in its turn can be interpreted as the question whether E_2 and E_1 have common vectors $a\,\xi_1 + b\,\xi_2 = -c\,\xi_3$, to which question (see above $a = b = c = 0$) the answer is "*NO*."

Generally speaking, for a matrix $\mathscr{A} \in \mathbb{R}^n$, its eigenspaces (if all geometric multiplicities are equal to the corresponding algebraic ones and if their sum is equal to n) form a direct decomposition of $\mathbb{R}^n$.

Theorem (on eigenbasis and direct eigendecomposition) Let λ_1, λ_2, ..., λ_k, $k \le n$, be distinct eigenvalues of an $n \times n$ real matrix $\mathscr{A}$ (with $n_{\lambda_1} + n_{\lambda_2} + \cdots + n_{\lambda_k} = n$) and let the Gauss–Jordan-modified (GJM) method provide n_{λ_1} unique linearly independent eigenvectors to span E_{λ_1}, and so on, then, since the eigenvectors from different eigenspaces are also linearly independent, the totality of eigenvectors from all eigenspaces E_{λ_1}, E_{λ_2}, ..., E_{λ_k} will be a set of n linearly independent vectors in $\mathbb{R}^n$— the eigenbasis of $\mathscr{A}$. In the basis, the matrix $\mathscr{A}$ will be diagonal with λ_1, λ_2, ..., λ_k, on its diagonal, each $\lambda_i, i = 1, 2, \ldots, k$, appearing n_{λ_i} times.

The space $\mathbb{R}^n$ is the direct sum of these eigenspaces:

$$\mathbb{R}^n = E_{\lambda_1} + E_{\lambda_2} + \cdots + E_{\lambda_k}.$$

5.2 Diagonalization of Square Matrices

Going back to example 2, we can verify the obtained results:

$$\mathscr{A}\,\xi_1 = \begin{bmatrix} 0 & 0 & -2 \\ 1 & 2 & 1 \\ 1 & 0 & 3 \end{bmatrix} \begin{bmatrix} 0 \\ -1 \\ 0 \end{bmatrix} = \begin{bmatrix} 0 \\ -2 \\ 0 \end{bmatrix} = 2\,\xi_1;$$

$$\mathscr{A}\,\xi_2 = \begin{bmatrix} 0 & 0 & -2 \\ 1 & 2 & 1 \\ 1 & 0 & 3 \end{bmatrix} \begin{bmatrix} 1 \\ 0 \\ -1 \end{bmatrix} = \begin{bmatrix} 2 \\ 0 \\ -2 \end{bmatrix} = 2\,\xi_2;$$

$$\mathscr{A}\,\xi_3 = \begin{bmatrix} 0 & 0 & -2 \\ 1 & 2 & 1 \\ 1 & 0 & 3 \end{bmatrix} \begin{bmatrix} 2 \\ -1 \\ -1 \end{bmatrix} = \begin{bmatrix} 2 \\ -1 \\ -1 \end{bmatrix} = \xi_3.$$

If we form the matrix, whose columns are linearly independent eigenvectors

$$\xi_1, \xi_2, \xi_3 : \mathscr{T} = \begin{bmatrix} 0 & 1 & 2 \\ -1 & 0 & -1 \\ 0 & -1 & -1 \end{bmatrix},$$

this matrix will be invertible (*rank* = 3), and the three above verifications can be written in a compact matrix form $\mathscr{A}\mathscr{T} = \mathscr{T}\mathscr{D}$, where $\mathscr{D}$ is the diagonal matrix.

$$\mathscr{D} = \begin{bmatrix} 2 & 0 & 0 \\ 0 & 2 & 0 \\ 0 & 0 & 1 \end{bmatrix}.$$

Indeed,

$$\overset{\mathscr{A}}{\begin{bmatrix} 0 & 0 & -2 \\ 1 & 2 & 1 \\ 1 & 0 & 3 \end{bmatrix}} \overset{\mathscr{T}}{\begin{bmatrix} 0 & 1 & 2 \\ -1 & 0 & -1 \\ 0 & -1 & -1 \end{bmatrix}} = \overset{\mathscr{T}}{\begin{bmatrix} 0 & 1 & 2 \\ -1 & 0 & -1 \\ 0 & -1 & -1 \end{bmatrix}} \overset{\mathscr{D}}{\begin{bmatrix} 2 & 0 & 0 \\ 0 & 2 & 0 \\ 0 & 0 & 1 \end{bmatrix}}$$

since both products are equal to $\begin{bmatrix} 0 & 2 & 2 \\ -2 & 0 & -1 \\ 0 & -2 & -1 \end{bmatrix}$.

A more explicit formulation is

$$\mathscr{D} = \mathscr{T}^{-1}\mathscr{A}\mathscr{T},$$

which means that the matrix $\mathscr{A}$ can be diagonalized by a similarity transformation with the invertible matrix $\mathscr{T}$ whose columns are the eigenvectors of $\mathscr{A}$, so that the diagonal matrix $\mathscr{D}$ has the corresponding eigenvalues of $\mathscr{A}$ on its diagonal. Obviously, the general prerequisite for this is that for all real eigenvalues $\lambda_1, \lambda_2, \dots, \lambda_k, k \leq n$, of $\mathscr{A}$ in $\mathbb{R}^n$ their algebraic and geometric multiplicities are the same numbers $n_{\lambda_1}, n_{\lambda_2}, \dots, n_{\lambda_k}$ and also $n_{\lambda_1} + n_{\lambda_2} + \cdots + n_{\lambda_k} = n$.

The matrix $\mathscr{T}$ is the transition matrix from the standard basis e_1, e_2, e_3 in $\mathbb{R}^3$ in which the matrix $\mathscr{A}$ is given, to the eigenbasis ξ_1, ξ_2, ξ_3 in which it takes on the diagonal form $\mathscr{D}$. This can be seen immediately since the multiplication of $\mathscr{T}$ with vectors from the standard basis gives the corresponding columns of $\mathscr{T}$, i.e., ξ_1, ξ_2, ξ_3: e.g.,

$$\mathscr{T} e_1 = \begin{bmatrix} 0 & 1 & 2 \\ -1 & 0 & -1 \\ 0 & -1 & -1 \end{bmatrix} \begin{bmatrix} 1 \\ 0 \\ 0 \end{bmatrix} = \begin{bmatrix} 0 \\ -1 \\ 0 \end{bmatrix} = \xi_1.$$

Remark If $\mathscr{A}$ in $\mathbb{R}^n$ has n distinct eigenvalues, then the above conditions are obviously satisfied, and $\mathscr{A}$ is diagonalizable.

Example To show when a 2×2 matrix $\mathscr{A} = \begin{bmatrix} a & b \\ c & d \end{bmatrix}$ is diagonalizable, we shall calculate its characteristic equation

$$\begin{vmatrix} a-\lambda & b \\ c & d-\lambda \end{vmatrix} = 0 \Rightarrow (a-\lambda)(d-\lambda) - bc = 0 \Rightarrow$$

$$\Rightarrow \lambda^2 - (a+d)\lambda + (ad-bc) = 0 \Rightarrow$$

$$\Rightarrow \lambda^2 - \text{tr}\mathscr{A}\,\lambda + \det\mathscr{A} = 0 \Rightarrow$$

$$\Rightarrow \lambda_{1,2} = \frac{(a+d) \pm \sqrt{(a+d)^2 - 4ad + 4bc}}{2}.$$

To have two distinct solutions of this equation, the discriminant D must be greater than zero:

$$D = a^2 + 2ad + d^2 - 4ad + 4bc = (a-d)^2 + 4bc > 0. \;\; \Delta$$

Theorem (on diagonalization of a complex matrix $\mathscr{A}$) A complex $n \times n$ matrix $\mathscr{A}$ (as a linear operator in $\mathbb{C}^n$) can be diagonalized by a similarity transformation if and only if each of its eigenvalues has equal algebraic and geometric multiplicity. (Note that $n_{\lambda_1} + n_{\lambda_2} + \cdots + n_{\lambda_k} = n$ is automatically satisfied for complex matrices.) Then, $\mathscr{A}$ is diagonalized by a similarity transformation with the invertible matrix $\mathscr{T}$ which has as its columns n linearly independent eigenvectors $\{\xi_1, \xi_2, \ldots, \xi_n\}$ of $\mathscr{A} : \mathscr{T}^{-1}\mathscr{A}\mathscr{T} = \mathscr{D}$, where $\mathscr{D}$ is the diagonal matrix with the corresponding eigenvalues of $\mathscr{A}$:

$$\mathscr{A}\mathscr{T} = \mathscr{T}\mathscr{D}, \text{ is}$$

$$\mathscr{A}[\xi_1 \; \xi_2 \; \ldots \; \xi_n] = [\xi_1 \; \xi_2 \; \ldots \; \xi_n] \begin{bmatrix} \lambda_1 & \cdots & \cdots & 0 \\ 0 & \lambda_2 & \cdots & 0 \\ \vdots & \vdots & \ddots & \vdots \\ 0 & \cdots & \cdots & \lambda_n \end{bmatrix}.$$

A short formulation of the condition in the above theorem is that $\mathscr{A}$ is diagonalizable (similar to a diagonal matrix) iff in $\mathbb{C}^n$ there exists a basis consisting of eigenvectors of $\mathscr{A}$ (the so-called *eigenbasis* of $\mathscr{A}$).

Remark In $\mathbb{R}^3$, the characteristic equation of $\mathscr{A}$ is cubic, so it can have either three real solutions or one real and a pair of complex conjugate solutions. We shall show later (Sect. 5.6) that to such a pair of c.c. solutions corresponds an invariant plane which is not an eigenspace of $\mathscr{A}$. So, to have an eigenbasis, a matrix $\mathscr{A}$ is not allowed to have c.c. solutions of its characteristic equation. Nevertheless, some very important matrices such as orthogonal ones $\mathscr{A}^T = \mathscr{A}^{-1}$ are just like that, and we shall investigate (Sect. 5.7) their canonical forms which are different from the diagonal form.

Furthermore, we shall show that the necessary and sufficient condition for a matrix $\mathscr{A}$ in $\mathbb{R}^3$ to have an orthonormal eigenbasis, in which it is represented by the diagonal matrix $\mathscr{D}$ with real eigenvalues on its diagonal, is that the matrix is symmetric $\mathscr{A}^T = \mathscr{A}$ (Sect. 5.6).

However, we are going first to investigate the diagonalization procedure in unitary spaces U_n.

5.3 Diagonalization of an Operator in U_n

We shall show now that a linear operator A in U_n can be represented in an ON basis $\{\bar{u}_1, \bar{u}_2, \dots, \bar{u}_n\}$ in that space by the matrix $\mathscr{A} = [a_{ij}]_{n \times n}$ where the matrix elements a_{ij}, $i, j = 1, 2, \dots, n$, are easily calculated by means of the inner product in $U_n : a_{ij} = (\bar{u}_i, A\bar{u}_j)$.

Proof The basic formula for representation of a linear operator in any vector space reads as

$$A\bar{u}_j = \sum_{k=1}^{n} a_{kj} \bar{u}_k, \ i = 1, 2, \dots, n.$$

Multiplying it from the left by the vector $\bar{u}_i$, we get

$$(\bar{u}_i, A\bar{u}_j) = \sum_{k=1}^{n} a_{kj}(\bar{u}_i, \bar{u}_k) = \sum_{k=1}^{n} a_{kj} \delta_{ik} = a_{ij}. \ \ \Delta$$

Going over to the new ON basis $\{\bar{u}'_1, \bar{u}'_2, \dots, \bar{u}'_n\}$ by the unitary replacement matrix $\mathscr{R}$ ($\mathscr{R}^{-1} = \mathscr{R}^{\dagger}$), we get the new representation matrix (Sect. 4.4)

$$\mathscr{A}' = \mathscr{R}^* \mathscr{A} \mathscr{R}^T = \mathscr{S} \mathscr{A} \mathscr{S}^{-1}, \text{ where}$$
$$\mathscr{S} = \left(\mathscr{R}^{-1}\right)^T = \mathscr{R}^* \text{ and } \mathscr{S}^{-1} = \mathscr{R}^T.$$

The matrix $\mathscr{S}$ is obviously also a unitary matrix $[\mathscr{S}^{-1} = \mathscr{R}^T = (\mathscr{S}^*)^T = \mathscr{S}^{\dagger}]$. So, all the matrices that represent the operator A in all ON bases in U_n form an equivalence class of the unitary similarity as an equivalence relation. [That the unitary similarity is an equivalence relation is proved easily since $U(n)$ is a group—the proof is analogous to that for the group $GL(n, F)$].

The important question now is which linear operators in U_n have a diagonal matrix (with complex diagonal elements) in the equivalence class of its representing matrices. In other words, which linear operators can be represented by complex diagonal matrixes by means of a unitary similarity transformation. More precisely, which is the most general class of linear operators in U_n which have an ON eigenbasis.

Definition A normal operator A in U_n is characterized by the property that it commutes with its adjoint $A^\dagger$ (Sect. 4.2):

$$AA^\dagger = A^\dagger A$$

Theorem (on diagonalizability of normal operators) A linear operator A in U_n has an orthonormal (ON) eigenbasis in U_n (in which it is represented by a diagonal matrix $\mathscr{D}$ with complex numbers on its diagonal) if and only if it is normal.

So, normal operators form the most general class of linear operators in U_n which are diagonalizable by unitary similarity.

Proof The normal nature of an operator A is a sufficient condition for the existence of an ON eigenbasis.

For this part of the proof (normal nature $\Rightarrow$ diagonalizability), we need several auxiliary statements (Lemmas).

Lemma 1 *If we have two operators A and B from $\hat{L}(U_n, U_n)$ which commute, then each eigenspace E_λ of A is an invariant subspace for B, i.e., B reduces in E_λ in the sense that its action on any vector from E_λ results in a vector which also belongs to E_λ.*

Proof Let $AB = BA$. Consider any eigenspace E_λ of A. (The operator A must have at least one eigenspace, the extreme case being $n_\lambda = n$ and $\dim E_\lambda = 1$). It is easy to verify that E_λ is an invariant subspace of B, i.e., for any $\bar{x} \in E_\lambda$ we have $B\bar{x} \in E_\lambda$:

$$\bar{x} \in E_\lambda \Leftrightarrow A\bar{x} = \lambda\bar{x},$$
$$A(B\bar{x}) = B(A\bar{x}) = B(\lambda\bar{x}) = \lambda(B\bar{x}) \Rightarrow B\bar{x} \in E_\lambda.$$

Since E_λ is invariant for B, it means that B reduces in E_λ to $B^{(\lambda)}$, which acts as B in the reduced domain E_λ. This new operator has in E_λ at least one eigenspace, so that vectors from this subspace are eigenvectors for both A and B, but generally with different eigenvalues.

Now, we start with A and $A^\dagger$ which commute. According to the above Lemma, they have at least one common eigenvector, which we denote as $\bar{u}_1$, i.e., $A\bar{u}_1 = \lambda_1\bar{u}_1$ and $A^\dagger\bar{u}_1 = \lambda_1{}^*\bar{u}_1$. The fact that the eigenvalues of A and $A^\dagger$ are the complex conjugates of each other, can be immediately proved, but this proof requires two more lemmas:

Lemma 2 *For a normal operator $ker A = ker A^\dagger$.*

Lemma 3 *$(A - \lambda I_U)$ is also a normal operator.*

Proofs

2. $$\bar{x} \in ker A \Leftrightarrow A\bar{x} = \bar{0}_u \Leftrightarrow (A\bar{x}, A\bar{x}) = 0.$$
$$\text{However, } (A\bar{x}, A\bar{x}) = (\bar{x}, A^\dagger A\bar{x}) = (\bar{x}, AA^\dagger\bar{x}) = (A^\dagger\bar{x}, A^\dagger\bar{x}),$$
$$\text{so that } (A\bar{x}, A\bar{x}) = 0 \Leftrightarrow (A^\dagger\bar{x}, A^\dagger\bar{x}) = 0 \Leftrightarrow A^\dagger\bar{x} = \bar{0}_u \Leftrightarrow \bar{x} \in ker A^\dagger.$$

3. We shall prove that $(A - \lambda I_U)$ commutes with its adjoint:

$$(A - \lambda I_U)(A - \lambda I_U)^\dagger = (A - \lambda I_U)(A^\dagger - \lambda^* I_U) =$$
$$= AA^\dagger - \lambda A^\dagger - \lambda^* A + \lambda\lambda^* I_U = A^\dagger A - \lambda^* A - \lambda A^\dagger + \lambda^* \lambda I_U =$$
$$= (A^\dagger - \lambda^* I_U)(A - \lambda I_U) = (A - \lambda I_U)^\dagger (A - \lambda I_U).$$

Now, we are ready to prove that the eigenvalues of A and $A^\dagger$ for the common eigenvector are the complex conjugates of each other. If $A\bar{u}_1 = \lambda\bar{u}_1$, then $(A - \lambda I_U)\bar{u}_1 = \bar{0}_U$. Since $(A - \lambda I_U)$ is a normal operator, each vector from its kernel is in the kernel of $(A - \lambda I_U)^\dagger = A^\dagger - \lambda^* I_U$, i.e., $(A^\dagger - \lambda^* I_U)\bar{u}_1 = \bar{0}_U$ or $A^\dagger \bar{u}_1 = \lambda^* \bar{u}_1$.

Observe the eigenspace $L(\bar{u}_1)$ spanned by $\bar{u}_1$ alone. Its orthocomplement $L(\bar{u}_1)^\perp$ is an invariant subspace for both A and $A^\dagger$. This is the essential part of the proof that a normal operator has an ON eigenbasis in U_n.

To prove this statement, consider an arbitrary vector $\bar{x} \in L(\bar{u}_1)^\perp$, i.e., $(\bar{x}, \bar{u}_1) = 0$. Then,

$$(A\bar{x}, \bar{u}_1) = (\bar{x}, A^\dagger \bar{u}_1) = (\bar{x}, \lambda^* \bar{u}_1) = \lambda^*(\bar{x}, \bar{u}_1) = 0 \Leftrightarrow A\bar{x} \in L(\bar{u}_1)^\perp.$$

So, $L(\bar{u}_1)^\perp$ is invariant under A. In an analogous manner, one proves that $L(\bar{u}_1)^\perp$ is also invariant under $A^\dagger$.

Now, we shall deal with the reduced parts of both A and $A^\dagger$ in $L(\bar{u}_1)^\perp$. These parts naturally commute with each other. Because of their commuting, there exists a common eigenvector $\bar{u}_2$ in $L(\bar{u}_1)^\perp$ with the eigenvalues λ_2 and λ_2^*, respectively. This common eigenvector spans the one-dimensional eigenspace $L(\bar{u}_2)$, which is orthogonal to $L(\bar{u}_1)$: $L(\bar{u}_1) \perp L(\bar{u}_2)$.

We continue the above procedure in the orthocomplement of $L(\bar{u}_2)$ with respect to the subspace $L(\bar{u}_1)^\perp$:

$$L(\bar{u}_1)^\perp \ominus L(\bar{u}_2).$$

(this notation is logical since $L(\bar{u}_1)^\perp = L(\bar{u}_2) \oplus [L(\bar{u}_1)^\perp \ominus L(\bar{u}_2)]$). This new orthocomplement is also invariant under both A and $A^\dagger$, and in it, there is a common eigenvector $\bar{u}_3$ of the reduced parts with eigenvalues λ_3 and λ_3^*, respectively.

Obviously, the eigenspace $L(\bar{u}_3)$ is orthogonal to both $L(\bar{u}_1)$ and $L(\bar{u}_2)$

$$L(\bar{u}_1) \perp L(\bar{u}_2) \perp L(\bar{u}_3).$$

In the same manner (due to the normal property of A), we get n $(\dim U_n = n)$ mutually orthogonal one-dimensional eigenspaces $L(\bar{u}_1), L(\bar{u}_2), L(\bar{u}_3), \ldots, L(\bar{u}_n)$, which in the orthogonal sum make the whole U_n:

$$L(\bar{u}_1) \oplus L(\bar{u}_2) \oplus L(\bar{u}_3) \oplus \cdots \oplus L(\bar{u}_n) = U_n.$$

We now divide each $\bar{u}_1, \bar{u}_2, \ldots, \bar{u}_n$, with its norm and thus we get an orthonormal (ON) eigenbasis of A. In this basis, the representing matrix of A is the diagonal matrix $\mathscr{D}$ with complex eigenvalues $\lambda_1, \lambda_2, \ldots, \lambda_n$, on its diagonal (note: these λs are not necessarily distinct).

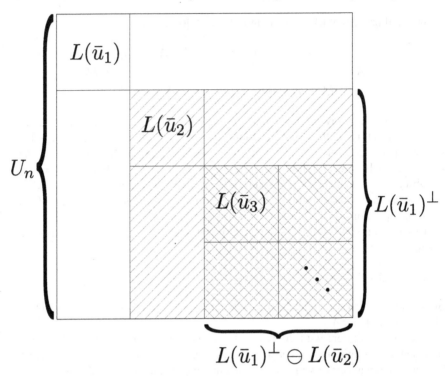

In conclusion, we can say that commutation of an operator A with its adjoint $A^\dagger$ in U_n $(AA^\dagger = A^\dagger A)$ implies ($\Rightarrow$) the existence of an ON eigenbasis of A in which A is represented by the diagonal matrix $\mathscr{D}$ with the corresponding complex eigenvalues on its diagonal.

Thus, the spectrum (the set of all eigenvalues) of the normal operator A is in the complex plane. Also, the geometric multiplicity of each eigenvalue of a normal operator A is equal to its algebraic multiplicity. (Note that the sum of algebraic multiplicities is necessarily equal to n, since U_n is a complex vector space.)

The second part of the proof (of the theorem on diagonalizability of normal operators) is to show that a diagonalizable operator A necessarily commutes with its adjoint $A^\dagger$. Let $\mathscr{D}$ be the diagonal matrix that represents such an operator. The adjoint matrix $\mathscr{D}^\dagger$ is simply the complex conjugate $\mathscr{D}^*$, since the operation of transposition does not change a diagonal matrix. But $\mathscr{D}$ and $\mathscr{D}^*$ obviously commute:

$$\mathscr{D}\mathscr{D}^* = \mathscr{D}^*\mathscr{D}.$$

Since the matrix $\mathscr{D}^\dagger = \mathscr{D}^*$ represents the adjoint operator $A^\dagger$, the commutation of $\mathscr{D}$ and $\mathscr{D}^*$ implies the commutation of operators which they represent: $\mathscr{A}\mathscr{A}^\dagger = \mathscr{A}^\dagger\mathscr{A}$, because the representation of operators by matrices is an isomorphism between the algebra (with unity) of operators and that of $n \times n$ matrices. Δ

Remark as a

Theorem (on isomorphism of algebras) The representation of operators acting in $V_n(F)$ by matrices from $F_{n \times n}$ [caused by the choice of a basis in $V_n(F)$] is an isomorphism between the algebras (with unity) $\hat{L}(V_n(F), V_n(F))$ of operators and $F_{n \times n} = \hat{L}(F^n, F^n)$ of matrices.

Proof

$V_n(F)$ F^n
basis   standard basis
$v = \{\bar{v}_1, \bar{v}_2, \ldots, \bar{v}_n\}$ $e = \{e_1, e_2, \ldots, e_n\}$

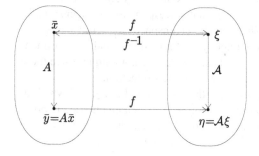

The isomorphism f between $V_n(F)$ and F^n is established through the bijection that connects the bases v and e:

$$\bar{x} = \sum_{i=1}^{n} \xi_i \bar{v}_i \overset{f}{\leftrightarrow} \xi = \sum_{i=1}^{n} \xi_i e_i = \begin{bmatrix} \xi_1 \\ \xi_2 \\ \vdots \\ \xi_n \end{bmatrix},$$

so that $\bar{v}_1 \overset{f}{\leftrightarrow} e_1, \bar{v}_2 \overset{f}{\leftrightarrow} e_2, \ldots, \bar{v}_n \overset{f}{\leftrightarrow} e_n.$

Applying f to both sides of $\bar{y} = A\bar{x}$, we get

$$f(\bar{y}) = f(A\bar{x}) = (fAf^{-1})f(\bar{x}), \text{ so that } \eta = (fAf^{-1})\xi.$$

From this follows, by comparison with $\eta = \mathscr{A}\xi$ that

$$\boxed{\mathscr{A} = fAf^{-1}}$$

which is also obvious from the figure. Thus, the operator A and its representing matrix $\mathscr{A}$ are equivalent operators with regard to the isomorphism f.

This is a bijection g between the algebras

$$\hat{L}(V_n(F), V_n(F)) \text{ and } \hat{L}(F^n, F^n):$$
$$g(A) = fAf^{-1} = \mathscr{A} \text{ and } g^{-1} = f^{-1}\mathscr{A}f = A.$$

This bijection g is, furthermore, an isomorphism:

1. $g(A+C) = f(A+C)f^{-1} = fAf^{-1} + fCf^{-1} = g(A) + g(C)$;
2. $g(aA) = f(aA)f^{-1} = a(fAf^{-1}) = a\,g(A)$;
3. $g(AC) = f(AC)f^{-1} = (fAf^{-1})(fCf^{-1}) = g(A)g(C)$;
4. $g(I_v) = fI_v f^{-1} = I_n$.

Remember also the basic formula for the representation matrix

$$\mathscr{A} = [a_{ij}]_{n \times n} \quad : \quad A\bar{v}_j = \sum_{i=1}^{n} a_{ij}\bar{v}_i \,, \; j = 1, 2, \ldots, n,$$

which means that we expand the images $\{A\bar{v}_1, A\bar{v}_2, \ldots, A\bar{v}_n\}$ in the basis $\{\bar{v}_1, \bar{v}_2, \ldots, \bar{v}_n\}$ and make the columns of $\mathscr{A}$ from the expansion coefficients. Δ

If we now collect all the eigenspaces $L(\bar{u}_i)$ that correspond to one eigenvalue λ_1 and make their orthogonal sum, we shall get the eigenspace $E_{\lambda_1} = \ker(A - \lambda_1 I_U)$, which is unique in the sense that it is completely determined by A and λ_1. Its dimension is n_{λ_1}. The elements of the above ON basis in E_{λ_1} are not unique, since there was a lot of arbitrariness in the described process of their selection.

Now, we continue with collecting all $L(\bar{u}_i)$ that correspond to λ_2, and which in the orthogonal sum give E_{λ_2} (with the dimension n_{λ_2}), which eigenspace is obviously orthogonal to E_{λ_1}. At the end of this procedure, we shall have $k \leq n$ mutually orthogonal eigenspaces $E_{\lambda_1}, E_{\lambda_2}, \ldots, E_{\lambda_k}$, with dimensions $n_{\lambda_1}, n_{\lambda_2}, \ldots, n_{\lambda_k}$, respectively, so that

$$E_{\lambda_1} \oplus E_{\lambda_2} \oplus \ldots \oplus E_{\lambda_k} = U_n$$

and, of course,

$$n_{\lambda_1} + n_{\lambda_2} + \ldots + n_{\lambda_k} = n.$$

Theorem (on the orthogonal decomposition of U_n into the eigenspaces of a normal operator A) If A is a normal operator in U_n, then its eigenspaces E_{λ_i}, $i = 1, 2, \ldots, k \leq n$, which correspond to distinct eigenvalues $\lambda_1, \lambda_2, \ldots, \lambda_k$, are uniquely determined as the kernels of the operators $(A - \lambda_i I_U)$ with dimensions n_{λ_i}, $i = 1, 2, \ldots k$ and $n_{\lambda_1} + n_{\lambda_2} + \ldots + n_{\lambda_k} = n$. Furthermore, they are mutually orthogonal, and in the orthogonal sum they give the whole U_n:

$$E_{\lambda_1} \oplus E_{\lambda_2} \oplus \ldots \oplus E_{\lambda_k} = U_n$$

The proof is given in the preceding text. Δ

5.3.1 Two Examples of Normal Matrices

1. Verify that $\mathscr{A} = \begin{bmatrix} 2 & i \\ i & 2 \end{bmatrix}$ is a normal matrix. Find a unitary matrix $\mathscr{T}$ such that $\mathscr{T}^{\dagger} \mathscr{A} \mathscr{T}$ is a diagonal matrix, and find that diagonal matrix.

We first verify that $\mathscr{A}$ is a normal matrix:

$$\mathscr{A}\mathscr{A}^\dagger = \begin{bmatrix} 2 & i \\ i & 2 \end{bmatrix} \begin{bmatrix} 2 & -i \\ -i & 2 \end{bmatrix} = \begin{bmatrix} 4+1 & 0 \\ 0 & 1+4 \end{bmatrix} = \begin{bmatrix} 5 & 0 \\ 0 & 5 \end{bmatrix},$$

$$\mathscr{A}^\dagger\mathscr{A} = \begin{bmatrix} 2 & -i \\ -i & 2 \end{bmatrix} \begin{bmatrix} 2 & i \\ i & 2 \end{bmatrix} = \begin{bmatrix} 4+1 & 0 \\ 0 & 1+4 \end{bmatrix} = \begin{bmatrix} 5 & 0 \\ 0 & 5 \end{bmatrix} \Rightarrow \mathscr{A}\mathscr{A}^\dagger = \mathscr{A}^\dagger\mathscr{A}.$$

Then, we find the characteristic equation $\det[\mathscr{A} - \lambda I_2] = 0$. This is

$$\begin{vmatrix} 2-\lambda & i \\ i & 2-\lambda \end{vmatrix} = 0 \text{ or } (2-\lambda)^2 - i^2 = 0 \text{ or}$$

$$(4 - 4\lambda + \lambda^2) + 1 = 0 \text{ or } \lambda^2 - 4\lambda + 5 = 0 \text{ with solutions}$$

$$\lambda_{1,2} = 2 \pm \sqrt{4-5} = 2 \pm i.$$

So, there are two distinct (complex conjugate) eigenvalues $\lambda_1 = 2 + i$ and $\lambda_2 = 2 - i$.

To find $\mathscr{T}$, we need to compute only one normalized eigenvector for each of these two distinct eigenvalues. (The eigenvectors will be automatically orthogonal). For $\lambda_1 = 2 + i$, we have to reduce the matrix.

$$\begin{bmatrix} 2-2-i & i & \bigm| & 0 \\ i & 2-2-i & \bigm| & 0 \end{bmatrix} = \begin{bmatrix} -i & i & \bigm| & 0 \\ i & -i & \bigm| & 0 \end{bmatrix} \sim \begin{bmatrix} 1 & -1 & \bigm| & 0 \\ 1 & -1 & \bigm| & 0 \end{bmatrix} \sim \begin{bmatrix} 1 & -1 & \bigm| & 0 \\ 0 & 0 & \bigm| & 0 \end{bmatrix} \Rightarrow$$

$$c_2 = (-1)c_1 \text{ or}$$

$$\Rightarrow$$

$$\bar{0} = (-1)c_1 + (-1)c_2$$

The eigenvector that corresponds to $\lambda_1 = 2 + i$ is $\begin{bmatrix} -1 \\ -1 \end{bmatrix}$, and after normalization it becomes $\bar{u}_1 = \frac{1}{\sqrt{2}} \begin{bmatrix} 1 \\ 1 \end{bmatrix}$. Similarly, for $\lambda_2 = 2 - i$, we find the normalized eigenvector $\bar{u}_2 = \frac{1}{\sqrt{2}} \begin{bmatrix} -1 \\ 1 \end{bmatrix}$. The matrix $\mathscr{T} = \begin{bmatrix} \bar{u}_1 & \bar{u}_2 \end{bmatrix} = \frac{1}{\sqrt{2}} \begin{bmatrix} 1 & -1 \\ 1 & 1 \end{bmatrix}$ is unitary since $\bar{u}_1$ and $\bar{u}_2$ are orthonormal vectors.

The matrix $\mathscr{T}$ diagonalizes $\mathscr{A}$:

$$\mathscr{T}^\dagger \mathscr{A} \mathscr{T} = \frac{1}{2} \begin{bmatrix} 1 & 1 \\ -1 & 1 \end{bmatrix} \begin{bmatrix} 2 & i \\ i & 2 \end{bmatrix} \begin{bmatrix} 1 & -1 \\ 1 & 1 \end{bmatrix} = \frac{1}{2} \begin{bmatrix} 2+i & 2+i \\ -2+i & 2-i \end{bmatrix} \begin{bmatrix} 1 & -1 \\ 1 & 1 \end{bmatrix} =$$

$$= \frac{1}{2} \begin{bmatrix} 4+2i & 0 \\ 0 & 4-2i \end{bmatrix} = \begin{bmatrix} 2+i & 0 \\ 0 & 2-i \end{bmatrix}. \quad \Delta$$

2. Find the eigenvalues and eigenspaces of the complex 3×3 matrix $\mathscr{A} = \begin{bmatrix} 1 & 0 & i \\ 0 & 2 & 0 \\ -i & 0 & 1 \end{bmatrix}$.

Find also the matrix $\mathscr{T}$ that diagonalizes $\mathscr{A}$ by unitary similarity.

Since $\mathscr{A} = \mathscr{A}^\dagger$ (it is self-adjoint), it is obviously a normal matrix ($\mathscr{A}\mathscr{A}^\dagger = \mathscr{A}^\dagger\mathscr{A}$).

The characteristic equation is

$$\det[\mathscr{A} - \lambda I_3] =$$

$$= \begin{vmatrix} 1-\lambda & 0 & i \\ 0 & 2-\lambda & 0 \\ -i & 0 & 1-\lambda \end{vmatrix} = (2-\lambda)\begin{vmatrix} 1-\lambda & i \\ -i & 1-\lambda \end{vmatrix} = (2-\lambda)\left[(1-\lambda)^2 + i^2\right] =$$

$$= (2-\lambda)(\cancel{1} - 2\lambda + \lambda^2 - \cancel{1}) = \lambda(2-\lambda)(\lambda - 2) = -\lambda(\lambda - 2)^2.$$

So, there are two distinct eigenvalues (both real because $\mathscr{A} = \mathscr{A}^\dagger$) $\lambda_1 = 0$ and $\lambda_2 = 2$, with algebraic multiplicities $n_{\lambda_1} = 1$ and $n_{\lambda_2} = 2$. Since $\mathscr{A}$ is normal, the eigenspace E_{λ_1} is one-dimensional, and E_{λ_2} two-dimensional. (The geometric multiplicities must be equal to the algebraic ones.)

Indeed, for $\lambda_1 = 0$, we have

$$\begin{bmatrix} 1 & 0 & i & | & 0 \\ 0 & 2 & 0 & | & 0 \\ -i & 0 & 1 & | & 0 \end{bmatrix} \sim \begin{bmatrix} 1 & 0 & i & | & 0 \\ 0 & 1 & 0 & | & 0 \\ 0 & 0 & 0 & | & 0 \end{bmatrix} \Rightarrow$$

$$\begin{array}{c} c_3 = i c_1 + 0 c_2 \text{ or} \\ \\ \Rightarrow \\ \bar{0} = i c_1 + 0 c_2 + (-1) c_3 \end{array} \quad \text{so } E_{\lambda_1} = \text{LIN}\left(\begin{bmatrix} i \\ 0 \\ -1 \end{bmatrix}\right).$$

For $\lambda_2 = 2$, $n_{\lambda_2} = 2$,

$$\begin{bmatrix} -1 & 0 & i & | & 0 \\ 0 & 0 & 0 & | & 0 \\ -i & 0 & -1 & | & 0 \end{bmatrix} \sim \begin{bmatrix} -1 & 0 & i & | & 0 \\ -i & 0 & -1 & | & 0 \\ 0 & 0 & 0 & | & 0 \end{bmatrix} \sim \begin{bmatrix} 1 & 0 & -i & | & 0 \\ 0 & 0 & 0 & | & 0 \\ 0 & 0 & 0 & | & 0 \end{bmatrix} \Rightarrow$$

$$c_2 = 0 c_1 \text{ or } \bar{0} = 0 c_1 + (-1) c_2 + 0 c_3$$

$$\Rightarrow$$

$$\text{and } c_3 = (-i) c_1 \text{ or } \bar{0} = (-i) c_1 + 0 c_2 + (-1) c_3.$$

Thus

$$E_{\lambda_2} = \text{LIN}\left(\begin{bmatrix} 0 \\ -1 \\ 0 \end{bmatrix}, \begin{bmatrix} -i \\ 0 \\ -1 \end{bmatrix}\right).$$

The vector $\bar{v}_1 = \begin{bmatrix} i \\ 0 \\ -1 \end{bmatrix}$ is necessarily orthogonal to $\bar{v}_2 = \begin{bmatrix} 0 \\ -1 \\ 0 \end{bmatrix}$ and

$\bar{v}_3 = \begin{bmatrix} -i \\ 0 \\ -1 \end{bmatrix}$, but these two are orthogonal by chance (no need for Gram–Schmidt procedure). We get the ON eigenbasis in $\mathbb{C}^3$ by dividing each of $\bar{v}_1$, $\bar{v}_2$, $\bar{v}_3$ by its norm. The unitary matrix $\mathscr{T}$ has as its columns these ON vectors:

$$\mathcal{T} = \frac{1}{\sqrt{2}} \begin{bmatrix} i & 0 & -i \\ 0 & -\sqrt{2} & 0 \\ -1 & 0 & -1 \end{bmatrix}.$$

The matrix $\mathcal{T}$ will diagonalize $\mathcal{A}$ by the unitary similarity

$$\mathcal{T}^{-1}\mathcal{A}\mathcal{T} = \mathcal{T}^{\dagger}\mathcal{A}\mathcal{T} = \mathcal{D} = \begin{bmatrix} 0 & 0 & 0 \\ 0 & 2 & 0 \\ 0 & 0 & 2 \end{bmatrix}.$$

This can be easily verified by $\mathcal{T}\mathcal{D} = \mathcal{A}\mathcal{T}$. Δ

5.4 The Actual Method for Diagonalization of a Normal Operator

A normal operator A $(AA^{\dagger} = A^{\dagger}A)$ in U_n has an orthonormal (ON) eigenbasis in which A is represented by the diagonal matrix $\mathcal{D}$ with complex eigenvalues on the diagonal. To find $\mathcal{D}$ (which is unique up to the order of these eigenvalues), we first represent A in any ON basis $u = \{\bar{u}_1, \bar{u}_2, \dots, \bar{u}_n\}$ in U_n by the matrix $\mathcal{A} = [a_{ij}]_{n \times n}$, where $a_{ij} = (\bar{u}_i, A\bar{u}_j)$.

The next step is to calculate the characteristic equation of the matrix $\mathcal{A}$

$$\det(\mathcal{A} - \lambda I_n) = (-\lambda)^n + p_1(-\lambda)^{n-1} + p_2(-\lambda)^{n-2} + \dots + p_{n-1}(-\lambda) + p_n = 0,$$

$$\text{where } p_1 = \text{tr}\mathcal{A}, \ p_n = \det\mathcal{A}, \text{ and } p_k, \ k = 2, 3, \dots, n-1,$$

are the sums of the principal minors of $\mathcal{A}$ of order k. [We have already proved that $\det(\mathcal{A} - \lambda I_n)$ does not depend on the choice of the basis u].

Solving the characteristic equation $\det(\mathcal{A} - \lambda I_n) = 0$ is usually the most difficult task if n is greater than 3. Various approximation methods are used to obtain the solutions (i.e., the eigenvalues).

Due to the fact that the field $\mathbb{C}$ is algebraically closed, we must get n solutions of this equation if we calculate every solution λ_i as many times as its algebraic multiplicity n_{λ_i}.

There are usually $k \leq n$ different solutions $\lambda_1, \lambda_2, \dots, \lambda_k$ with

$$n_{\lambda_1} + n_{\lambda_2} + \dots + n_{\lambda_k} = n.$$

Furthermore, we have to solve for nontrivial solutions the homogeneous linear system $(\mathcal{A} - \lambda_i I_n)\xi = \bar{0}_n$ for every λ_i, $i = 1, 2, \dots k$, where ξ is the matrix-column of unknowns

$$\xi = [\xi_1 \ \xi_2 \ \dots \ \xi_n]^T.$$

Since $\mathcal{A}$ is a normal matrix, the GJM reduction will provide n_{λ_i} (geometric multiplicity) linearly independent basis vectors for the eigenspace $E_{\lambda_i} = \ker(\mathcal{A} - \lambda_i I_n)$, $i = 1, 2, \dots, k$.

We have to use now the Gram–Schmidt procedure for orthonormalization to get n_{λ_i} ON basis vectors in the eigenspace E_{λ_i}.

When we do this procedure for every E_{λ_i}, $i = 1, 2, \ldots, k$, then we have almost finished our job since eigenvectors from two different eigenspaces are automatically orthogonal.

Proof (that eigenvectors from two different eigenspaces are orthogonal)

We shall use the fact that for normal matrices a common eigenvector of $\mathscr{A}$ and $\mathscr{A}^\dagger$ corresponds to the complex conjugate eigenvalues λ and λ^*.

Let $\mathscr{A}v = \lambda_1 v$ and $\mathscr{A}w = \lambda_2 w$ with $\lambda_1 \neq \lambda_2$. Then, $\lambda_1(v, w) = (\lambda_1^* v, w) = (\mathscr{A}^\dagger v, w) = (v, \mathscr{A}w) = (v, \lambda_2 w) = \lambda_2(v, w)$, which gives $(\lambda_1 - \lambda_2)(v, w) = 0 \Rightarrow (v, w) = 0$ since $\lambda_1 \neq \lambda_2$. Δ

Thus, we have n orthonormal eigenvectors $\{c_1, c_2, \ldots c_n\}$ of $\mathscr{A}$ in $\mathbb{C}^n$ forming an ON eigenbasis of $\mathscr{A}$.

We form a new $n \times n$ matrix $\mathscr{T}$ whose columns are these ON vectors

$$\mathscr{T} = \begin{bmatrix} | & | & & | \\ c_1 & c_2 & \cdots & c_n \\ | & | & & | \end{bmatrix}.$$

This matrix is unitary, since its columns are ON vectors. This is one of the characteristic criteria for unitary matrices.

Proof If $\mathscr{T}^{-1} = \mathscr{T}^\dagger$, then this is equivalent to $\mathscr{T}^\dagger \mathscr{T} = I_n$. The (i, j)-element of this matrix equality is

$$c_i^\dagger c_j = \delta_{ij} \text{ or } (c_i, c_j) = \delta_{ij}, \text{ where } c_i, \; i = 1, 2, \ldots, n \text{ are the columns of } \mathscr{T}.$$

$$\text{So, } \mathscr{T}^\dagger \mathscr{T} = I_n \Leftrightarrow (c_i, c_j) = \delta_{ij}, \; i, j = 1, 2, \ldots, n. \quad \Delta$$

What we have by now is n matrix equalities $\mathscr{A}c_1 = \lambda_1 c_1$, $\mathscr{A}c_2 = \lambda_2 c_2$, $\ldots$, $\mathscr{A}c_n = \lambda_n c_n$, where $\lambda_1, \lambda_2, \ldots, \lambda_n$ are the corresponding eigenvalues, which are not necessarily all distinct.

Notice that this set of equalities can be written concisely as

$$\begin{bmatrix} | & | & & | \\ \mathscr{A}c_1 & \mathscr{A}c_2 & \cdots & \mathscr{A}c_n \\ | & | & & | \end{bmatrix}_{n \times n} = \begin{bmatrix} | & | & & | \\ \lambda c_1 & \lambda c_2 & \cdots & \lambda c_n \\ | & | & & | \end{bmatrix}_{n \times n} \quad \text{or}$$

$$\mathscr{A} \begin{bmatrix} | & | & & | \\ c_1 & c_2 & \cdots & c_n \\ | & | & & | \end{bmatrix} = \begin{bmatrix} | & | & & | \\ c_1 & c_2 & \cdots & c_n \\ | & | & & | \end{bmatrix} \begin{bmatrix} \lambda_1 & 0 & \cdots & 0 \\ 0 & \lambda_2 & \cdots & 0 \\ \vdots & \vdots & \ddots & \vdots \\ 0 & 0 & 0 & \lambda_n \end{bmatrix} \quad \text{or}$$

$$\mathscr{A}\mathscr{T} = \mathscr{T}\mathscr{D}.$$

Since $\mathscr{T}^{-1} = \mathscr{T}^\dagger$, it means that $\mathscr{T}$ is an invertible matrix, and we finally have

$$\mathscr{D} = \mathscr{T}^{-1} \mathscr{A} \mathscr{T}.$$

This demonstrates that the normal $n \times n$ matrix $\mathscr{A}$ is diagonalizable by the unitary similarity.

The matrix $\mathscr{T}$ is in fact the unitary transition matrix $\mathscr{T} e_1 = c_1$, $\mathscr{T} e_2 = c_2$, ..., $\mathscr{T} e_n = c_n$, from the standard ON basis $\{e_1, e_2, \ldots, e_n\}$ in $\mathbb{C}^n$ in which $\mathscr{A}$ is represented by itself to the ON eigenbasis $\{c_1, c_2, \ldots, c_n\}$ in which $\mathscr{A}$ changes to the diagonal matrix $\mathscr{D}$.

As far as the initial normal operator A is concerned, it is represented by the diagonal matrix $\mathscr{D}$ in the ON eigenbasis which is obtained from the initial ON basis $u = \{\bar{u}_1, \bar{u}_2, \ldots, \bar{u}_n\}$ (in which A was represented by the normal matrix $\mathscr{A}$) by the unitary transition operator T, i.e., $\{T\bar{u}_1, T\bar{u}_2, \ldots, T\bar{u}_n\}$.

Remember that the transition operator T is represented by the same transition matrix $\mathscr{T}$ both in the initial basis u and in the (new) eigenbasis, and as such, it can be reproduced by one of them and the matrix $\mathscr{T}$:

$$T\bar{u}_i = \sum_{j=1}^{n} t_{ji}\bar{u}_j, \; i = 1,2,\ldots,n.$$

(An operator is given if we know how it acts on any basis.)

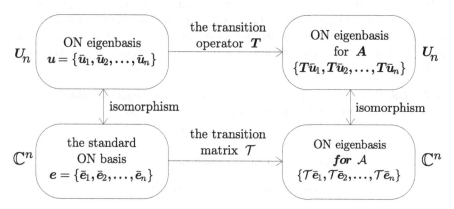

Therefore, the elements of the diagonal matrix

$$\mathscr{D} = \begin{bmatrix} \lambda_1 & 0 & \cdots & 0 \\ 0 & \lambda_2 & \cdots & 0 \\ \vdots & \vdots & \ddots & \vdots \\ 0 & 0 & \cdots & \lambda_n \end{bmatrix}$$

are calculated by means of the ON eigenbasis for A

$$(T\bar{u}_i, AT\bar{u}_j) = \lambda_i \delta_{ij}, \; i, j = 1,2,\ldots,n.$$

The last formula can be written as $(\bar{u}_i, (T^\dagger AT)\bar{u}_j) = \lambda_i \delta_{ij}$, which is the (i, j)-element of the matrix

$$\mathscr{T}^\dagger \mathscr{A} \mathscr{T} = \mathscr{T}^{-1} \mathscr{A} \mathscr{T} = \mathscr{D}.$$

Thus, we are back to the statement that the normal $n \times n$ matrix $\mathscr{A}$ can be brought to a diagonal form $\mathscr{D}$ by a unitary similarity transformation with the transition matrix $\mathscr{T}$ whose columns are the vectors from the ON eigenbasis of $\mathscr{A}$ in $\mathbb{C}^n$.

Remark The transition matrix $\mathscr{T}$ is the transposed matrix of the unitary replacement matrix $\mathscr{R}$ ($\mathscr{T} = \mathscr{R}^T$), and it is the inverse of the unitary contragredient matrix $\mathscr{S}$ ($\mathscr{S} = (\mathscr{R}^T)^{-1} = \mathscr{T}^{-1}$), so that we also have $\mathscr{S}\mathscr{A}\mathscr{S}^{-1} = \mathscr{S}\mathscr{A}\mathscr{S}^\dagger = \mathscr{D}$, which is a frequently used expression.

5.5 The Most Important Subsets of Normal Operators in U_n

5.5.1 The Unitary Operators $A^\dagger = A^{-1}$

We have found that the spectrum (the set of all eigenvalues) of a normal operator ($AA^\dagger = A^\dagger A$) lies in the complex plane. In that plane, there are two important subsets—the unit circle ($z^* = z^{-1}$) $\Leftrightarrow |z|^2 = 1$ and the real axis ($z^* = z$). It is an interesting fact that the normal operators whose spectra are limited to one of these subsets of the complex plane are the most important from the quantum mechanical point of view—namely, the operators that represent symmetries and observables, respectively.

Definition Unitary operators in U_n form a subset of normal operators which is characterized by the fact that their adjoints are equal to their inverses:

$$A^\dagger = A^{-1}$$

There is one immediate consequence of this definition:

every unitary operator is invertible.

Theorem (on five equivalent definitions of unitary operators) A linear operator $A \in \hat{L}(U_n, U_n)$ is called a unitary operator if it satisfies one of the following five properties:

(1) $A^\dagger = A^{-1}$;
(2) $AA^\dagger = A^\dagger A = I_U$;
(3) $(A\bar{x}, A\bar{y}) = (\bar{x}, \bar{y})$ for all $\bar{x}, \bar{y} \in U_n$ (a unitary operator preserves the inner product in U_n);
(4) $||A\bar{x}|| = ||\bar{x}||$ for all $\bar{x} \in U_n$ (every unitary operator preserves the norm of vectors, it is an isometric operator);
(5) The eigenvalue spectrum of A lies on the unit circle in the complex plane.

Proof The properties (1) and (2) obviously imply each other: (1) $\Leftrightarrow$ (2). Furthermore, we shall use a circular scheme of proof:

$$(2) \;\Rightarrow\; (3)$$
$$\nwarrow \qquad \nearrow$$
$$(4)$$

From (2) follows (3):

$$(A\bar{x}, A\bar{y}) = (\bar{x}, A^\dagger A\bar{y}) \overset{(2)}{=} (\bar{x}, \bar{y}), \; \forall \bar{x}, \bar{y} \in U_n.$$

From (3) we have (4):

$$(A\bar{x}, A\bar{x}) = ||A\bar{x}||^2 \overset{(3)}{=} ||\bar{x}||^2 \text{ or } ||A\bar{x}|| = ||\bar{x}||, \; \forall \bar{x} \in U_n$$

And finally, (4) gives (2):

$$(A^\dagger A\bar{x}, \bar{x}) \overset{ADJOINT}{=} (A\bar{x}, A\bar{x}) \overset{(4)}{=} (\bar{x}, \bar{x}), \; \forall \bar{x} \in U_n.$$

So that $((A^\dagger A - I_U)\bar{x}, \bar{x}) = 0$ for all $\bar{x} \in U_n$. The conclusion $A^\dagger A = I_U$ follows from the Lemma statement:

$$(A\bar{x}, \bar{x}) = 0 \text{ for all } \bar{x} \in U_n \Leftrightarrow A \text{ is the } zero \text{ operator.}$$

Proof If A is the zero operator, then obviously $(A\bar{x}, \bar{x}) = 0$, $\forall x \in U_n$. On the other hand, if $(A\bar{x}, \bar{x}) = 0$ for all $\bar{x} \in U_n$, then it is also valid for the sum $\bar{u} + \bar{v}$:

$$(A(\bar{u} + \bar{v}), \bar{u} + \bar{v}) = 0 \Rightarrow (A\bar{u}, \bar{u}) + (A\bar{v}, \bar{v}) + (A\bar{u}, \bar{v}) + (A\bar{v}, \bar{u}) = 0$$

for all $\bar{u}, \bar{v} \in U_n$.

The same is true for the sum $\bar{u} + i\bar{v}$:

$$(A(\bar{u} + i\bar{v}), \bar{u} + i\bar{v}) = -i(A\bar{v}, \bar{u}) + i(A\bar{u}, \bar{v}) = 0 \text{ or}$$
$$(A\bar{u}, \bar{v}) - (A\bar{v}, \bar{u}) = 0.$$

Summing up these two expressions, we get $(A\bar{u}, \bar{v}) = 0$ for all $\bar{u}, \bar{v} \in U_n$. We can now choose $\bar{v} = A\bar{u}$, and get

$$||A\bar{u}||^2 = 0 \overset{\substack{POSITIVE \\ DEFINITENESS}}{\Rightarrow} A\bar{u} = \bar{0} \text{ for all } \bar{u} \in U_n.$$

So, A must be the zero operator, since its kernel is the whole U_n. $\triangle$ (Lemma)

The last part of the proof of the theorem is to show that (1) and (5) are equivalent: (1) $\Leftrightarrow$ (5). The diagonal matrix $\mathscr{D}$ that represents A in an ON eigenbasis satisfies also (1) $\mathscr{D}^\dagger = \mathscr{D}^{-1}$. But

$$\mathscr{D}^\dagger = \mathscr{D}^* = \begin{bmatrix} z_1^* & 0 & \cdots & 0 \\ 0 & z_2^* & \cdots & 0 \\ \vdots & \vdots & \ddots & \vdots \\ 0 & 0 & \cdots & z_n^* \end{bmatrix},$$

and $\mathscr{D}^{-1}$ has on its diagonal the inverse eigenvalues

$$\mathscr{D}^{-1} = \begin{bmatrix} z_1^{-1} & 0 & \cdots & 0 \\ 0 & z_2^{-1} & \cdots & 0 \\ \vdots & \vdots & \ddots & \vdots \\ 0 & 0 & \cdots & z_n^{-1} \end{bmatrix}.$$

Consequently, $z_i^* = z_i^{-1}$, $i = 1, 2, \ldots, n$, i.e., All the eigenvalues lie on the unit circle.

If on the other hand, $\mathscr{D}^\dagger = \mathscr{D}^* = \mathscr{D}^{-1}$ (the eigenspectrum is on the unit circle), this implies the unitarity of A: $A^\dagger = A^{-1}$, since the representation of operators by matrices induced by an ON eigenbasis of A is an isomorphism of algebras with unity in which $A^\dagger$ and A^{-1} are represented just by $\mathscr{D}^\dagger$ and $\mathscr{D}^{-1}$. Δ

Conclusion: unitary operators in U_n are the widest class of operators in U_n that are diagonalizable by the unitary similarity with the eigenvalue spectrum on the unit circle in the complex plane.

With regard to the $n \times n$ matrix $\mathscr{A}$ that represents a unitary operator A in an ON basis in U_n, we see from the previous argument that $\mathscr{A}$ must be a unitary matrix: $\mathscr{A}^\dagger = \mathscr{A}^{-1}$. Such a unitary matrix can be characterized by two additional properties:

(6) The rows of $\mathscr{A}$ are orthonormal vectors;
(7) The columns are also orthonormal vectors.

Proof We have already proved (7) when discussing the unitary transition matrix $\mathscr{T}$:

$$\mathscr{T}^\dagger \mathscr{T} = I_n \Leftrightarrow (c_i, c_j) = \delta_{ij}, \quad i, j = 1, 2, \ldots, n, \quad \text{where } c_i \text{ are the columns of } \mathscr{T}.$$

Similarly, for (6), $\mathscr{A}\mathscr{A}^\dagger = I_n$ is equivalent to $(r_i, r_j) = \delta_{ij}$ $i = 1, 2, \ldots, n$, where r_i are the rows of $\mathscr{A}$. In more detail, the (i, j) element of $\mathscr{A}\mathscr{A}^\dagger = I_n$ is $\sum_{k=1}^{n} a_{ik} a_{jk}^* = \delta_{ij}$. Taking the complex conjugate we get $(r_i, r_j) = \delta_{ij}$. Δ

In Sect. 4.4, we stated (without proof) that $n \times n$ unitary matrices form a group $U(n)$ in the algebra $\mathbb{C}_{n \times n}$ of all complex square matrices, and as a subgroup of the group $GL(n, \mathbb{C})$ of all invertible matrices. This group $U(n)$ is the group of all automorphisms in $\mathbb{C}^n$, since unitary matrices map bijectively $\mathbb{C}^n$ onto $\mathbb{C}^n$ preserving the complete inner-product complex vector space structure of $\mathbb{C}^n$. (Note that in $\mathbb{C}^n$ there are no linmaps which are not matrices.)

Proof [that all the $n \times n$ unitary matrices form a group $U(n)$]

1. $I_n \in U(n)$;

2. $\mathscr{A} \in U(n) \Rightarrow (\mathscr{A}^{-1})^{-1} = \mathscr{A} = (\mathscr{A}^{-1})^\dagger \Rightarrow \mathscr{A}^{-1} \in U(n);$
3. $\mathscr{A} \in U(n)$ and $\mathscr{B} \in U(n) \Rightarrow (\mathscr{A}\mathscr{B})^{-1} = \mathscr{B}^{-1}\mathscr{A}^{-1} = \mathscr{B}^\dagger \mathscr{A}^\dagger = (\mathscr{A}\mathscr{B})^\dagger \Rightarrow \mathscr{A}\mathscr{B} \in U(n)$. $\triangle$

Unitary operators in U_n exhibit one property which is important especially in the theory of representations of groups by unitary operators.

Theorem (on reducibility of unitary operators) Let A be a unitary operator in U_n and let W be a subspace of U_n which is invariant under A (operator A reduces in W as A_1—restriction of the domain), then the orthocomplement $W^\perp$ is also invariant under A (operator A reduces in $W^\perp$ as A_2). The space U_n is the orthogonal sum of W and $W^\perp$: $U_n = W \oplus W^\perp$. So, the action of A in U_n breaks up into components

$$A\bar{x} = A_1\bar{x}_1 + A_2\bar{x}_2,$$

since every vector $\bar{x} \in U_n$ is uniquely expressible as the sum

$$\bar{x} = \bar{x}_1 + \bar{x}_2, \ \bar{x}_1 \in W, \ \bar{x}_2 \in W^\perp, \ (\bar{x}_1, \bar{x}_2) = 0.$$

Proof Since A is an invertible operator, it maps the invariant subspace W onto itself, because as such it cannot reduce the dimensionality of W. This means that every vector $\bar{x} \in W$ is the image of another vector $\bar{x}'$, so that $A\bar{x}' = \bar{x}$.

Take now an arbitrary vector $\bar{y}$ from $W^\perp$, $\bar{y} \in W^\perp \Leftrightarrow (\bar{y}, \bar{x}) = 0, \ \forall \bar{x} \in W$. We shall show that $A\bar{y} \in W^\perp \Leftrightarrow (A\bar{y}, \bar{x}) = 0, \ \forall \bar{x} \in W$, i.e., that $A\bar{y}$ is also orthogonal to any $\bar{x} \in W$:

$$(A\bar{y}, \bar{x}) = (A\bar{y}, A\bar{x}') = (\bar{y}, \bar{x}) = 0; \ \forall \bar{x} \in W.$$

So, $W^\perp$ is invariant under A, and A reduces in $W^\perp$ as A_2 (it acts as A in the restricted domain $W^\perp$). Since

$$U_n = W \oplus W^\perp \Rightarrow \bar{x} \in U_n, \ \text{is uniquely written as}$$
$$\bar{x} = \bar{x}_1 + \bar{x}_2, \ \bar{x}_1 \in W, \ \bar{x}_2 \in W^\perp, \ (\bar{x}_1, \bar{x}_2) = 0, \ \text{so}$$
$$A\bar{x} = A(\bar{x}_1 + \bar{x}_2) = A\bar{x}_1 + A\bar{x}_2 = A_1\bar{x}_1 + A_2\bar{x}_2,$$

and as U_n breaks up into the orthogonal sum of W and $W^\perp$, so does A into the direct sum of two reduced operators A_1 and A_2. $\triangle$

If we choose an ON basis in U_n which is adapted to the orthogonal decomposition $U_n = W \oplus W^\perp$ [if $\dim W = k < n$, then the first k vectors of this basis span W, and the rest of $(n-k)$ vectors span $W^\perp$], then the operator A is represented in such a basis by a block diagonal matrix

$$\mathscr{A} = \begin{bmatrix} \mathscr{A}_1 & 0 \\ 0 & \mathscr{A}_2 \end{bmatrix} \begin{matrix} \}k \\ \}n-k \end{matrix}$$
$$\underbrace{}_{k} \ \underbrace{}_{n-k}$$

We usually say that the matrix $\mathscr{A}$ is the direct sum of matrices $\mathscr{A}_1$ and $\mathscr{A}_2$:

$$\mathscr{A} = \mathscr{A}_1 + \mathscr{A}_2.$$

This process of breaking $\mathscr{A}$ into the direct sum of smaller matrices $\mathscr{A}_1$ and $\mathscr{A}_2$ is of fundamental importance in reducing unitary representations of groups into direct sums of irreducible representations, i.e., those which have no invariant subspaces.

5.5.2 The Hermitian Operators $A^\dagger = A$

Definition Hermitian operators in U_n form a subset of normal operators which are self-adjoint, i.e., their adjoints are equal to themselves:

$$A^\dagger = A.$$

Theorem (on four equivalent definitions of Hermitian operators) A linear operator $A \in \hat{L}(U_n, U_n)$ is called a Hermitian operator if it satisfies one of the following four properties:

(1) $A^\dagger = A$ (it is self-adjoint);
(2) $(\bar{x}, A\bar{y}) = (A\bar{x}, \bar{y})$, $\forall \bar{x}, \bar{y} \in U_n$ (it "jumps" from one to the other factor in the inner product without change);
(3) $(\bar{x}, A\bar{x}) \in \mathbb{R}$, $\forall \bar{x} \in U_n$ (The "expectation value" of A (a quantum mechanical term) is always real);
(4) The eigenvalue spectrum of A lies on the *real axis* in the complex plane.

Proof The properties (1) and (2) obviously imply each other: (1) $\Leftrightarrow$ (2). We shall now prove (1) $\Leftrightarrow$ (3).

If $A^\dagger = A$, then $(\bar{x}, A\bar{x}) = (A\bar{x}, \bar{x}) = (\bar{x}, A\bar{x})^*$, which means that $(\bar{x}, A\bar{x})$ is a real number for all $\bar{x} \in U_n$.

If, on the other hand, $(\bar{x}, A\bar{x})$ is always real $(\forall \bar{x} \in U_n)$, then we get $(\bar{x}, A\bar{x}) = (\bar{x}, A\bar{x})^* = (A\bar{x}, \bar{x}) = (\bar{x}, A^\dagger \bar{x})$ or $(\bar{x}, (A - A^\dagger)\bar{x}) = 0$ for all $\bar{x} \in U_n$. The Lemma from Sect. 5.5.1 gives immediately $A = A^\dagger$.

Finally, we have to prove (1) $\Leftrightarrow$ (4). The property $A^\dagger = A$ implies the analogous one for the representing diagonal matrix $\mathscr{D}$ in any ON eigenbasis of A, i.e., $\mathscr{D}^\dagger = \mathscr{D}$, so that $\mathscr{D}$ is a Hermitian matrix. But, $\mathscr{D}^\dagger = \mathscr{D}^*$, so that $\mathscr{D}^* = \mathscr{D}$ is equivalent to saying that all the eigenvalues of A are real numbers $(z_i^* = z_i, i = 1, 2, \ldots, n)$, so (1) $\Rightarrow$ (4). The reasoning in the reverse order gives (4) $\Rightarrow$ (1). $\triangle$

The Hermitian operators play in $\hat{L}(U_n, U_n)$ the role analogous to the role of real numbers in $\mathbb{C}$, so we can uniquely write any linear operator B in U_n in the Descartes (Cartesian) form $B = H_1 + iH_2$, where H_1 and H_2 are Hermitian operators, in analogy with $z = x + iy$ for $z \in \mathbb{C}$ and $x, y \in \mathbb{R}$. Indeed,

$$H_1 = \frac{B + B^\dagger}{2}, \ H_2 = \frac{B - B^\dagger}{2i}, \ \text{and} \ H_1^\dagger = H_1, \ H_2^\dagger = H_2.$$

For a linear operator $C \in \hat{L}(U_n, U_n)$, the products $CC^\dagger$ and $C^\dagger C$ are always Hermitian operators $[(CC^\dagger) = CC^\dagger$ and $(C^\dagger C)^\dagger = C^\dagger C]$, but in the general case, they are not equal (they are equal only if C is a normal operator).

Hermitian operators in U_n form a real vector space, since $(A + B)^\dagger = A^\dagger + B^\dagger = A + B$ and $(aA)^\dagger = a^* A^\dagger = aA$, $a \in \mathbb{R}$. So, the subset of Hermitian operators in the superspace $\hat{L}(U_n, U_n)$ is closed under addition and multiplication with a real number.

But, the product AB of two Hermitian operators A and B is again a Hermitian operator if and only if A and B commute: we have $(AB)^\dagger = B^\dagger A^\dagger = BA$, so commuting $(BA = AB)$ guarantees the Hermitian nature of AB. Similarly, the Hermitian nature of AB, i.e., $(AB)^\dagger = AB$, implies the commuting of A and B.

Note that the zero operator $\hat{0}$ and the identity operator I_U are Hermitian: $(\bar{x}, \hat{0}\bar{y}) = (\hat{0}\bar{x}, \bar{y}) = 0$ and $(\bar{x}, I_U \bar{y}) = (I_U \bar{x}, \bar{y}) = (\bar{x}, \bar{y})$, $\forall \bar{x}, \bar{y} \in U_n$.

The quantity $(\bar{x}, A\bar{x})$ is called the expectation value in Quantum Mechanics. It is the average value of many measurements of the physical observable A which is measured in the state described by the unit vector $\bar{x}$. It is natural that such a number must be real. So, the operators in Quantum Mechanics which correspond to the observables are Hermitian operators.

An important subset of Hermitian operators consists of *positive operators*.

Definition A Hermitian operator A is called a *positive operator* if $(\bar{x}, A\bar{x}) \geq 0$ for all $\bar{x} \in U_n$. The eigenvalue spectrum of a positive operator consists of real numbers which are greater or equal to 0. The short notation for these operators is $A \geq 0$.

Positive operators are essential in Quantum Statistical Physics where the state of the system is not described by vectors in infinite-dimensional unitary space (as is done in Quantum Mechanics), but by positive operators (with the trace equal to 1), which are called *statistical operators*.

Having described unitary and Hermitian operators in U_n in detail, we can go back to normal operators and present two equivalent definitions of these operators in terms of Hermitian and unitary ones.

Theorem (on two equivalent definitions of normal operators) A normal operator A in U_n which is basically defined by commutation with its adjoint, $AA^\dagger = A^\dagger A$, can be equivalently defined in two more ways:

(1) for a normal operator A it is characteristic that in its Descartes form $A = H_1 + iH_2$, these Hermitian operators commute: $H_1 H_2 = H_2 H_1$;
(2) Every normal operator A in U_n can be written in polar form (in analogy with the polar form of a complex number $z = |z|e^{i\theta}$) as a product $A = HU$, where H is a positive (Hermitian) operator $(H \geq 0)$ and U is a unitary operator $(U^\dagger = U^{-1})$. Furthermore, it is characteristic for the normal operator that its polar factors commute:

$$A = HU = UH.$$

Proof (1) Since $H_1 = \frac{A + A^\dagger}{2}$ and $H_2 = \frac{A - A^\dagger}{2i}$, we can easily obtain

$$H_1 H_2 - H_2 H_1 = \frac{1}{2i}(A^\dagger A - AA^\dagger),$$

so that the zero on one side implies the zero on the other side.

(2) If operator A is normal, it can be represented in an ON eigenbasis by the diagonal matrix $\mathscr{D}$ whose diagonal consists of complex eigenvalues z_j, $j = 1, 2, \ldots, n$. (Note that $\mathscr{D}$ is unique up to the order of the eigenvalues). We can write every eigenvalue in polar form: $z_j = |z_j| e^{i\theta_j}$, $j = 1, 2, \ldots, n$ (for $z_j = 0$ we have $|z_j| = 0$, and the phase factor can be any number whose modulus is one). After this factorization of eigenvalues, we can write $\mathscr{D}$ as the product of diagonal matrices $\mathscr{D} = \mathscr{H}\mathscr{U}$, where the first factor $\mathscr{H}$ is a diagonal positive (Hermitian) matrix, and the second factor $\mathscr{U}$ is a diagonal unitary matrix. Being diagonal, these matrices commute $\mathscr{H}\mathscr{U} = \mathscr{U}\mathscr{H}$.

Using now the isomorphism between algebras $\hat{L}(U_n, U_n)$ and $\hat{L}(C^n, C^n)$ induced by the original choice of the ON eigenbasis, we can write

$$A = HU = UH.$$

On the other hand, if an operator A can be factorized into the product of a positive and a unitary operator, so that these factors commute $A = HU = UH$, then $A^\dagger = U^\dagger H = HU^\dagger$, which implies that A is a normal operator:

$$AA^\dagger = HUU^\dagger H = H^2 \text{ and } A^\dagger A = HU^\dagger UH = H^2,$$

$$\text{so that } AA^\dagger = A^\dagger A, \text{ since } U^\dagger U = UU^\dagger = I. \ \Delta$$

5.5.3 The Projection Operators $P^\dagger = P = P^2$

The projection operators form a subset of the Hermitian operators (more precisely of the positive ones) and they play an important role in the formalism of Quantum Mechanics.

Definition A linear operator $P \in \hat{L}(U_n, U_n)$ which is Hermitian $(P^\dagger = P)$ and idempotent $(P^2 = P)$ is called a projection operator.

Theorem (on the eigenvalue spectrum of a projection operator and the corresponding eigen decomposition of U_n) The range ran $P = \{P\bar{x} | \bar{x} \in U_n\}$ of every projection operator P is equal (due to the idempotent property of P) to the eigenspace E_1 of P which corresponds to the eigenvalue 1:

$$\text{ran } P = \{\bar{x} | \bar{x} \in U_n \text{ and } P\bar{x} = 1 \cdot \bar{x}\}.$$

The kernel ker $P = \{\bar{x} | \bar{x} \in U_n \text{ and } P\bar{x} = \bar{0}\} = E_0$ (The eigenspace corresponding to the eigenvalue 0) is equal (due to the Hermitian property of P) to the orthocomplement of the range:

$$\text{ker } P = (\text{ran } P)^\perp.$$

This means that the whole U_n is decomposed into the orthogonal sum of ran P and ker P:

$$U_n = \text{ran } P \oplus \text{ker } P.$$

Note that in these subspaces P acts as a multiplicative constant (eigenvalue) 1 and 0, respectively, and that there are no more eigenvalues of P.

An arbitrary $\bar{x} \in U_n$ is uniquely decomposed into orthogonal components—one from ran $P = E_1$ and the other from ker $P = E_0$: $\bar{x} = \bar{y} + \bar{z}$, $\bar{y} \in$ ran P, $\bar{z} \in$ ker P. The action of P on $\bar{x}$ gives the component $\bar{y}$:

$$P\bar{x} = P\bar{y} + P\bar{z} = 1 \cdot \bar{y} + 0 \cdot \bar{z} = \bar{y}, \text{ which is called}$$

the *projection* of $\bar{x}$ on E_1.

Proof We shall start with E_1, which is, in fact, the subspace in U_n of vectors that are invariant under P:

$$E_1 = \{\bar{x} | \bar{x} \in U_n \text{ and } P\bar{x} = \bar{x}\}.$$

We now want to prove ran $P = E_1$ and ker $P = E_1^{\perp}$.

Since ran $P = \{P\bar{x} | \bar{x} \in U_n\}$, then, due to the idempotency $PP = P$, it follows that $P(P\bar{x}) = P\bar{x}$, $\forall \bar{x} \in U_n$, i.e., an arbitrary vector $P\bar{x}$ from ran P is invariant under P or, in set language, ran $P \subseteq E_1$. Furthermore, any vector $\bar{x}$ from E_1 is the image of itself $(P\bar{x} = \bar{x})$, so it belongs to the set ran P of all images, $E_1 \subseteq$ ran P. Thus, ran $P = E_1$.

From the theorem about adjoint operators (Sect. 4.2), we have the general statement ran $A^{\dagger} = (\ker A)^{\perp}$ for any operator A in U_n and its adjoint. For Hermitian operators $A^{\dagger} = A$, this equality becomes simpler:

$$\text{ran } A = (\ker A)^{\perp} \text{ or } \ker A = (\text{ran } A)^{\perp}.$$

For the projection operator P, this gives immediately

$$\ker P = (\text{ran } P)^{\perp} = E_1^{\perp}.$$

It follows that $U_n = $ ran $P \oplus \ker P = E_1 \oplus E_0$. Thus, the eigenvalue spectrum of P is just $\{0, 1\}$. The rest of the theorem follows straightforwardly from the above decomposition of U_n. $\triangle$

We call P the projection operator onto its eigenspace E_1 (the subspace of invariant vectors), since its range is E_1.

There is one very practical formula when one works with projection operators, the formula that is based on two properties which define projection operators:

$$(P\bar{x}, \bar{y}) = (P^2\bar{x}, \bar{y}) = (P\bar{x}, P\bar{y}) = (\bar{x}, P^2\bar{y}) = (\bar{x}, P\bar{y}),$$

$$\text{thus } \boxed{(P\bar{x}, \bar{y}) = (P\bar{x}, P\bar{y}) = (\bar{x}, P\bar{y})},$$

and P can act on either factor in the inner product or on both factors without any change.

The matrix $\mathscr{P}$ of the projection operator P has an extremely simple form in any ON basis which is adapted to the orthogonal decomposition $U_n = $ ran $P \oplus \ker P$.

This form of $\mathscr{P}$ is characteristic for projection operators, and it is responsible for their important applications in Quantum Mechanics. If the rank, i.e., $\dim \mathrm{ran}\, P$, is m, then $\mathscr{P}$ is an $n \times n$ diagonal matrix, so that the first m elements on the diagonal are 1, and the rest of the $(n-m)$ elements on the diagonal are 0. It should be noted that the trace of $\mathscr{P}$ is equal to its $rank(m)$. Since the trace does not depend on the chosen basis, we can say that the trace is a feature of the operator itself:

$$\mathrm{tr}\, P = \dim \mathrm{ran}\, P.$$

This is a very useful equality.

Examples of Projection Operators

(1) The projection operator on the whole U_n is just the identity operator $I_U : I_U \bar{x} = \bar{x}$ for all $\bar{x} \in U_n$;

(2) The projection operator on the null subspace $\{\bar{0}\}$ is the zero operator $\hat{0}$ (it maps every $\bar{x} \in U_n$ onto $\bar{0}$);

(3) The projection operator on the one-dimensional subspace (the line) $L(\bar{v})$, which is spanned by the normalized vector $\bar{v}$ ($||\bar{v}|| = 1$) is given by its action on $\forall \bar{x} \in U_n$: $P_{L(\bar{v})}\bar{x} = (\bar{v},\bar{x})\,\bar{v}$, i.e., it is the projection of $\bar{x}$ along $\bar{v}$;

(4) The projection operator on the subspace V, which is spanned by an ON set of vectors $\{\bar{v}_1,\bar{v}_2,\dots,\bar{v}_m\}$, $m < n$ (these vectors are an ON basis in V), is given as the sum of the projections of $\bar{x}$ along all vectors of the basis: $P_V\bar{x} = \sum_{i=1}^{m}(\bar{v}_i,\bar{x})\,\bar{v}_i$. See later (Sect. 5.5.4) the theorem on the sum of orthogonal projection operators;

(5) The projection operator on a line in $\mathbb{C}^n$. Every normalized vector $x = [x_1\, x_2\, \dots\, x_n]^T \in \mathbb{C}^n$ defines one projection operator (a matrix)

$$\mathscr{P} = xx^{\dagger} = \begin{bmatrix} x_1 \\ x_2 \\ \vdots \\ x_n \end{bmatrix} \begin{bmatrix} x_1^* & x_2^* & \dots & x_n^* \end{bmatrix} =$$

$$= \begin{bmatrix} x_1x_1^* & x_1x_2^* & \dots & x_1x_n^* \\ x_2x_1^* & x_2x_2^* & \dots & x_2x_n^* \\ \vdots & \vdots & \ddots & \vdots \\ x_nx_1^* & x_nx_2^* & \dots & x_nx_n^* \end{bmatrix}.$$

This matrix is Hermitian since $\mathscr{P}^{\dagger} = (xx^{\dagger})^{\dagger} = xx^{\dagger} = \mathscr{P}$ and idempotent since $\mathscr{P}^2 = (xx^{\dagger})(xx^{\dagger}) = x(x^{\dagger}x)x^{\dagger} = xx^{\dagger} = \mathscr{P}$, because x is a normalized vector $||x||^2 = x^{\dagger}x = 1$. The projection operator $\mathscr{P}$ has its trace equal to 1: $\mathrm{tr}(xx^{\dagger}) = x^{\dagger}x = 1$, so that ran $\mathscr{P}$ is a one-dimensional subspace in $\mathbb{C}^n$. We can verify that x is invariant under $\mathscr{P}$: $\mathscr{P}x = (xx^{\dagger})x = x(x^{\dagger}x) = x$, thus ran $\mathscr{P} = L(x)$, i.e., $\mathscr{P} = xx^{\dagger}$ projects on the subspace (the line) which is spanned by x. The action of $\mathscr{P}$ on an arbitrary vector $y \in \mathbb{C}^n$ is [like in example (3)]

$$\mathscr{P}y = (xx^\dagger)y = x(x^\dagger y) = (x,y)x.$$

5.5.4 Operations with Projection Operators

The following three theorems about the addition, multiplication, and subtraction of two projection operators are important for most of the applications.

Definition We say that two projection operators P_1 and P_2 are orthogonal if their ranges are orthogonal: ran $P_1 \perp$ ran P_2.

This property can be algebraically characterized by $P_1 P_2 = P_2 P_1 = \hat{0}$ [where $\hat{0}$ is the zero operator in $\hat{L}(U_n, U_n)$]:

$$P_1 P_2 = P_2 P_1 = \hat{0} \Leftrightarrow \text{ran } P_1 \perp \text{ran } P_2.$$

This can be shown very easily: if $P_1 P_2 = P_2 P_1 = \hat{0}$, then for any $\bar{x}_1 \in$ ran P_1 $(P_1 \bar{x}_1 = \bar{x}_1)$ and $\bar{x}_2 \in$ ran P_2 $(P_2 \bar{x}_2 = \bar{x}_2)$, we obtain $(\bar{x}_1, \bar{x}_2) = (P_1 \bar{x}_1, P_2 \bar{x}_2) = (\bar{x}_1, P_1 P_2 \bar{x}_2) = (\bar{x}_1, \hat{0}\bar{x}_2) = 0 \Rightarrow$ ran $P_1 \perp$ ran P_2. On the other hand, if ran $P_1 \perp$ ran P_2 then for any $\bar{x}, \bar{y} \in U_n$ we have $(P_1 \bar{x}, P_2 \bar{x}) = 0$, since $P_1 \bar{x} \in$ ran P_1 and $P_2 \bar{y} \in$ ran P_2. It gives further $(\bar{x}, P_1 P_2 \bar{y}) = 0$ for any $\bar{x}, \bar{y} \in U_n$, so that the Lemma in Sect. 5.5.1 enables us to conclude $P_1 P_2 = \hat{0}$. Taking the adjoint of both sides (remember that $\hat{0}$ is also a Hermitian operator) finalizes the proof $P_1 P_2 = P_2 P_1 = \hat{0}$. Δ

Theorem (on the sum of two orthogonal projection operators) Let P_1 and P_2 be two projection operators: $P_1^2 = P_1 = P_1^\dagger$ and $P_2^2 = P_2 = P_2^\dagger$. The operator $P_1 + P_2$ is a projection operator if and only if P_1 and P_2 are orthogonal:

$$(P_1 + P_2)^2 = P_1 + P_2 = (P_1 + P_2)^\dagger \Leftrightarrow P_1 P_2 = P_2 P_1 = \hat{0}.$$

In that case, ran $(P_1 + P_2) =$ ran $P_1 \oplus$ ran P_2, meaning that the sum $P_1 + P_2$ projects on the orthogonal sum of the ranges ran $P_1 \oplus$ ran P_2.

Proof It is sufficient for the projection operators P_1 and P_2 to be orthogonal, that their sum is also such an operator:

$$(P_1 + P_2)^2 = P_1^2 + \cancel{P_1 P_2} + \cancel{P_2 P_1} + P_2^2 = P_1 + P_2 \quad (P_1 + P_2 \text{ is idempotent}),$$
$$(P_1 + P_2)^\dagger = P_1^\dagger + P_2^\dagger = P_1 + P_2 \quad (P_1 + P_2 \text{ is Hermitian}).$$

If $P_1 + P_2$ is a projection operator, then it follows that P_1 and P_2 must be orthogonal:

$$(P_1 + P_2)^2 = P_1 + P_2$$
$$\cancel{P_1^2} + P_1 P_2 + P_2 P_1 + \cancel{P_2^2} = \cancel{P_1} + \cancel{P_2} \Rightarrow P_1 P_2 + P_2 P_1 = \hat{0} \quad (*)$$

Multiplying $(*)$ first from the left with P_1, and then from the right also with P_1, we have

$$P_1 P_2 + P_1 P_2 P_1 = \hat{0}$$
$$P_1 P_2 P_1 + P_2 P_1 = \hat{0}.$$

Subtracting these equalities results in

$$P_1 P_2 - P_2 P_1 = \hat{0}. \quad (**)$$

The two expressions $(*)$ and $(**)$ give together

$$P_1 P_2 = \hat{0} \text{ implying also } P_2 P_1 = \hat{0}.$$

Now, we shall prove that ran $(P_1 + P_2) \subseteq$ ran $P_1 \oplus$ ran P_2. The vector $(P_1 + P_2)\bar{x}$ is an arbitrary vector in ran $(P_1 + P_2)$ when $\bar{x}$ runs over the whole U_n. However, $(P_1 + P_2)\bar{x} = P_1\bar{x} + P_2\bar{x} \in$ ran $P_1 \oplus$ ran P_2.

Furthermore, we shall demonstrate that ran $P_1 \oplus$ ran $P_2 \subseteq$ ran $(P_1 + P_2)$. If $\bar{x}$ is an arbitrary vector from ran $P_1 \oplus$ ran P_2 it means that $\bar{x} = \bar{y} + \bar{z}$, $\bar{y} \in$ ran P_1 $(P_1 \bar{y} = \bar{y})$ and $\bar{z} \in$ ran P_2 $(P_2 \bar{z} = \bar{z})$. We can now show that $\bar{x} = \bar{y} + \bar{z}$ is invariant under $P_1 + P_2$, i.e., that it belongs to ran $(P_1 + P_2)$:

$$(P_1 + P_2)(\bar{y} + \bar{z}) = P_1 \bar{y} + P_2 \bar{y} + P_1 \bar{z} + P_2 \bar{z} = \bar{y} + P_2(P_1\bar{y}) + P_1(P_2\bar{z}) + \bar{z} = \bar{y} + \bar{z}. \quad \Delta$$

This theorem can be immediately generalized: Let $\{P_1, P_2, \dots, P_k\}$ be a set of projection operators in U_n. The operator $\{P_1 + P_2 + \cdots + P_k\}$ is a projection operator iff all these operators are mutually orthogonal, i.e.,

$$P_m P_n = \hat{0}, \ m \neq n, \ m,n = 1,2,\dots,k.$$

Then, $\{P_1 + P_2 + \cdots + P_k\}$ projects on the orthogonal sum of the corresponding ranges ran $P_1 \oplus$ ran $P_2 \oplus \dots \oplus$ ran P_k.

Definition The set of nonzero projection operators $\{P_1, P_2, \dots, P_k\}$ which are mutually orthogonal and are such that their sum is the identity operator I_U (the projection operator on U_n) $\sum_{i=1}^{k} P_i = I_U$, is called the decomposition of the unity.

The obvious consequence of this decomposition is that U_n is also decomposed into the orthogonal sum of the corresponding ranges $U_n = \sum_{i=1}^{k} \oplus$ran P_i and vice versa.

A typical example of the decomposition of the unity is given if one takes an ON basis $\{\bar{u}_1, \bar{u}_2, \dots, \bar{u}_n\}$ in U_n and follows the above procedure:

$$I_U \bar{x} = \sum_{i=1}^{n} (\bar{u}_i, \bar{x}) \bar{u}_i \text{ and } U_n = \sum_{i=1}^{n} \oplus L(\bar{u}_i). \text{ [cf. example (4) in Sect. 5.5.3]}$$

Now is a good opportunity to discuss the relation between the projection operator P and the operator $I_U - P$. We shall first show that $I_U - P$ is also a projection operator: $(I_U - P)^\dagger = I_U^\dagger - P^\dagger = I_U - P$ (it is a Hermitian operator, since both I_U and P are Hermitian); $(I_U - P)^2 = I_U^2 - 2P + P^2 = I_U - P$ (it is idempotent since both I_U

and P are idempotent). It is worth mentioning that the fact that $I_U - P$ is a projection operator implies $P^2 = P = P^\dagger$; If $I_U - P$ is a projection operator, then we have just demonstrated that so is $I_U - (I_U - P)$ which is equal to P.

Furthermore, P and $I_U - P$ are orthogonal:

$$(I_U - P)P = P(I_U - P) = P - P^2 = \hat{0},$$

so that their sum is also a projection operator $P + (I_U - P) = I_U$, implying

$$\text{ran } P \perp \text{ran } (I_U - P) \text{ and } U_n = \text{ran } P \oplus \text{ran } (I_U - P) \text{ or}$$
$$\text{ran } (I_U - P) = U_n \ominus \text{ran } P = (\text{ran } P)^\perp = \ker P.$$

Thus, $I_U - P$ projects on the orthocomplement of ran P.

Theorem (on the product of two commuting projection operators) If P_1 and P_2 are two projection operators, then their product $P_1 P_2$ is also a projection operator iff P_1 and P_2 commute:

$$(P_1 P_2)^2 = P_1 P_2 = (P_1 P_2)^\dagger \Leftrightarrow P_1 P_2 = P_2 P_1.$$

Then $P_1 P_2$ projects on the intersection of the ranges of P_1 and P_2:

$$\text{ran } (P_1 P_2) = \text{ran } P_1 \cap \text{ran } P_2.$$

Proof The commuting condition is sufficient for $P_1 P_2$ to be a projection operator:

$$(P_1 P_2)^2 = P_1 P_2 P_1 P_2 = P_1^2 P_2^2 = P_1 P_2;$$
$$(P_1 P_2)^\dagger = P_2^\dagger P_1^\dagger = P_2 P_1 = P_1 P_2.$$

If $P_1 P_2$ is a projection operator (Hermitian and idempotent), then it is obvious (from the Hermitian nature of P_1 and P_2) that they necessarily commute:

$$\left.\begin{matrix} (P_1 P_2)^\dagger = P_1 P_2 \\ (P_1 P_2)^\dagger = P_2^\dagger P_1^\dagger = P_2 P_1 \end{matrix}\right\} \Rightarrow P_1 P_2 = P_2 P_1.$$

The vector $P_1 P_2 \bar{x}$ is an arbitrary element in ran $(P_1 P_2)$ when $\bar{x}$ runs through the whole U_n. Due to the commuting property of P_1 and P_2, this vector can be written in two forms: $P_1 (P_2 \bar{x}) \in \text{ran } P_1$ and $P_2 (P_1 \bar{x}) \in \text{ran } P_2$. This means that $P_1 P_2 \bar{x} \in$ ran $P_1 \cap \text{ran } P_2$ or

$$\text{ran } (P_1, P_2) \subseteq \text{ran } P_1 \cap \text{ran } P_2.$$

On the other hand, if $\bar{y}$ is any vector from ran $P_1 \cap \text{ran } P_2$, this implies both $\bar{y} \in$ ran P_1 $(P_1 \bar{y} = \bar{y})$ and $\bar{y} \in \text{ran } P_2$ $(P_2 \bar{y} = \bar{y})$. Placing one expression into the other, we get $P_1 P_2 \bar{y} = \bar{y}$ or $\bar{y} \in \text{ran } (P_1 P_2) \Rightarrow \text{ran } P_1 \cap \text{ran } P_2 \subseteq \text{ran } (P_1 P_2)$. $\triangle$

Theorem (on the difference of two projection operators) The difference $P_1 - P_2$ of two projection operators P_1 and P_2 is a projection operator if and only if the range ran P_2 is a subspace of ran P_1:

$$(P_1 - P_2)^2 = P_1 - P_2 = (P_1 - P_2)^\dagger \Leftrightarrow \text{ran } P_2 \subseteq \text{ran } P_1.$$

The algebraic form of this condition is $P_1 P_2 = P_2 P_1 = P_2$.

In that case the difference $P_1 - P_2$ projects on the orthogonal difference of the corresponding ranges

$$\text{ran } (P_1 - P_2) = \text{ran } P_1 \ominus \text{ran } P_2.$$

(This is the subspace of ran P_1 consisting of all vectors orthogonal to the whole ran P_2 or the orthocomplement of ran P_2 with respect to ran P_1).

Proof We shall first find an equivalent algebraic form of the statement that $P_1 - P_2$ is a projection operator:

$$[P_1 - P_2 \text{ is a projection operator}] \Leftrightarrow$$
$$\Leftrightarrow [I_U - (P_1 - P_2) = (I_U - P_1) + P_2 \text{ is a projection operator}] \Leftrightarrow$$
$$\Leftrightarrow [(I_U - P_1) \text{ and } P_2 \text{ must be orthogonal } (I_U - P_1)P_2 = P_2(I_U - P_1) = \hat{0}] \Leftrightarrow$$
$$\Leftrightarrow [P_1 P_2 = P_2 P_1 = P_2] \ (*)$$

But, $P_2(I_U - P_1) = \hat{0}$ is equivalent to ran $P_2 \perp$ ran $(I_U - P_1) = $ ran $P_2 \perp (\text{ran } P_1)^\perp$. Since ran P_2 is orthogonal to $(\text{ran } P_1)^\perp$, it is a subspace of ran P_1:

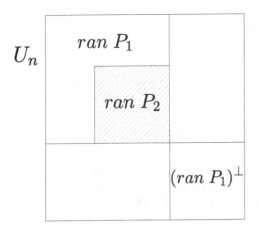

ran $P_2 \subseteq$ ran P_1.

Furthermore, since $P_1 - P_2 \overset{(*)}{=} P_1 - P_1 P_2 = P_1(I_U - P_2)$, this means that the product P_1 and $(I_U - P_2)$ projects on the intersection of ran P_1 and ran $(I_U - P_2) = U_n \ominus$ ran P_2 or

$$\text{ran } P_1 \cap (U_n \ominus \text{ran } P_2) \overset{\text{DISTRIBUTIVITY}}{=} \text{ran } P_1 \cap U_n \ominus \text{ran } P_1 \cap \text{ran } P_2 =$$
$$\text{ran } P_1 \ominus \text{ran } P_2. \ \triangle$$

5.5.5 The Spectral Form of a Normal Operator A

Going back to the theorem on the orthogonal decomposition of U_n into the eigenspaces of a normal operator:

$$E_{\lambda_1} \oplus E_{\lambda_2} \oplus \cdots \oplus E_{\lambda_k} = U_n,$$

we remember that the eigenspaces E_{λ_i}, $i = 1, 2, \ldots, k$, are uniquely determined as kernels $\ker (A - \lambda_i I_U)$, where λ_i are the distinct eigenvalues of A, that E_{λ_i} are mutually orthogonal subspaces and that in the orthogonal sum they give the whole U_n.

This theorem can be given another form. In the language of projection operators, it reads as

$$\sum_{i=1}^{k} P_{\lambda_i} = I_U \text{ with } P_{\lambda_i} P_{\lambda_j} = \hat{0} \text{ for } i, j = 1, 2, \ldots, k, \ i \neq j,$$

where P_{λ_i} is the projection operator on the eigenspace E_{λ_i} and is called the *eigen projection operator* corresponding to the eigenvalue λ_i.

This means that to every normal operator A there corresponds one and only one eigen decomposition of the unity.

For an arbitrary $\bar{x} \in U_n$, this decomposition gives $\bar{x} = \sum_{i=1}^{k} P_{\lambda_i} \bar{x}$, and applying A on both sides, we have

$$A\bar{x} = \sum_{i=1}^{k} A(P_{\lambda_i} \bar{x}) = \sum_{i=1}^{k} \lambda_i (P_{\lambda_i} \bar{x}) = (\sum_{i=1}^{k} \lambda_i P_{\lambda_i}) \bar{x}, \text{ since } P_{\lambda_i} \bar{x} \in E_{\lambda_i}.$$

Thus, we obtain the operator equality

$$\boxed{A = \sum_{i=1}^{k} \lambda_i P_{\lambda i}}$$

This is the *spectral form* of the normal operator A in terms of its distinct eigenvalues and the corresponding projection operators. This is the most characteristic formula for expressing the eigen problem of the normal operator A.

Now, we have achieved the main theorem in the theory of unitary spaces:

Theorem (on the spectral form of a normal operator A) To every normal operator A $(AA^\dagger = A^\dagger A)$ in U_n, there corresponds a unique set of complex numbers (the eigenvalue spectrum of A) $\{\lambda_1, \lambda_2, \ldots, \lambda_k\}$, $k \leq n$, and the corresponding set of eigen projection operators $\{P_{\lambda_1}, P_{\lambda_2}, \ldots, P_{\lambda_k}\}$, so that

1. The eigenvalues λ_i, $i = 1, 2, \ldots k$, are distinct from each other;
2. The projection operators P_{λ_i} are mutually orthogonal;
3. $\sum_{i=1}^{k} P_{\lambda_i} = I_U$—the eigen decomposition of the unity;
4. $A = \sum_{i=1}^{k} \lambda_i P_{\lambda_i}$—the spectral form of A.

The proof is given in the previous text. Δ

We shall conclude our discussion on normal operators in U_n by a theorem which is fundamental for simultaneous measurement of two commuting observables in Quantum Mechanics.

Theorem (on simultaneous diagonalization of two commuting normal operators) Two normal operators A and B in U_n have a common orthonormal eigenbasis if and only if they commute ($AB = BA$).

Proof If A and B commute, then every eigenspace E_{λ_i}, $i = 1, 2, \ldots, k$, of A is an invariant subspace for B (cf. Lemma 1, Sect. 5.3), i.e., B reduces to B_i by restricting the domain of B to E_{λ_i}. Of course, B_i is a normal operator in E_{λ_i}, and we can find an ON eigenbasis of B_i in E_{λ_i}. Collecting the eigenbases of all B_i, $i = 1, 2, \ldots, k$, we get a common ON eigenbasis in U_n for both A and B.

If, on the other hand, A and B have such a basis, then they are represented in that basis by diagonal matrices $\mathscr{D}^A$ and $\mathscr{D}^B$, which obviously commute, implying, through the isomorphism of algebras, that $AB = BA$. Δ

5.6 Diagonalization of a Symmetric Operator in $\mathbb{E}_3$

In Sect. 4.4, we have seen that a representing matrix $\mathscr{A}$ of an operator A in the Euclidean space $\mathbb{E}_3$ changes by the orthogonal similarity $\mathscr{A}' = \mathscr{R}\mathscr{A}\mathscr{R}^{-1} = \mathscr{R}\mathscr{A}\mathscr{R}^T$, when the ON basis $v = \{\bar{v}_1, \bar{v}_2, \bar{v}_3\}$, in which A was represented by $\mathscr{A}$, changes to the new ON basis $v' = \{\bar{v}'_1, \bar{v}'_2, \bar{v}'_3\}$ (in which A is represented by $\mathscr{A}'$) by the orthogonal replacement matrix $\mathscr{R}$ ($\mathscr{R}^{-1} = \mathscr{R}^T$):

$$\bar{v}'_i = \sum_{j=1}^{3} r_{ij}\bar{v}_j, \ i = 1, 2, 3, \ \text{or} \ \begin{bmatrix} \bar{v}'_1 \\ \bar{v}'_2 \\ \bar{v}'_3 \end{bmatrix} = \mathscr{R} \begin{bmatrix} \bar{v}_1 \\ \bar{v}_2 \\ \bar{v}_3 \end{bmatrix}.$$

When the matrix $\mathscr{R}$ goes through the whole orthogonal group $O(3)$, we get in this way all ON bases in $\mathbb{E}_3$ and the class $\{\mathscr{R}\mathscr{A}\mathscr{R}^{-1} | \mathscr{R} \in O(3)\}$ of all representing matrices of A in these ON bases. So, the matrix $\mathscr{A} = [a_{ij}]_{3\times3}$ and the ON basis $\{\bar{v}_1, \bar{v}_2, \bar{v}_3\}$ determine on one side the operator A ($A\bar{v}_j = \sum_{i=1}^{3} a_{ij}\bar{v}_i$, $i = 1, 2, 3$ or $A\bar{x} = \sum_{i,j=1}^{3} a_{ij}x_j\bar{v}_i$, $\bar{x} \in \mathbb{E}_3$), and on the other side the class $\{\mathscr{R}\mathscr{A}\mathscr{R}^{-1} | \mathscr{R} \in O(3)\}$. This class is in fact an orthogonal (Cartesian) tensor of the second order

$$\mathscr{A}' = \mathscr{R}\mathscr{A}\mathscr{R}^{-1} \Leftrightarrow a'_{km} = \sum_{i,j=1}^{3} r_{ki}r_{mj}a_{ij}, \; k,m = 1,2,3,$$

where $\mathscr{R}^{-1} = \mathscr{R}^T$ or $\mathscr{R} \in O(3)$.

An orthogonal tensor of the second order.

an ON basis in $\mathbb{E}_3$ $v = \{\bar{v}_1, \bar{v}_2, \bar{v}_3\}$ $\downarrow$ $\begin{bmatrix} \bar{v}'_1 \\ \bar{v}'_2 \\ \bar{v}'_3 \end{bmatrix} = \mathscr{R} \begin{bmatrix} \bar{v}_1 \\ \bar{v}_2 \\ \bar{v}_3 \end{bmatrix}$ $\mathscr{R} \in O(3)$	a bijection $\longrightarrow$	the real matrix representing the operator A in v $\mathscr{A} = [a_{ij}]_{3\times3};$ $a_{ij} = (\bar{v}_i, A\bar{v}_j), \; i,j=1,2,3$ $\downarrow$ $\mathscr{A}' = \mathscr{R}\mathscr{A}\mathscr{R}^T = \mathscr{R}\mathscr{A}\mathscr{R}^{-1}$ $\mathscr{R} \in O(3)$

The set of all ON bases in $\mathbb{E}_3$.   The set of representing matrices of A in all ON bases in $\mathbb{E}_3$.

This tensor is the bijective map that connects the set of all ON bases in $\mathbb{E}_3$ with the class of representing matrices of A, so that this map is determined by the pair $[\{\bar{v}_1, \bar{v}_2, \bar{v}_3\}, \mathscr{A}]$. Therefore, the operator $\mathscr{A}$ and its orthogonal tensor are in fact the same entity.

(It can be easily shown that the operator A will be given by the same formula $A\bar{x} = \sum_{i,j=1}^{3} a_{ij}x_j\bar{v}_i, \; \bar{x} \in \mathbb{E}_3$, if we use any other pair $[\{\bar{v}'_1, \bar{v}'_2, \bar{v}'_3\}, \mathscr{A}']$ from its tensor instead of $[\{\bar{v}_1, \bar{v}_2, \bar{v}_3\}, \mathscr{A}]$. Indeed,

$$A\bar{x} = A\sum_{m=1}^{3} x'_m\bar{v}'_m = \sum_{m=1}^{3} x'_m A\bar{v}'_m = \sum_{m=1}^{3} x'_m \sum_{k=1}^{3} a'_{km}\bar{v}'_k =$$

$$= \sum_{k,m=1}^{3} \left(\sum_{i,j=1}^{3} r_{ki}r_{mj}a_{ij} \right) \sum_{n=1}^{3} r_{mn}x_n \sum_{p=1}^{3} r_{kp}\bar{v}_p =$$

$$= \sum_{k,m,i,j,n,p=1}^{3} \underbrace{\{\mathscr{R}^T\}_{pk}\{\mathscr{R}\}_{ki}}_{\delta_{pi}}\underbrace{\{\mathscr{R}^T\}_{nm}\{\mathscr{R}\}_{mj}}_{\delta_{nj}} a_{ij}x_n\bar{v}_p = \sum_{i,j=1}^{3} a_{ij}x_j\bar{v}_i.)$$

Now we ask one of the most important questions in the theory of Euclidean spaces: which operators A in $\mathbb{E}_3$ have in their tensors a pair that consists of an ON eigenbasis of A and the corresponding diagonal matrix $\mathscr{D}$ with real eigenvalues of A on its diagonal?

Remember that in unitary spaces U_n the answer to a similar question was that these operators are the normal ones $(A^+A = AA^+)$, the only difference was that the diagonal elements were complex numbers, and the tensor was a unitary mixed tensor of the second order.

We shall answer this question by starting, as is quite natural, with the characteristic cubic equation of A and $\mathbb{E}_3$ (cf. the remark at the end of Sect. 5.2):

$$\det[\mathscr{A} - \lambda I_3] = -\lambda^3 + p_1\lambda^2 - p_2\lambda + p_3 = 0,$$

where $p_1 = \text{tr}\mathscr{A}$, $p_3 = \det\mathscr{A}$, and p_2 is the sum of the principal minors of the second order in any matrix $\mathscr{A}$ that represents A. (In other words, p_1, p_2, p_3 are the sums of the principal minors of the first, second, and third order in any matrix $\mathscr{A}$ that represents A).

It is worthwhile pointing out that p_1, p_2, p_3 are the three basic invariants of every orthogonal tensor of the second order—these are the combinations of elements in a representing matrix $\mathscr{A}$ that remain the same for all matrices of its class.

This cubic equation with real coefficients can have three real solutions or one real and two complex conjugate solutions (remember that the field $\mathbb{R}$ of real numbers is not algebraically closed—Sect. 5.1. the Fundamental Theorem of Algebra). Therefore, every linear operator in $\mathbb{E}_3$ has at least one one-dimensional eigenspace (an eigenline) which corresponds to that always existing one real solution of the characteristic cubic equation.

Let us investigate what corresponds geometrically to a possible pair of complex conjugate solutions $\lambda = \alpha + i\beta$ and $\lambda^* = \alpha - i\beta$. We have to get out of $\mathbb{E}_3$ (the so-called *complexification* of $\mathbb{E}_3$) and consider the homogeneous system of linear equations

$$[\mathscr{A} - \lambda I_3]\zeta = \bar{0}$$

where now the vector ζ of unknowns is a complex vector $\zeta = [\xi_1 + i\eta_1, \ \xi_2 + i\eta_2, \ \xi_3 + i\eta_3]^T$:

$$\mathscr{A}\zeta = \lambda\zeta, \text{ or in detail}$$

$$\begin{bmatrix} a_{11} & a_{12} & a_{13} \\ a_{21} & a_{22} & a_{23} \\ a_{31} & a_{32} & a_{33} \end{bmatrix} \begin{bmatrix} \xi_1 + i\eta_1 \\ \xi_2 + i\eta_2 \\ \xi_3 + i\eta_3 \end{bmatrix} = (\alpha + i\beta) \begin{bmatrix} \xi_1 + i\eta_1 \\ \xi_2 + i\eta_2 \\ \xi_3 + i\eta_3 \end{bmatrix}.$$

If we separate the real and imaginary parts of this linear system, and introduce two real vectors

$$\xi = [\xi_1\ \xi_2\ \xi_3]^T \text{ and } \eta = [\eta_1\ \eta_2\ \eta_3]^T,$$

we get immediately

$$\mathscr{A}\xi = \alpha\xi - \beta\eta \text{ and } \mathscr{A}\eta = \alpha\eta + \beta\xi.$$

Notice that we get the same result for the complex conjugate solution λ^*: $\mathscr{A}\zeta^* = \lambda^*\zeta^*$, since the matrix $\mathscr{A}$ is a real one.

We have obtained a two-dimensional subspace (a plane) $L(\xi, \eta)$ in $\mathbb{R}^3$ which is spanned by two linearly independent vectors ξ and η and which is an invariant subspace under $\mathscr{A}$. The vectors ξ and η are linearly independent, since if they were dependent, i.e., $\eta = c\xi$, that would mean

$$\mathscr{A}\xi = \alpha\xi - c\beta\xi = (\alpha - c\beta)\xi,$$

and $(\alpha - c\beta)$ would be a real eigenvalue of $\mathscr{A}$, contrary to the assumed existence of two complex conjugate solutions.

Due to the isomorphism between $\mathbb{E}_3$ and $\mathbb{R}^3$, which is established by the choice of the ON basis $\{\bar{v}_1, \bar{v}_2, \bar{v}_3\}$ in which the operator A is represented by the matrix $\mathscr{A}$, the analogous conclusion is valid for the operator A and the invariant plane $L(\bar{x}, \bar{y})$ which is spanned by linearly independent vectors

$$\bar{x} = \sum_{i=1}^{3} \xi_i \bar{v}_i \text{ and } \bar{y} = \sum_{i=1}^{3} \eta_i \bar{v}_i \text{ in } \mathbb{E}_3 :$$
$$A\bar{x} = \alpha\bar{x} - \beta\bar{y} \text{ and } A\bar{y} = \alpha\bar{y} + \beta\bar{x}.$$

Therefore, to the complex conjugate pair $\lambda = \alpha + i\beta$, $\lambda^* = \alpha - i\beta$ of solutions of the characteristic cubic equation of A there corresponds in $\mathbb{E}_3$ a two-dimensional invariant subspace (an invariant plane) $L(\bar{x}, \bar{y})$ of the operator A, which is not (for $\beta \neq 0$) an eigenspace for A. We shall need this result for the treatment of the canonical form of orthogonal operators (Sect. 5.7).

We can conclude that an operator A in $\mathbb{E}_3$ is diagonalizable (i.e., that there exists an eigenbasis in which A is represented by a diagonal matrix $\mathscr{D}$ with real elements) only if all three solutions of its characteristic cubic equation are real. This is a necessary, but not sufficient, condition:

$$\text{real spectrum} \overset{\not\Rightarrow}{\underset{\Leftarrow}{}} \text{operator } A \text{ diagonalizable,}$$

since one can easily find operators with a real spectrum, but without an eigenbasis (see Sect. 5.1, the first example $\mathscr{A} = \begin{bmatrix} 2 & 1 & 0 \\ -1 & 0 & 1 \\ 1 & 3 & 1 \end{bmatrix}$).

Theorem (on orthogonally diagonalizable operators in $\mathbb{E}_3$) The necessary and sufficient condition that a linear operator A in $\mathbb{E}_3$ has an orthonormal (ON) eigenbaisis, i.e., that it is orthogonally diagonalizable, is that this operator is symmetric

$$(A\bar{x}, \bar{y}) = (\bar{x}, A\bar{y}), \forall \bar{x}, \bar{y} \in \mathbb{E}_3.$$

Proof We shall first show that the matrices which represent such an operator in all ON bases are symmetric: $\mathscr{A}^T = \mathscr{A}$, i.e., symmetric with respect to the main diagonal. This is obvious since

$$a_{ij} = (\bar{v}_i, A\bar{v}_j) \overset{sym.}{=} (A\bar{v}_i, \bar{v}_j) = (v_j, A v_i) = a_{ji} \Leftrightarrow \mathscr{A}^T = \mathscr{A},$$

for any ON basis $\{\bar{v}_1, \bar{v}_2, \bar{v}_3\}$.

Next, we shall prove that all eigenvalues of a symmetric operator A are real, which is necessary for A to be diagonalizable. Suppose the opposite, that a symmetric operator A has a complex conjugate pair $\lambda = \alpha + i\beta$ and $\lambda^* = \alpha - i\beta$ as solutions of its characteristic equation. Then in $\mathbb{E}_3$ there exist linearly independent vectors $\bar{x}$ and $\bar{y}$ for which

$$A\bar{x} = \alpha\bar{x} - \beta\bar{y} \text{ and } A\bar{y} = \alpha\bar{y} + \beta\bar{x}.$$

Multiplying the first expression by $\bar{y}$ from the right, and the second one by $\bar{x}$ from the left, we get

$$(A\bar{x}, \bar{y}) = \alpha(\bar{x}, \bar{y}) - \beta(\bar{y}, \bar{y}) \text{ and } (\bar{x}, A\bar{y}) = \alpha(\bar{x}, \bar{y}) + \beta(\bar{x}, \bar{x}).$$

For the symmetric operator A the left hand sides are equal; so are the righthand ones, implying

$$\beta[(\bar{x}, \bar{x}) + (\bar{y}, \bar{y})] = 0.$$

Since both $\bar{x}$ and $\bar{y}$ are nonzero vectors (being linearly independent), this means $\beta = 0$, and the assumption about the complex conjugate pair of solutions is proven wrong. So, if the operator A is symmetric its eigenvalues are real.

We shall now prove the sufficiency of the condition: when an operator A is symmetric, it always has an ON eigenbasis in $\mathbb{E}_3$ (operator A symmetric $\Rightarrow$ there exists an ON eigenbasis). We start the proof by choosing a real eigenvalue λ_1 of A and find one eigenvector $\bar{x}_1$ which corresponds to λ_1. The line $L(\bar{x}_1)$ spanned by $\bar{x}_1$ is a one-dimensional eigenspace of A (it is equal to E_{λ_1} or it is a subspace of E_{λ_1} depending an whether the algebraic multiplicity of λ_1 is 1 or greater). But the two-dimensional orthocomplement of $L(\bar{x}_1)$ is always invariant under A: Let $\bar{y} \in L(\bar{x}_1)^\perp$, i.e., $(\bar{y}, \bar{x}_1) = 0$. We want to prove that $A\bar{y} \in L(\bar{x}_1)^\perp$, i.e., $(A\bar{y}, \bar{x}_1) = 0$. Indeed, $(A\bar{y}, \bar{x}_1) \overset{sym.}{=} (\bar{y}, A\bar{x}_1) = (\bar{y}, \lambda_1\bar{x}_1) = \lambda_1(\bar{y}, \bar{x}_1) = 0$. We can now consider the reduced operator A_1 obtained when we limit the domain of A to its invariant subspace $L(\bar{x}_1)^\perp$. Operator A_1 is of course symmetric, and it has in $L(\bar{x}_1)^\perp$ one real eigenvalue λ_2 and the corresponding one-dimensional eigenspace $L(\bar{x}_2)$, which is obviously orthogonal to $L(\bar{x}_1)$. The subspace $L(\bar{x}_1)^\perp \ominus L(\bar{x}_2)$ [the orthocomplement of $L(\bar{x}_2)$ with respect to $L(\bar{x}_1)^\perp$] is also invariant under A.

[If $\bar{z} \in L(\bar{x}_1)^\perp \ominus L(\bar{x}_2)$, this means $(\bar{z}, \bar{x}_1) = 0$ and $(\bar{z}, \bar{x}_2) = 0$. *Then* $(A\bar{z}, \bar{x}_1) \overset{sym.}{=} (\bar{z}, A\bar{x}_1) = (\bar{z}, \lambda_1\bar{x}_1) = \lambda_1(\bar{z}, \bar{x}_1) = 0$ and $(A\bar{z}, \bar{x}_2) \overset{sym.}{=} (\bar{z}, A\bar{x}_2) = (\bar{z}, \lambda_2\bar{x}_2) = \lambda_2(\bar{z}, \bar{x}_2) = 0$, so that $A\bar{z} \in L(\bar{x}_1)^\perp \ominus L(\bar{x}_2)$].

But, $L(\bar{x}_1)^\perp \ominus L(\bar{x}_2)$ is necessarily a one-dimensional subspace, and as such, it is an eigenline spanned by an eigenvector $\bar{x}_3$ which is orthogonal to both $\bar{x}_1$ and $\bar{x}_2$.

[A one-dimensional subspace W in $\mathbb{E}_3$ is always spanned by a nonzero vector, say $\bar{x}$:

$$W = L(\bar{x}) = \{\alpha\bar{x} | \alpha \in \mathbb{R}\}.$$

To be invariant under A, it must be $A(\alpha\bar{x}) \in W$, $\forall \alpha \in \mathbb{R}$. For this, it is necessary and sufficient that $A\bar{x} \in W$ $(A(\alpha\bar{x}) \in W, \forall \alpha \in \mathbb{R} \Leftrightarrow A\bar{x} \in W)$. But, $A\bar{x} \in W$ means that $A\bar{x}$ is a product of $\bar{x}$ with some real constant λ, i.e., $A\bar{x} = \lambda\bar{x}$, so that $\bar{x}$ is an eigenvector of A and W an eigenspace (an eigenline).]

So, we have an orthogonal decomposition of $\mathbb{E}_3$:

$$\mathbb{E}_3 = L(\bar{x}_1) \oplus L(\bar{x}_2) \oplus L(\bar{x}_3).$$

We can now normalize the eigenvectors $\bar{x}_1$, $\bar{x}_2$, $\bar{x}_3$, and so doing get an orthonormal eigenbasis in $\mathbb{E}_3$ in which the operator A is represented by a real diagonal matrix $\mathscr{D}$. Thus, we have proved the sufficient condition: A is a symmetric operator $\Rightarrow$ there exists an ON eigenbasis (A is orthogonally diagonalizable).

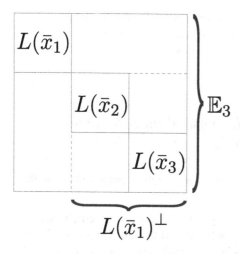

The implication in the inverse sense (i.e., the necessary condition) is rather trivial: if $\mathscr{D}$ is a representing diagonal matrix of an operator A in the given ON eigenbasis, then it is obviously a symmetric matrix ($\mathscr{D}^T = \mathscr{D}$), which means that A is also a symmetric operator $(A\bar{x}, \bar{y}) = (\bar{x}, A\bar{y})$, $\forall \bar{x}, \bar{y} \in \mathbb{E}_3$.
[An operator is symmetric if and only if its representing matrix in any ON basis is symmetric:

$$(A\bar{x}, \bar{y}) = (\bar{x}, A\bar{y}), \ \forall \bar{x}, \bar{y} \in \mathbb{E}_3 \Leftrightarrow$$
$$\Leftrightarrow (A\bar{v}_i, \bar{v}_j) = (\bar{v}_i, A\bar{v}_j), \ i, j = 1, 2, 3, \text{ and}$$
$$\{\bar{v}_1, \bar{v}_2, \bar{v}_3\} \text{ in an ON basis in } \mathbb{E}_3.$$

The above $\Rightarrow$ implication we discussed at the beginning of the proof, but now we need the $\Leftarrow$ implication:

$$(A\bar{x}, \bar{y}) = (A \sum_{i=1}^{3} \xi_i \bar{v}_i), \ \sum_{j=1}^{3} \eta_j \bar{v}_j) =$$
$$= \sum_{i,j=1}^{3} \xi_i \eta_j (A\bar{v}_i, \bar{v}_j)$$

and similarly

$$(\bar{x}, A\bar{y}) = \sum_{i,j=1}^{3} \xi_i \eta_j (\bar{v}_i, A\bar{v}_j),$$

so that

$$(A\bar{v}_i, \bar{v}_j) = (\bar{v}_i, A\bar{v}_j)$$

implies

$$(A\bar{x}, \bar{y}) = (\bar{x}, A\bar{y}).] \ \Delta$$

Note that the above theorem deals with orthonormal eigenbases in $\mathbb{E}_3$. This suggests that there are 3×3 real matrices in R_3 (see Sect. 5.1, the second example) which possesses eigenbases which are not ON, because the matrices are not symmetric.

5.6.1 The Actual Procedure for Orthogonal Diagonalization of a Symmetric Operator in $\mathbb{E}_3$

This procedure is completely analogous to that for a normal operator in U_n, so that there is no need to repeat the proofs.

(i) We represent the symmetric operator A given in $\mathbb{E}_3$ by a symmetric matrix $\mathscr{A}$ in $\mathbb{R}^3$ in some ON basis $\{\bar{v}_1, \bar{v}_2, \bar{v}_3\}$ in $\mathbb{E}_3$. But, if the initial space is $\mathbb{R}^3$, then the operator is from the start given as the symmetric matrix $\mathscr{A}$ which represents itself in the standard ON basis $\{e_1, e_2, e_3\}$ in $\mathbb{R}^3$.

(ii) Then, we solve the eigen problem for the matrix $\mathscr{A}$: we find different real eigenvalues λ_i, $i = 1, \ldots, k \leq 3$ with the algebraic multiplicities n_i ($\sum_{i=1}^{k} n_i = 3$) by solving the characteristic equation $\det[\mathscr{A} - \lambda I_3] = 0$ of $\mathscr{A}$. Furthermore, for each λ_i, we calculate n_i eigenvectors matrix-columns (remember that the geometric multiplicity must be equal to the algebraic one for each λ_i) by solving for nontrivial solutions the corresponding homogeneous linear system $\mathscr{A}\xi = \lambda_i \xi$. When some n_i are greater than 1, we have to use the Gram–Schmidt procedure to get an ON basis in the corresponding eigenspace. When $n_i = 1$, we need only normalize the corresponding eigenvector. Remember that eigenspaces for different λ_i are automatically orthogonal.

Proof Let $\mathscr{A}$ be a 3×3 symmetric matrix, and let $\bar{v}_1$ and $\bar{v}_2$ be two eigenvectors corresponding to distinct eigenvalues λ_1 and λ_2. We want to show that $\bar{v}_1$ and $\bar{v}_2$ are orthogonal. We shall first prove $\lambda_1(\bar{v}_1, \bar{v}_2) = \lambda_2(\bar{v}_1, \bar{v}_2)$. So $\lambda_1(\bar{v}_1, \bar{v}_2) = (\lambda_1 \bar{v}_1, \bar{v}_2) = (\mathscr{A}\bar{v}_1, \bar{v}_2) = (\mathscr{A}\bar{v}_1)^T \bar{v}_2 = \bar{v}_1^T \mathscr{A}^T \bar{v}_2$. On the other hand, $\lambda_2(\bar{v}_1, \bar{v}_2) = (\bar{v}_1, \lambda_2 \bar{v}_2) = (\bar{v}_1, \mathscr{A}\bar{v}_2) = \bar{v}_1^T \mathscr{A} \bar{v}_2$. Taking into account that $\mathscr{A}^T = \mathscr{A}$, we prove the above statement, which can be written as $(\lambda_1 - \lambda_2)(\bar{v}_1, \bar{v}_2) = 0 \Rightarrow (\bar{v}_1, \bar{v}_2) = 0$ since $(\lambda_1 - \lambda_2) \neq 0$. Δ

Thus, we have obtained an ON eigenbasis of $\mathscr{A}$ in $\mathbb{R}^3$: $\{\bar{c}_1, \bar{c}_3, \bar{c}_3\}$.

(iii) We form a 3×3 real matrix $\mathscr{T}$ whose columns are the ON eigenvectors $\{\bar{c}_1, \bar{c}_3, \bar{c}_3\}$:

$$\mathscr{T} = \begin{bmatrix} | & | & | \\ c_1 & c_2 & c_3 \\ | & | & | \end{bmatrix} = \begin{bmatrix} c_{11} & c_{12} & c_{13} \\ c_{21} & c_{22} & c_{23} \\ c_{31} & c_{32} & c_{33} \end{bmatrix}.$$

This new matrix $\mathscr{T}$ is orthogonal $\mathscr{T}^{-1} = \mathscr{T}^T$ since its columns are orthonormal: $(c_i, c_j) = \delta_{ij}$, $i, j = 1, 2, 3 \Leftrightarrow \mathscr{T}^T \mathscr{T} = I_3 \Leftrightarrow \mathscr{T}^{-1} = \mathscr{T}^T$. The orthogonal matrix $\mathscr{T}$ maps $\mathscr{A}$ by similarity transformation into the diagonal matrix $\mathscr{D}$ which has real λ_i on its diagonal (each eigenvalue appearing n_i times):

$$\mathscr{D} = \mathscr{T}^{-1} \mathscr{A} \mathscr{T} = \mathscr{T}^T \mathscr{A} \mathscr{T}.$$

This demonstrates that the real 3×3 symmetric matrix $\mathscr{A}$ is orthogonally diagonalizable, i.e., by orthogonal similarity.

The orthogonal matrix $\mathscr{T}$ is the transition matrix from the standard ON basis $\{e_1, e_2, e_3\}$ in $\mathbb{R}^3$ to the ON eigenbasis $\{c_1, c_2, c_3\}$ of $\mathscr{A}$ in which $\mathscr{A}$ is represented by $\mathscr{D}$: $\mathscr{T} e_i = c_i$, $i = 1, 2, 3$.

Now is a very good moment to show in the most simple way that the matrix which is orthogonally diagonalizable must be symmetric (again the necessary condition from our theorem):

If $\mathscr{D} = \mathscr{T}^T \mathscr{A} \mathscr{T}$, where $\mathscr{D}$ is a diagonal matrix, and $\mathscr{T}$ is an orthogonal matrix $(\mathscr{T} \mathscr{T}^T = I_3)$, then $(\mathscr{T} \mathscr{T}^T) \mathscr{A} (\mathscr{T} \mathscr{T}^T) = \mathscr{T} \mathscr{D} \mathscr{T}^T$ or $\mathscr{A} = \mathscr{T} \mathscr{D} \mathscr{T}^T$. Transposing both sides, we get $\mathscr{A}^T = (\mathscr{T}^T)^T \mathscr{D}^T \mathscr{T}^T = \mathscr{T} \mathscr{D} \mathscr{T}^T$, so that $\mathscr{A}^T = \mathscr{A}$.

(iv) The ON eigenbasis $\{\vec{v}_1', \vec{v}_2', \vec{v}_3'\}$ in which the initial operator A in $\mathbb{E}_3$ is represented by the diagonal matrix $\mathscr{D}$ is obtained from the ON basis $\{\bar{v}_1, \bar{v}_2, \bar{v}_3\}$ in which A was represented by $\mathscr{A}$ by making use of the orthogonal replacement matrix $\mathscr{R} = \mathscr{T}^T$:

$$\vec{v}_i' = \sum_{j=1}^{3} r_{ij} \bar{v}_j, \quad i = 1, 2, 3.$$

Remark If one has studied the eigen problem in the unitary space U_n (Sects. 5.3, 5.4, and 5.5), one can use that knowledge to formulate the orthogonal diagonalization of symmetric matrices in $\mathbb{R}^3$. Obviously, every real $(\mathscr{A}^* = \mathscr{A})$ symmetric $(\mathscr{A}^T = \mathscr{A})$ matrix is also Hermitian $[\mathscr{A}^+ = (\mathscr{A}^*)^T = \mathscr{A}]$. We use property (4) in the first theorem in Sect. 5.5.2 which says that all the eigenvalues of a Hermitian matrix are real numbers, so that the eigen spectrum of a symmetric matrix consists only of real numbers. Hermitian matrices as a subset of normal $(\mathscr{A}^+ \mathscr{A} = \mathscr{A} \mathscr{A}^+)$ ones have three real eigenvalues, counting them with their algebraic multiplicity. Now the eigenvectors of a real symmetric $\mathscr{A}$ can be computed by GJM procedure applied to $\mathscr{A} - \lambda_i I_3$, where the λ_i are the three eigenvalues of $\mathscr{A}$. Because $\mathscr{A}$ and λ_i are real, the GJM reduced echelon form will be a real matrix also. The dimension of each kernel (the geometric multiplicity of λ_i) must be equal to its algebraic multiplicity. The eigenspaces E_{λ_i} for different λ_i are orthogonal (Sect. 5.4), and we only have to apply Gram–Schmidt orthonormalization procedure in each E_{λ_i}. Thus, we get a real orthonormal eigenbasis of $\mathscr{A}$ in $\mathbb{R}^3$. The transition matrix $\mathscr{T}$ whose columns are these ON eigenvectors is a real orthogonal 3×3 matrix $(\mathscr{T}^{-1} = \mathscr{T}^T)$ which diagonalizes $\mathscr{A}$ by orthogonal similarity:

$$\mathscr{T}^{-1} \mathscr{A} \mathscr{T} = \mathscr{T}^T \mathscr{A} \mathscr{T} = \mathscr{D}.$$

The diagonal matrix $\mathscr{D}$ has the real eigenvalues of $\mathscr{A}$ on its diagonal. It should be emphasized once again that real symmetric matrices are the only linear operators in $\mathbb{R}^3$ that can be orthogonally diagonalized.

Example Perform the orthogonal diagonalization of the real symmetric 3×3 matrix

$$\mathscr{A} = \begin{bmatrix} 4 & 2 & 2 \\ 2 & 4 & 2 \\ 2 & 2 & 4 \end{bmatrix}.$$

(1) the characteristic cubic equation for the matrix $\mathscr{A}$ is

$$\det[\mathscr{A} - \lambda I_3] = \begin{vmatrix} 4-\lambda & 2 & 2 \\ 2 & 4-\lambda & 2 \\ 2 & 2 & 4-\lambda \end{vmatrix} =$$

$$= -\lambda^3 + (4+4+4)\lambda^2 - (12+12+12)$$
$$+(64+8+8-16-16-16) = -\lambda^3 + 12\lambda^2 - 36\lambda + 32 = 0.$$

Since the free term 32 is 2^5 and we know that it is the product of three real eigenvalues, we can conclude that there are only two possibilities: $2^1, 2^1, 2^3$ or $2^1, 2^2, 2^2$, for the spectrum (also any two can be negative). Since 2 is necessarily one eigenvalue, we shall divide the characteristic polynomial by $\lambda - 2$:

$$(\lambda^3 - 12\lambda^2 + 36\lambda - 32) : (\lambda - 2) = \lambda^2 - 10\lambda + 16$$
$$\underline{\lambda^3 + 2\lambda^2}$$
$$-10\lambda^2 + 36\lambda$$
$$\underline{+10\lambda^2 - 20\lambda}$$
$$+16\lambda - 32$$
$$\underline{-16\lambda + 32}$$
$$0$$

So $\lambda_1 = 2$, and the other two eigenvalues are the solutions of the quadratic equation

$$\lambda^2 - 10\lambda + 16 = 0 \text{ or}$$

$$\lambda_{2,3} = \frac{10 \pm \sqrt{100 - 64}}{2} = \frac{10 \pm 6}{2} \text{ or } \lambda_2 = 2 \text{ and } \lambda_3 = 8.$$

Solving the homogeneous linear system $\mathscr{A}\,\xi = \lambda\xi$ or $(\mathscr{A} - \lambda I_3)\,\xi = \bar{0}_3$ for the double eigenvalue $\lambda = 2$, we get the augmented matrix

$$\begin{bmatrix} 2 & 2 & 2 & | & 0 \\ 2 & 2 & 2 & | & 0 \\ 2 & 2 & 2 & | & 0 \end{bmatrix}$$

and the Gauss–Jordan modified reduced echelon form can be obtained immediately:

$$\begin{bmatrix} 1 & 1 & 1 & | & 0 \\ 0 & -1 & 0 & | & 0 \\ 0 & 0 & -1 & | & 0 \end{bmatrix}.$$

The two basis vectors of the kernel of $\mathscr{A} - 2I_3$ are

$$\bar{v}_1 = \begin{bmatrix} 1 \\ -1 \\ 0 \end{bmatrix} \text{ and } \bar{v}_2 = \begin{bmatrix} 1 \\ 0 \\ -1 \end{bmatrix}.$$

They are obviously linearly independent, but we have to apply Gram–Schmidt procedure to get the ON basis in E_2. The first vector in the ON basis is the normalized $\bar{v}_1$:

$$\bar{c}_1 = \bar{v}_1 / \|\bar{v}_1\| = \frac{1}{\sqrt{2}} \begin{bmatrix} 1 \\ -1 \\ 0 \end{bmatrix}.$$

The second vector is obtained as the normalized normal from $\bar{v}_2$ on $\bar{c}_1$:

$$\bar{c}_2 = [\bar{v}_2 - (\bar{c}_1, \bar{v}_2)\,\bar{c}_1] / \|\bar{v}_2 - (\bar{c}_1, \bar{v}_2)\,\bar{c}_1\| =$$

$$\left(\begin{bmatrix} 1 \\ 0 \\ -1 \end{bmatrix} - \frac{1}{2} \times 1 \times \begin{bmatrix} 1 \\ -1 \\ 0 \end{bmatrix} \right) / \left\| \begin{bmatrix} \frac{1}{2} \\ \frac{1}{2} \\ -1 \end{bmatrix} \right\| = \sqrt{\frac{2}{3}} \begin{bmatrix} \frac{1}{2} \\ \frac{1}{2} \\ -1 \end{bmatrix} = \frac{1}{\sqrt{6}} \begin{bmatrix} 1 \\ 1 \\ -2 \end{bmatrix}.$$

The only basis vector in the eigenspace E_8 is obtained by solving the homogeneous linear system $(\mathscr{A} - 8I_3)\xi = \bar{0}_3$ for a nontrivial solution. The corresponding augmented matrix

$$\begin{bmatrix} -4 & 2 & 2 & | & 0 \\ 2 & -4 & 2 & | & 0 \\ 2 & 2 & -4 & | & 0 \end{bmatrix} \text{ is GJM reduced to}$$

$$\begin{bmatrix} -4 & 2 & 2 & | & 0 \\ 2 & -4 & 2 & | & 0 \\ 2 & 2 & -4 & | & 0 \end{bmatrix} \sim \begin{bmatrix} -2 & 1 & 1 & | & 0 \\ 2 & -4 & 2 & | & 0 \\ 2 & 2 & -4 & | & 0 \end{bmatrix} \sim \begin{bmatrix} -2 & 1 & 1 & | & 0 \\ 0 & -3 & 3 & | & 0 \\ 0 & 3 & -3 & | & 0 \end{bmatrix} \sim$$

$$\sim \begin{bmatrix} -2 & 1 & 1 & | & 0 \\ 0 & -3 & 3 & | & 0 \\ 0 & 0 & 0 & | & 0 \end{bmatrix} \sim \begin{bmatrix} 1 & -\frac{1}{2} & -\frac{1}{2} & | & 0 \\ 0 & 1 & -1 & | & 0 \\ 0 & 0 & 0 & | & 0 \end{bmatrix} \sim$$

$$\sim \begin{bmatrix} 1 & 0 & -1 & | & 0 \\ 0 & 1 & -1 & | & 0 \\ 0 & 0 & 0 & | & 0 \end{bmatrix} \begin{matrix} Gauss-Jordan \\ modification \end{matrix} \begin{bmatrix} 1 & 0 & -1 & | & 0 \\ 0 & 1 & -1 & | & 0 \\ 0 & 0 & -1 & | & 0 \end{bmatrix}.$$

So that $\bar{v}_3 = \begin{bmatrix} -1 \\ -1 \\ -1 \end{bmatrix}$, and $\bar{c}_3 = \frac{1}{\sqrt{3}} \begin{bmatrix} 1 \\ 1 \\ 1 \end{bmatrix}$.

The orthogonal transition matrix

$$\mathcal{T} = \begin{bmatrix} | & | & | \\ c_1 & c_2 & c_2 \\ | & | & | \end{bmatrix} = \frac{1}{\sqrt{6}} \begin{bmatrix} \sqrt{3} & 1 & \sqrt{2} \\ -\sqrt{3} & 1 & \sqrt{2} \\ 0 & -2 & \sqrt{2} \end{bmatrix}.$$

Finally, diagonalize $\mathscr{A}$ by orthogonal similarity $\mathscr{D} = \mathcal{T}^T \mathscr{A} \mathcal{T}$:

$$\frac{1}{6} \begin{bmatrix} \sqrt{3} & -\sqrt{3} & 0 \\ 1 & 1 & -2 \\ \sqrt{2} & \sqrt{2} & \sqrt{2} \end{bmatrix} \begin{bmatrix} 4 & 2 & 2 \\ 2 & 4 & 2 \\ 2 & 2 & 4 \end{bmatrix} \begin{bmatrix} \sqrt{3} & 1 & \sqrt{2} \\ -\sqrt{3} & 1 & \sqrt{2} \\ 0 & -2 & \sqrt{2} \end{bmatrix} =$$

$$= \frac{1}{6} \begin{bmatrix} 2\sqrt{3} & -2\sqrt{3} & 0 \\ 2 & 2 & -4 \\ 8\sqrt{2} & 8\sqrt{2} & 8\sqrt{2} \end{bmatrix} \begin{bmatrix} \sqrt{3} & 1 & \sqrt{2} \\ -\sqrt{3} & 1 & \sqrt{2} \\ 0 & -2 & \sqrt{2} \end{bmatrix} =$$

$$= \frac{2}{6} \begin{bmatrix} \sqrt{3} & -\sqrt{3} & 0 \\ 1 & 1 & -2 \\ 4\sqrt{2} & 4\sqrt{2} & 4\sqrt{2} \end{bmatrix} \begin{bmatrix} \sqrt{3} & 1 & \sqrt{2} \\ -\sqrt{3} & 1 & \sqrt{2} \\ 0 & -2 & \sqrt{2} \end{bmatrix} =$$

$$= \frac{1}{3} \begin{bmatrix} 6 & 0 & 0 \\ 0 & 6 & 0 \\ 0 & 0 & 24 \end{bmatrix} = \begin{bmatrix} 2 & 0 & 0 \\ 0 & 2 & 0 \\ 0 & 0 & 8 \end{bmatrix} = \mathscr{D}. \ \Delta$$

5.6.2 Diagonalization of Quadratic Forms

A quadratic form in two real variables x and y is a sum of squares of the variables and products of the two variables:

$$ax^2 + bxy + byx + cy^2 = ax^2 + 2bxy + cy^2.$$

This form can be written as a matrix product:

$$[x \ y] \begin{bmatrix} a & b \\ b & c \end{bmatrix} \begin{bmatrix} x \\ y \end{bmatrix} = [ax + by \ bx + cy] \begin{bmatrix} x \\ y \end{bmatrix} =$$

$$= (ax + by)x + (bx + cy)y = ax^2 + 2bxy + cy^2.$$

It is important to notice that the matrix $\begin{bmatrix} a & b \\ b & c \end{bmatrix}$ is symmetric.

Definition A quadratic form in the n real variables $\{x_1, x_2, \ldots, x_n\}$ is an expression

that can be written as $\begin{bmatrix} x_1 & x_2 & \ldots & x_n \end{bmatrix} \mathscr{A} \begin{bmatrix} x_1 \\ x_2 \\ \vdots \\ x_n \end{bmatrix}$, where $\mathscr{A}$ is a real symmetric $n \times n$

matrix.

If we introduce the n-vector $\mathscr{X} = \begin{bmatrix} x_1 \\ x_2 \\ \vdots \\ x_n \end{bmatrix}$, then this form is a compact matrix product

$$\mathscr{X}^T \mathscr{A} \mathscr{X} = \sum_{i=1}^{n} a_{ii} x_i^2 + \sum_{i \neq j=1}^{n} a_{ij} x_i x_j, \ a_{ij} = a_{ji}.$$

Example In the three variables $\{x, y, z\}$, a quadratic form is

$$\begin{bmatrix} x & y & z \end{bmatrix} \begin{bmatrix} a & d & e \\ d & b & f \\ e & f & c \end{bmatrix} \begin{bmatrix} x \\ y \\ z \end{bmatrix} = ax^2 + by^2 + cz^2 + dxy + dyx + exz +$$

$$+ ezx + fyz + fzy = ax^2 + by^2 + cz^2 + 2dxy + 2exz + 2fyz. \ \triangle$$

Quadratic forms have numerous applications in geometry (in particular in analytic geometry), in vibrations of mechanical systems (elastic bodies), in statistics, in electrical engineering (electric circuits), and in quantum mechanics.

For this reason, it is important that we know how to diagonalize them, which means how to remove the cross-product terms by changing variables.

The procedure for orthogonal diagonalization of a symmetric matrix $\mathscr{A}$ in $\mathbb{R}^3$ (described in Sect. 5.6.1) can be immediately generalized to $\mathbb{R}^n$, so we have:

Theorem (on diagonalization of quadratic forms in $\mathbb{R}^n$) Let a quadratic form in $\mathbb{R}^n$ be given as

$$\mathscr{X}^T \mathscr{A} \mathscr{X} = \begin{bmatrix} x_1 & x_2 & \ldots & x_n \end{bmatrix} \begin{bmatrix} a_{11} & a_{12} & \cdots & a_{1n} \\ a_{12} & a_{22} & \cdots & a_{2n} \\ \vdots & \vdots & \ddots & \vdots \\ a_{1n} & a_{2n} & \cdots & a_{nn} \end{bmatrix} \begin{bmatrix} x_1 \\ x_2 \\ \vdots \\ x_n \end{bmatrix},$$

where $\mathscr{A}$ is obviously a symmetric matrix: $\mathscr{A}^T = \mathscr{A}$ or $a_{ij} = a_{ji}$, $i \neq j$, $i, j = 1, 2, \ldots, n$.

Then there exists an orthogonal matrix $\mathscr{T}$ whose columns are the ON vectors of the eigenbasis of $\mathscr{A}$, which diagonalizes A by orthogonal similarity

$$\mathscr{T}^{-1}\mathscr{A}\mathscr{T} = \mathscr{T}^T\mathscr{A}\mathscr{T} = \mathscr{D} = \begin{bmatrix} \lambda_1 & 0 & \cdots & 0 \\ 0 & \lambda_2 & \cdots & 0 \\ \vdots & \vdots & \ddots & \vdots \\ 0 & 0 & \cdots & \lambda_n \end{bmatrix},$$

where $\lambda_1, \lambda_2, \ldots, \lambda_n$, are the corresponding real eigenvalues of $\mathscr{A}$. If we make the substitution $\mathscr{X} = \mathscr{T}\mathscr{Y}$, $\mathscr{Y} = \begin{bmatrix} y_1 & y_2 & \cdots & y_n \end{bmatrix}^T$, then

$$\mathscr{X}^T\mathscr{A}\mathscr{X} = (\mathscr{T}\mathscr{Y})^T\mathscr{A}\mathscr{T}\mathscr{Y} = \mathscr{Y}^T(\mathscr{T}^T\mathscr{A}\mathscr{T})\mathscr{Y} = \mathscr{Y}^T\mathscr{D}\mathscr{Y} =$$

$$= \begin{bmatrix} y_1 & y_2 & \cdots & y_n \end{bmatrix} \begin{bmatrix} \lambda_1 & 0 & \cdots & 0 \\ 0 & \lambda_2 & \cdots & 0 \\ \vdots & \vdots & \ddots & \vdots \\ 0 & 0 & \cdots & \lambda_n \end{bmatrix} \begin{bmatrix} y_1 \\ y_2 \\ \vdots \\ y_n \end{bmatrix} = \lambda_1 y_1^2 + \lambda_2 y_2^2 + \cdots + \lambda_n y_n^2,$$

which is a quadratic form without cross-product terms. So, the orthogonal matrix $\mathscr{T}$ reduces the quadratic form $\mathscr{X}^T\mathscr{A}\mathscr{X}$ to a sum of squares of the new variables $\{y_1, y_2, \ldots, y_n\}$.

The proof is obvious. $\quad\Delta$

Example Find a change of variables that reduces the given quadratic form to a sum of squares.

$$3x^2 + 4y^2 + 5z^2 + 4xy - 4yz \Rightarrow \mathscr{A} = \begin{bmatrix} 3 & 2 & 0 \\ 2 & 4 & -2 \\ 0 & -2 & 5 \end{bmatrix} \Rightarrow$$

$$\det[\mathscr{A} - \lambda I_3] = -\lambda^3 + 12\lambda^2 - 39\lambda + 28 = 0 \Rightarrow$$

$$\lambda_1 \lambda_2 \lambda_3 = 28, \quad \lambda_1 + \lambda_2 + \lambda_3 = 12 \Rightarrow$$

$$\lambda_1 = 7, \lambda_2 = 4, \lambda_3 = 1, \bar{v}_1 = \frac{1}{3}\begin{bmatrix} 1 \\ 2 \\ -2 \end{bmatrix}, \bar{v}_2 = \frac{1}{3}\begin{bmatrix} 2 \\ 1 \\ 2 \end{bmatrix}, \bar{v}_3 = \frac{1}{3}\begin{bmatrix} 2 \\ -2 \\ -1 \end{bmatrix} \Rightarrow$$

$$\mathscr{T} = \frac{1}{3}\begin{bmatrix} 1 & 2 & 2 \\ 2 & 1 & -2 \\ -2 & 2 & -1 \end{bmatrix}, \quad \mathscr{T}^T\mathscr{A}\mathscr{T} = \begin{bmatrix} 7 & 0 & 0 \\ 0 & 4 & 0 \\ 0 & 0 & 1 \end{bmatrix}, \quad \begin{bmatrix} x \\ y \\ z \end{bmatrix} = \mathscr{T}\begin{bmatrix} x' \\ y' \\ z' \end{bmatrix},$$

$$7x'^2 + 4y'^2 + z'^2.$$

5.6.3 Conic Sections in $\mathbb{R}^2$

The most important conic sections (conics) in $\mathbb{R}^2$ are ellipses, hyperbolas, and parabolas. They are nondegenerate conics. There are also degenerate conics which are single points (degenerate ellipses), two intersecting lines (degenerate hyperbolas) and two parallel lines (degenerate parabolas).

A nondegenerate conic is in standard position relative to the coordinate axes if its equation can be expressed as

1. $\frac{x^2}{a^2} + \frac{y^2}{b^2} = 1$, $a, b > 0$ for ellipses $(a > b, b > a)$ and circles $(a = b)$.

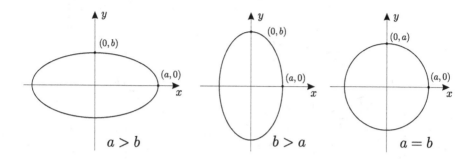

2. $\frac{x^2}{a^2} - \frac{y^2}{b^2} = 1$, $a, b > 0$ and $-\frac{x^2}{a^2} + \frac{y^2}{b^2} = 1$, $a, b > 0$ for hyperbolas.

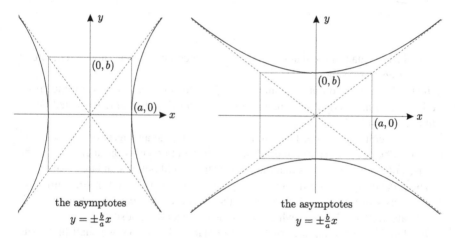

3. $y^2 = ax$ $(a > 0$ and $a < 0)$ and $x^2 = ay$ $(a > 0$ and $a < 0)$ for parabolas.

An example of a degenerate ellipse [a single point $(0,0)$] is $\frac{x^2}{a^2} + \frac{y^2}{b^2} = 0$, an example of a degenerate hyperbola (two intersecting lines $y = \pm\frac{b}{a}x$—asymptotes) is $\frac{x^2}{a^2} - \frac{y^2}{b^2} = 0$, and examples of degenerate parabolas (two parallel lines—horizontal or vertical) are $y^2 = a$ and $x^2 = a$.

We shall now turn our attention to the general quadratic equation in x and y

$$ax^2 + 2bxy + cy^2 + dx + ey + f = 0,$$

where $a, b, \ldots, f$, are real numbers, and at least one of the numbers a, b, c, is not zero. We call $ax^2 + 2bxy + cy^2$ *the associated quadratic form*. We shall find out

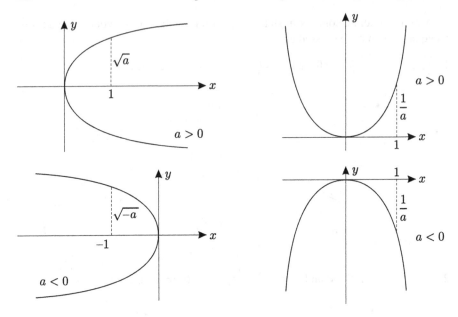

that graphs of quadratic equations in x and y are conic sections (nondegenerate or possibly degenerate ones).

It should be noticed that no conic in standard position has an xy-term (the cross-product term) in its equation. Such a term indicates that the conic is rotated out of standard position.

Also, no conic in standard position has both an x^2 and an x term or both a y^2 and a y term. The occurrence of either of these pairs (and no cross-product term) in the equation of a conic indicates that the conic is translated out of standard position.

The procedure for identifying the graph of a conic (that is not in standard position) consists in rotating and translating the xy-coordinate axes to obtain an $x''y''$-coordinate system relative to which the conic is in standard position.

To identify conics that are rotated out of standard position, we shall first write their quadratic equations

$$ax^2 + 2bxy + cy^2 + dx + ey + f = 0$$

in the matrix form

$$[x\ y]\begin{bmatrix} a & b \\ b & c \end{bmatrix}\begin{bmatrix} x \\ y \end{bmatrix} + [d\ e]\begin{bmatrix} x \\ y \end{bmatrix} + f = 0 \text{ or}$$

$$\mathscr{X}^T \mathscr{A} \mathscr{X} + \mathscr{K} \mathscr{X} + f = 0 \text{ where}$$

$$\mathscr{X} = \begin{bmatrix} x \\ y \end{bmatrix}, \ \mathscr{A} = \begin{bmatrix} a & b \\ b & c \end{bmatrix}, \ \mathscr{K} = [d\ e].$$

To eliminate the cross-product term $2bxy$, we shall find the 2×2 orthogonal matrix $\mathscr{T}$ that orthogonally diagonalizes the associated quadratic form $\mathscr{X}^T \mathscr{A} \mathscr{X}$ and makes the coordinate substitution

$$\begin{bmatrix} x \\ y \end{bmatrix} = \mathscr{T} \begin{bmatrix} x' \\ y' \end{bmatrix} = \begin{bmatrix} c_{11} & c_{12} \\ c_{21} & c_{22} \end{bmatrix} \begin{bmatrix} x' \\ y' \end{bmatrix} \text{ or } \mathscr{X} = \mathscr{T} \mathscr{X}'.$$

[Note that the determinant of $\mathscr{T}$ is either $+1$ or -1 (see later, Sect. 5.7.2, the discussion about rotations $\det \mathscr{T} = +1$ and reflections $\det \mathscr{T} = -1$ in $\mathbb{R}^2$), so that we have to arrange, if necessary, that $\det \mathscr{T} = 1$ by interchange of columns].
Finally,

$$(\mathscr{T} \mathscr{X}')^T \mathscr{A} (\mathscr{T} \mathscr{X}') + \mathscr{K} (\mathscr{T} \mathscr{X}') + f = 0$$

will give

$$\mathscr{X}'^T (\mathscr{T}^T \mathscr{A} \mathscr{T}) \mathscr{X}' + \mathscr{K} (\mathscr{T} \mathscr{X}') + f = 0$$

or

$$[x' \ y'] \begin{bmatrix} \lambda_1 & 0 \\ 0 & \lambda_2 \end{bmatrix} \begin{bmatrix} x' \\ y' \end{bmatrix} + [d \ e] \begin{bmatrix} c_{11} & c_{12} \\ c_{21} & c_{22} \end{bmatrix} \begin{bmatrix} x' \\ y' \end{bmatrix} + f = 0$$

or

$$\lambda_1 x'^2 + \lambda_2 y'^2 + d'x' + e'y' + f = 0,$$

where $d' = d c_{11} + e c_{21}$, $e' = d c_{12} + e c_{22}$.

Every equation of this form with at least one of λ_1 and λ_2 nonzero describes an (possibly degenerate or imaginary) ellipse, hyperbola, or parabola.

Now, we can say that the graph is an ellipse if $\lambda_1 \lambda_2 > 0$, a hyperbola if $\lambda_1 \lambda_2 < 0$, and a parabola if $\lambda_1 \lambda_2 = 0$.

Since this quadratic equation in the rotated $x'y'$-coordinate system contains x'^2, x', y'^2, y' terms, but no cross-product term, its graph is a conic that is translated with respect to the $x'y'$-coordinate system. So, we have to translate the coordinate system to the new one $x''y''$ in which the conic is in the standard position. To do this, we first collect x' and y' terms:

$$(\lambda_1 x'^2 + d'x') + (\lambda_2 y'^2 + e'y') + f = 0,$$

and then complete the squares on the two expressions in parentheses.

Finally, we translate the coordinate axes $x'y'$ by the translating equations $x'' = x' - p$, $y'' = y' - q$, where p and q are the coordinates of the center of the conic in the coordinate system x' and y'.

Note One useful method for determining which conic is represented by the quadratic equation

$$ax^2 + 2bxy + cy^2 + dx + ey + f = 0$$

is obtained before we diagonalize the symmetric matrix of the associated quadratic form $ax^2 + 2bxy + cy^2$, i.e., the matrix $\mathscr{A} = \begin{bmatrix} a & b \\ b & c \end{bmatrix}$. The characteristic equation is

$$\det(\mathscr{A} - \lambda I_2) = \begin{vmatrix} a - \lambda & b \\ b & c - \lambda \end{vmatrix} = 0 \text{ or}$$

$$(a-\lambda)(c-\lambda) - b^2 = \lambda^2 - (a+c)\lambda + ac - b^2 = 0.$$

The eigenvalues λ_1 and λ_2 are given by

$$\lambda_{1,2} = \frac{(a+c) \pm \sqrt{(a+c)^2 - 4(ac - b^2)}}{2} = \frac{(a+c) \pm \sqrt{(a+c)^2 + 4b^2 - 4ac}}{2}.$$

These eigenvalues must be real numbers because $\mathscr{A}$ is a symmetric matrix. Indeed, $(a-c)^2 + 4b^2 > 0$, which guarantees the reality of the roots. But, if $4b^2 - 4ac$ is a negative number, the square root will be smaller than $(a+c)$, and λ_1 and λ_2 will have the same sign $(\lambda_1\lambda_2 > 0)$; if $4b^2 - 4ac > 0$, we have $\lambda_1\lambda_2 < 0$, and if $4b^2 - 4ac = 0$, one of the eigenvalues will be zero $(\lambda_1\lambda_2 = 0)$.

$$\begin{aligned}
Conclusion : 4b^2 - 4ac &< 0 \Rightarrow \text{the conic is an ellipse,}\\
4b^2 - 4ac &> 0 \Rightarrow \text{the conic is a hyperbola,}\\
4b^2 - 4ac &= 0 \Rightarrow \text{the conic is a parabola.}
\end{aligned}$$

Two Examples

1. Describe and draw the conic whose equation is

$$5x^2 - 4xy + 8y^2 + \frac{20}{\sqrt{5}}x - \frac{80}{\sqrt{5}}y + 4 = 0 \text{ or}$$

$$\mathscr{X}^T \mathscr{A} \mathscr{X} + \mathscr{K}\mathscr{X} + 4 = 0, \text{ where}$$

$$\mathscr{X} = \begin{bmatrix} x \\ y \end{bmatrix}, \quad \mathscr{A} = \begin{bmatrix} 5 & -2 \\ -2 & 8 \end{bmatrix} \quad \mathscr{K} = \begin{bmatrix} \frac{20}{\sqrt{5}} & -\frac{80}{\sqrt{5}} \end{bmatrix}.$$

The associated quadratic form $5x^2 - 4xy + 8y^2$ has a cross-product term which should be eliminated by orthogonal diagonalization of the corresponding 2×2 symmetric matrix $\mathscr{A} = \begin{bmatrix} 5 & -2 \\ -2 & 8 \end{bmatrix}$. Its eigenvalues λ_1 and λ_2 are of the same sign since $4b^2 - 4ac = 16 - 160 = -144 < 0$, and this conic is an ellipse. The eigenvalues are

$$\lambda_{1,2} = \frac{(a+c) \pm \sqrt{(a+c)^2 - 4(ac - b^2)}}{2} = \frac{13 \pm \sqrt{169 - 144}}{2} = \frac{13 \pm 5}{2} \Rightarrow$$

$$\lambda_1 = 4, \quad \lambda_2 = 9.$$

The corresponding eigenvectors must be orthogonal, and they are obtained by the GJM reduction method of the homogeneous linear systems:
For $\lambda_1 = 4$,

$$\begin{bmatrix} 5-4 & -2 \\ -2 & 8-4 \end{bmatrix} = \begin{bmatrix} 1 & -2 \\ -2 & 4 \end{bmatrix} \sim \begin{bmatrix} 1 & -2 \\ 0 & 0 \end{bmatrix} \xrightarrow{GJM} \begin{bmatrix} 1 & -2 \\ 0 & -1 \end{bmatrix} \Rightarrow \bar{c}_1 = \frac{1}{\sqrt{5}} \begin{bmatrix} 2 \\ 1 \end{bmatrix};$$

For $\lambda_2 = 9$,

$$\begin{bmatrix} 5-9 & -2 \\ -2 & 8-9 \end{bmatrix} = \begin{bmatrix} -4 & -2 \\ -2 & -1 \end{bmatrix} \sim \begin{bmatrix} 1 & \frac{1}{2} \\ 0 & 0 \end{bmatrix} \xrightarrow{\text{GJM}} \begin{bmatrix} 1 & \frac{1}{2} \\ 0 & -1 \end{bmatrix} \Rightarrow \bar{c}_2 = \frac{1}{\sqrt{5}} \begin{bmatrix} -1 \\ 2 \end{bmatrix}.$$

The rotation matrix is $\mathscr{T} = \begin{bmatrix} \frac{2}{\sqrt{5}} & -\frac{1}{\sqrt{5}} \\ \frac{1}{\sqrt{5}} & \frac{2}{\sqrt{5}} \end{bmatrix}$, and it represents the rotation (det $\mathscr{T} =$ 1) in $\mathbb{R}^2$ through the angle α counter-clockwise $\mathscr{T} = \begin{bmatrix} \cos\alpha & -\sin\alpha \\ \sin\alpha & \cos\alpha \end{bmatrix}$, where $\alpha = 26,565°$.

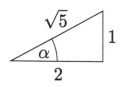

The quadratic equation after the orthogonal substitution

$$\mathscr{X} = \begin{bmatrix} x \\ y \end{bmatrix} = \mathscr{T} \begin{bmatrix} x' \\ y' \end{bmatrix} = \mathscr{T}\mathscr{X}' \text{ becomes}$$

$$[\mathscr{T}\mathscr{X}']^T \mathscr{A}(\mathscr{T}\mathscr{X}') = \mathscr{K}(\mathscr{T}\mathscr{X}') + 4 = 0 \Rightarrow$$

$$\Rightarrow \mathscr{X}'^T \begin{bmatrix} 4 & 0 \\ 0 & 9 \end{bmatrix} \mathscr{X}' + \begin{bmatrix} \frac{20}{\sqrt{5}} & -\frac{80}{\sqrt{5}} \end{bmatrix} \begin{bmatrix} \frac{2}{\sqrt{5}} & -\frac{1}{\sqrt{5}} \\ \frac{1}{\sqrt{5}} & \frac{2}{\sqrt{5}} \end{bmatrix} \mathscr{X}' + 4 = 0 \text{ or}$$

$$4x'^2 + 9y'^2 - 8x' - 36y' + 4 = 0.$$

To bring the conic into standard position we have to translate the $x'y'$-axes. We first collect x' and y' terms $4(x'^2 - 2x') + 9(y'^2 - 4y') = -4$, and then complete the squares

$$4(x'^2 - 2x' + 1) + 9(y'^2 - 4y' + 4) - 4 - 36 = -4 \text{ or}$$
$$4(x' - 1)^2 + 9(y' - 2)^2 = 36.$$

Then, we translate the coordinate axes

$$x'' = x' - 1 \text{ and } y'' = y' - 2.$$

Finally, the equation of our ellipse in standard position in the $x''y''$-coordinate system becomes

$$\boxed{\frac{x''^2}{9} + \frac{y''^2}{4} = 1}, \ a = 3, \ b = 2.$$

The sketch of this ellipse is as shown in the diagram, where the rotation of the coordinate system xy through the angle $\alpha = 26,565°$ is clearly indicated, as well as the subsequent translation of the $x'y'$ axes by $+1$ and $+2$, respectively. This is the ellipse $5x^2 - 4xy + 8y^2 + \frac{20}{\sqrt{5}}x - \frac{80}{\sqrt{5}}y + 4 = 0$ or $\frac{x''^2}{9} + \frac{y''^2}{4} = 1$.

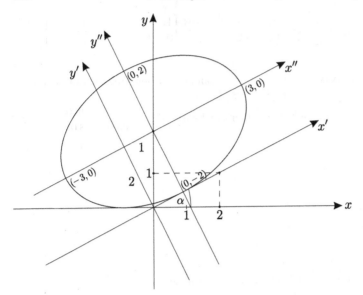

2. Use rotation and translation of axes to sketch the curve $2xy + 2\sqrt{2}x = 1$.
   The symmetric coefficient matrix of the associated quadratic form $2xy$ is $\mathscr{A} = \begin{bmatrix} 0 & 1 \\ 1 & 0 \end{bmatrix}$. Since $4b^2 - 4ac = 4 > 0$, we have a hyperbola. The eigenvalues of $\mathscr{A}$ are $\lambda_{1,2} = \frac{\pm\sqrt{4}}{2} = \pm 1$. The corresponding eigenvectors must be orthogonal. For $\lambda_1 = 1$,

$$\begin{bmatrix} -1 & 1 \\ 1 & -1 \end{bmatrix} \sim \begin{bmatrix} 1 & -1 \\ 0 & 0 \end{bmatrix} \xrightarrow{\text{GJM}} \begin{bmatrix} 1 & -1 \\ 0 & -1 \end{bmatrix} \Rightarrow \bar{c}_1 = \frac{1}{\sqrt{2}}\begin{bmatrix} 1 \\ 1 \end{bmatrix};$$

For $\lambda_2 = -1$,

$$\begin{bmatrix} 1 & 1 \\ 1 & 1 \end{bmatrix} \sim \begin{bmatrix} 1 & 1 \\ 0 & 0 \end{bmatrix} \xrightarrow{\text{GJM}} \begin{bmatrix} 1 & 1 \\ 0 & -1 \end{bmatrix} \Rightarrow \bar{c}_2 = \frac{1}{\sqrt{2}}\begin{bmatrix} -1 \\ 1 \end{bmatrix}.$$

Thus, $\mathscr{T} = \begin{bmatrix} \frac{1}{\sqrt{2}} & -\frac{1}{\sqrt{2}} \\ \frac{1}{\sqrt{2}} & \frac{1}{\sqrt{2}} \end{bmatrix} = \begin{bmatrix} \cos 45° & -\sin 45° \\ \sin 45° & \cos 45° \end{bmatrix}$

represents the rotation $(\det \mathscr{T} = 1)$ of the xy-coordinate system counter-clockwise through $45°$, and orthogonally diagonalizes $\mathscr{A}$:

$$\mathscr{T}^T \mathscr{A} \mathscr{T} = \frac{1}{2}\begin{bmatrix} 1 & 1 \\ -1 & 1 \end{bmatrix}\begin{bmatrix} 0 & 1 \\ 1 & 0 \end{bmatrix}\begin{bmatrix} 1 & -1 \\ 1 & 1 \end{bmatrix} = \begin{bmatrix} 1 & 0 \\ 0 & -1 \end{bmatrix}.$$

The substitution $\begin{bmatrix} x \\ y \end{bmatrix} = \mathscr{T} \begin{bmatrix} x' \\ y' \end{bmatrix}$ or $x = \frac{1}{\sqrt{2}}(x' - y')$, $y = \frac{1}{\sqrt{2}}(x' + y')$, eliminates the cross-product term, and leaves the quadratic equation in $x'y'$-coordinates as

$$x'^2 - y'^2 + 2x' - 2y' = 1.$$

Completing the squares, we get

$$(x'^2 + 2x' + 1) - (y'^2 + 2y' + 1) - 1 + 1 = 1 \Rightarrow (x' + 1)^2 - (y' + 1)^2 = 1.$$

Translating both axes x' and y' by -1, i.e., $x'' = x' + 1$, $y'' = y' + 1$, we obtain

$$\boxed{x''^2 - y''^2 = 1}, \ a = b = 1.$$

This quadratic equation in $x''y''$-coordinates represents a hyperbola in standard position obtained by a counter-clockwise rotation of the xy-coordinate system through $45°$ with subsequent translation of both coordinates x' and y' by -1.

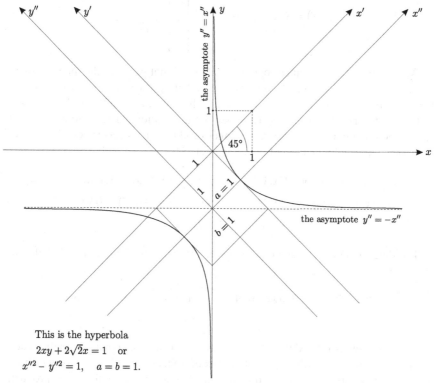

This is the hyperbola
$$2xy + 2\sqrt{2}x = 1 \quad \text{or}$$
$$x''^2 - y''^2 = 1, \quad a = b = 1.$$

Δ

5.7 Canonical Form of Orthogonal Matrices

5.7.1 Orthogonal Matrices in $\mathbb{R}^n$

Real orthogonal matrices of size $n \times n$ are defined as those invertible matrices ($\mathscr{A}^{-1}$ exists) for which the inverse matrix $\mathscr{A}^{-1}$ is equal to its transpose $\mathscr{A}^T$: $\mathscr{A}^{-1} = \mathscr{A}^T$ or $\mathscr{A}\,\mathscr{A}^T = \mathscr{A}^T \mathscr{A} = I_n$.

These are extremely important in many applications, since they are precisely those matrices (i.e., linear operators in $\mathbb{R}^n$) which preserve the inner product. To show this characteristic property of orthogonal matrices, let us first consider the inner product in $\mathbb{R}^n$

$$(\bar{x}, \bar{y}) = x_1 y_1 + x_2 y_2 + \cdots + x_n y_n = \sum_{i=1}^{n} x_i y_i, \quad \text{where}$$

$\bar{x} = [x_1\, x_2 \ldots x_n]^T$ and $\bar{y} = [y_1\, y_2 \ldots y_n]^T$ are vectors from $\mathbb{R}^n$.

This inner product can be redefined by means of matrix multiplication (Sect. 3.1)

$$(\bar{x}, \bar{y}) = \bar{x}^T \bar{y} = [x_1\, x_2 \ldots x_n] \begin{bmatrix} y_1 \\ y_2 \\ \vdots \\ y_n \end{bmatrix} = \sum_{i=1}^{n} x_i y_i.$$

We want to find those linear operators $\mathscr{A}$ in $\mathbb{R}^n$ that preserve the inner product: $(\mathscr{A}\bar{x}, \mathscr{A}\bar{y}) = (\bar{x}, \bar{y})$. They can be obtained immediately, $(\mathscr{A}\bar{x}, \mathscr{A}\bar{y}) = (\mathscr{A}\bar{x})^T (\mathscr{A}\bar{y}) = \bar{x}^T (\mathscr{A}^T \mathscr{A})\bar{y} = \bar{x}^T \bar{y}$ (if $\mathscr{A}^T \mathscr{A} = I_n$) $= (\bar{x}, \bar{y})$. Therefore, it is sufficient for an $n \times n$ matrix to preserve the inner product in $\mathbb{R}^n$ that it is orthogonal. But, to be sure that only orthogonal matrices preserve the inner product (the necessary condition), we have to use the analogue of the Lemma in Sect. 5.5.1:

$$(\mathscr{A}\bar{x}, \mathscr{A}\bar{y}) = (\bar{x}, \bar{y}) \text{ for all } \bar{x}, \bar{y} \in \mathbb{R}^n \Rightarrow (\mathscr{A}\bar{x}, \mathscr{A}\bar{x}) = (\bar{x}, \bar{x}) \text{ for all } \bar{x} \in \mathbb{R}^n \Rightarrow$$
$$\Rightarrow (\bar{x}, \mathscr{A}^T \mathscr{A}\bar{x}) = (\bar{x}, \bar{x}) \,\forall \bar{x} \in \mathbb{R}^n \Rightarrow (\bar{x}, (\mathscr{A}^T \mathscr{A} - I_n)\bar{x}) = 0 \,\forall \bar{x} \in \mathbb{R}^n$$
$$\Rightarrow \mathscr{A}^T \mathscr{A} = I_n. \,\Delta$$

The determinants of orthogonal matrices are equal to either $+1$ or -1 due to $\mathscr{A}^T \mathscr{A} = I_n$:

$$\det(\mathscr{A}^T \mathscr{A}) = \det(\mathscr{A}^T)\det(\mathscr{A}) = (\det \mathscr{A})^2 = \det I_n = 1 \text{ or}$$
$$\det \mathscr{A} = \pm 1.$$

We have already proved that the set of all $n \times n$ orthogonal matrices is the group $O(n)$ (Sect. 4.4). This is the group of all automorphisms in $\mathbb{R}^n$, since every orthogonal matrix is a bijective linear operator in $\mathbb{R}^n$ that preserves the inner product. Remember, an automorphism of a given algebraic structure (here $\mathbb{R}^n$) is a bijective map that preserves the complete set of operations that define this structure (here linear combinations and the inner product).

Obviously, the group $O(n)$ is a subset of the group $GL(n,\mathbb{R})$ of all real $n \times n$ invertible matrices. Since the group operations in $GL(n,\mathbb{R})$ and $O(n)$ are the same (the multiplication of $n \times n$ matrices), it follows that $O(n)$ is a subgroup of $GL(n,\mathbb{R})$.

On the other hand, the set of all orthogonal $n \times n$ matrices with the determinant $\det \mathscr{A} = +1$ forms a subgroup $SO(n)$ (special orthogonal matrices) of $O(n)$: this set is closed under matrix multiplication since $\det(\mathscr{A}\mathscr{B}) = \det \mathscr{A} \cdot \det \mathscr{B} = 1 \cdot 1 = 1$, as well as under inversion $\det(\mathscr{A}^{-1}) = (\det \mathscr{A})^{-1} = 1$. The groups $SO(n)$, $n = 2,3,\ldots$, play important roles in applications in mathematics, as well as in physics; for instance, the rotational symmetry of atoms [$SO(3)$ commutes with the energy operator] gives as a result the electron orbits in atoms.

We defined orthogonal matrices by the property $\mathscr{A}^{-1} = \mathscr{A}^T$ (or equivalently by $\mathscr{A}\mathscr{A}^T = \mathscr{A}^T\mathscr{A} = I_n$). It is very useful to know that orthogonal matrices can be characterized by two more properties which concern their columns and rows. They can be obtained straightforwardly from the above definitions: $\mathscr{A}^T\mathscr{A} = I_n \Leftrightarrow (c_i, c_j) = \delta_{ij}$, $i,j = 1,2,\ldots,n$ (already used in Sect. 5.6.1) and $\mathscr{A}\mathscr{A}^T = I_n \Leftrightarrow (r_i, r_j) = \delta_{ij}$. So, the columns c_i, $i = 1,2,\ldots,n$, as well as the rows r_i, $i = 1,2,\ldots,n$, of an orthogonal matrix are orthonormal (ON) sets of vectors.

5.7.2 Orthogonal Matrices in $\mathbb{R}^2$ (Rotations and Reflections)

Rotations

We shall first consider the rotation $\mathrm{Rot}(\alpha) : \mathbb{R}^2 \to \mathbb{R}^2$ in the plane $\mathbb{R}^2$ through an angle α counterclockwise about the origin 0. The rotation $\mathrm{Rot}(\alpha)$ is a linear operator in $\mathbb{R}^2$ (a 2×2 real matrix), since it preserves the addition of vectors $\mathrm{Rot}(\alpha)(\bar{x} + \bar{y}) = \mathrm{Rot}(\alpha)\bar{x} + \mathrm{Rot}(\alpha)\bar{y}$, as well as the multiplication of scalars with vectors $\mathrm{Rot}(\alpha)(c\bar{x}) = c\,\mathrm{Rot}(\alpha)\bar{x}$. It also preserves the length (the norm) of vectors $||\mathrm{Rot}(\alpha)\bar{x}|| = ||\bar{x}||$, so it is an orthogonal matrix.

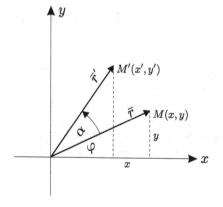

Now, consider the relation between an arbitrary vector $\vec{r} \in \mathbb{R}^2$ and its image $\vec{r}' = \text{Rot}(\alpha)\vec{r}$ after the rotation. The column matrix of $\vec{r}$ is

$$\vec{r} = \begin{bmatrix} x \\ y \end{bmatrix} = \begin{bmatrix} r\cos\varphi \\ r\sin\varphi \end{bmatrix}, \quad \text{where } r = ||\vec{r}||.$$

That of $\vec{r}'$ is

$$\vec{r}' = \begin{bmatrix} x' \\ y' \end{bmatrix} = \begin{bmatrix} r\cos(\varphi+\alpha) \\ r\sin(\varphi+\alpha) \end{bmatrix}, \quad \text{where } r = ||\vec{r}'||.$$

Furthermore,

$$\begin{bmatrix} r\cos(\varphi+\alpha) \\ r\sin(\varphi+\alpha) \end{bmatrix} = \begin{bmatrix} r(\cos\varphi\cos\alpha - \sin\varphi\sin\alpha) \\ r(\sin\varphi\cos\alpha + \cos\varphi\sin\alpha) \end{bmatrix} =$$
$$= \begin{bmatrix} x\cos\alpha - y\sin\alpha \\ x\sin\alpha + y\cos\alpha \end{bmatrix} = \begin{bmatrix} \cos\alpha & -\sin\alpha \\ \sin\alpha & \cos\alpha \end{bmatrix} \begin{bmatrix} x \\ y \end{bmatrix} = \begin{bmatrix} x' \\ y' \end{bmatrix}.$$

Therefore, $\vec{r}' = \text{Rot}(\alpha)\vec{r}$, where

$$\text{Rot}(\alpha) = \begin{bmatrix} \cos\alpha & -\sin\alpha \\ \sin\alpha & \cos\alpha \end{bmatrix}.$$

The determinant of this orthogonal matrix (the rows, as well as the columns are ON vectors) is obviously $+1$ $(\cos^2\alpha + \sin^2\alpha = 1)$. So, $\text{Rot}(\alpha) \in SO(2)$.

We shall now prove that every matrix $\mathscr{A} = \begin{bmatrix} a & b \\ c & d \end{bmatrix} \in SO(2)$ is a rotation in $\mathbb{R}^2$ counterclockwise through the angle $\alpha = \arccos a$ around the origin.

Consider $\mathscr{A} = \begin{bmatrix} a & b \\ c & d \end{bmatrix}$, $a,b,c,d \in \mathbb{R}$ with properties $\mathscr{A}^{-1} = \mathscr{A}^T$ and $\det\mathscr{A} = 1$, i.e., $\mathscr{A} \in SO(2)$. The inverse $\mathscr{A}^{-1}$ can be obtained by the general rule $\mathscr{A}^{-1} = \frac{1}{\det\mathscr{A}}\text{adj}\mathscr{A} = \frac{1}{+1}\begin{bmatrix} d & -b \\ -c & a \end{bmatrix}$, where $\text{adj}\mathscr{A}$ is the transposed matrix of cofactors of $\mathscr{A}$.

But since $\mathscr{A}^{-1} = \mathscr{A}^T$, we have also $\mathscr{A}^{-1} = \begin{bmatrix} a & c \\ b & d \end{bmatrix}$. Obviously, $\begin{bmatrix} d & -b \\ -c & a \end{bmatrix} = \begin{bmatrix} a & c \\ b & d \end{bmatrix}$ implies $a = d$ and $b = -c$. Thus, the two conditions $\mathscr{A}^{-1} = \mathscr{A}^T$ and $\det\mathscr{A} = +1$ give $\mathscr{A} = \begin{bmatrix} a & -c \\ c & a \end{bmatrix}$, so it seems that every 2×2 special orthogonal matrix is determined by two parameters. But, in addition, the condition $\det\mathscr{A} = 1$ for special orthogonal matrices gives $a^2 + c^2 = 1 \Rightarrow c = \sqrt{1-a^2}$ reducing the number of parameters to one, i.e., a for which $|a| \leq 1$. To be able to express $\mathscr{A} = \begin{bmatrix} a & -c \\ c & a \end{bmatrix}$, with the condition $a^2 + c^2 = 1$, by only one parameter, it is quite natural to use the trigonometric identity $\sin^2\alpha + \cos^2\alpha = 1$ for $a^2 + c^2 = 1$, replacing a with $\cos\alpha$ and c with $\sin\alpha = \sqrt{1-\cos^2\alpha}$.

Thus, we obtain for $\mathscr{A}$, which is now expressed by one parameter α, the rotation matrix for α

$$\mathscr{A} = \text{Rot}(\alpha) = \begin{bmatrix} \cos\alpha & -\sin\alpha \\ \sin\alpha & \cos\alpha \end{bmatrix}, \quad \text{where } \alpha = \arccos a, \ |a| \le 1.$$

Remark If we replaced a with $\sin\beta$ and c with $\cos\beta$ (which is another possibility), we would get

$$\mathscr{A} = \begin{bmatrix} \sin\beta & -\cos\beta \\ \cos\beta & \sin\beta \end{bmatrix} = \begin{bmatrix} \cos(\frac{\pi}{2}-\beta) & -\sin(\frac{\pi}{2}-\beta) \\ \sin(\frac{\pi}{2}-\beta) & \cos(\frac{\pi}{2}-\beta) \end{bmatrix} = \text{Rot}(\frac{\pi}{2}-\beta),$$

which is the rotation through the complementary angle of β, i.e., for $(\frac{\pi}{2}) - \beta$. Here $\beta = \arcsin a$. But, since always $\arcsin a + \arccos a = \frac{\pi}{2}$, we see that $\beta = \frac{\pi}{2} - \arccos a = \frac{\pi}{2} - \alpha$, so that $\alpha = \frac{\pi}{2} - \beta$, and $\mathscr{A} = \text{Rot}(\alpha)$, as before.

Therefore, $\mathscr{A} \in SO(2)$, determined by $\mathscr{A}^{-1} = \mathscr{A}^T$ and $\det\mathscr{A} = +1$, is a rotation in $\mathbb{R}^2$.

One more important property of the group $SO(2)$ is that it is a commutative (Abelian) group:

$$\text{Rot}(\alpha_1)\text{Rot}(\alpha_2) = \begin{bmatrix} \cos\alpha_1 & -\sin\alpha_1 \\ \sin\alpha_1 & \cos\alpha_1 \end{bmatrix} \begin{bmatrix} \cos\alpha_2 & -\sin\alpha_2 \\ \sin\alpha_2 & \cos\alpha_2 \end{bmatrix} =$$

$$= \begin{bmatrix} \cos\alpha_1\cos\alpha_2 - \sin\alpha_1\sin\alpha_2 & -\cos\alpha_1\sin\alpha_2 - \sin\alpha_1\cos\alpha_2 \\ \sin\alpha_1\cos\alpha_2 + \cos\alpha_1\sin\alpha_2 & -\sin\alpha_1\sin\alpha_2 + \cos\alpha_1\cos\alpha_2 \end{bmatrix} =$$

$$= \begin{bmatrix} \cos(\alpha_1+\alpha_2) & -\sin(\alpha_1+\alpha_2) \\ \sin(\alpha_1+\alpha_2) & \cos(\alpha_1+\alpha_2) \end{bmatrix} = \text{Rot}(\alpha_1+\alpha_2) =$$

$$= \text{Rot}(\alpha_2)\text{Rot}(\alpha_1).$$

On the other hand, orthogonal 2×2 matrices $\mathscr{B}$ with $\det\mathscr{B} = -1$ do not form a subgroup of $O(2)$, since this set is not closed under matrix multiplication: $\det(\mathscr{B}\mathscr{B}') = \det\mathscr{B} \cdot \det\mathscr{B}' = (-1) \cdot (-1) = 1 \ne -1$.

Reflections

We shall prove that a most general 2×2 orthogonal matrix $\mathscr{B} = \begin{bmatrix} a & b \\ c & d \end{bmatrix}$ with determinant equal to -1 represents a reflection in a line through the origin 0 in $\mathbb{R}^2$ that makes an angle α (counterclockwise) with the y-axis, where $\alpha = \frac{1}{2}\arccos(-a)$.

However, before embarking on this proof, we analyze reflections in $\mathbb{R}^2$, to find the general matrix expressions for them.

To represent by a 2×2 matrix the reflection $\text{Ref}(\alpha)$ in the line through the origin 0 in $\mathbb{R}^2$ that makes an angle α (counterclockwise) with the y-axis, we notice that it can be performed in three steps:

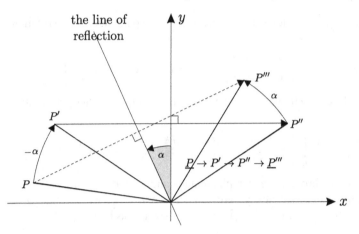

First, rotate the plane through the angle $-\alpha$ (clockwise), in fact perform $\mathrm{Rot}(-\alpha) = = \begin{bmatrix} \cos\alpha & \sin\alpha \\ -\sin\alpha & \cos\alpha \end{bmatrix} = = \mathrm{Rot}^{-1}(\alpha)$. The reflection axis now coincides with the y-axis. Second, perform the reflection in the y-axis $\mathrm{Ref}(0°) = \begin{bmatrix} -1 & 0 \\ 0 & 1 \end{bmatrix}$.

$$\left[\begin{array}{c} \text{The reflection in the } y\text{-axis is} \\ x' = -x, y' = y \text{ or } \begin{bmatrix} x' \\ y' \end{bmatrix} = \begin{bmatrix} -1 & 0 \\ 0 & 1 \end{bmatrix} \begin{bmatrix} x \\ y \end{bmatrix} \end{array} \right].$$

$(-x,y) \longleftarrow \overset{\displaystyle y}{\Big|} \quad (x,y)$

$\longrightarrow x$

Finally, rotate the plane back $\mathrm{Rot}(\alpha) = \begin{bmatrix} \cos\alpha & -\sin\alpha \\ \sin\alpha & \cos\alpha \end{bmatrix}$. So,

$$\mathscr{B} = \mathrm{Ref}(\alpha) = \mathrm{Rot}(\alpha)\mathrm{Ref}(0°)\mathrm{Rot}(-\alpha) =$$

$$= \begin{bmatrix} \cos\alpha & -\sin\alpha \\ \sin\alpha & \cos\alpha \end{bmatrix} \begin{bmatrix} -1 & 0 \\ 0 & 1 \end{bmatrix} \begin{bmatrix} \cos\alpha & \sin\alpha \\ -\sin\alpha & \cos\alpha \end{bmatrix} =$$

$$= \begin{bmatrix} -\cos^2\alpha + \sin^2\alpha & -2\sin\alpha\cos\alpha \\ -2\sin\alpha\cos\alpha & -\sin^2\alpha + \cos^2\alpha \end{bmatrix} = \begin{bmatrix} -\cos 2\alpha & -\sin 2\alpha \\ -\sin 2\alpha & \cos 2\alpha \end{bmatrix}.$$

It is obvious that $\mathrm{Ref}(\alpha)$ is an orthogonal matrix (its columns, as well as rows, are orthonormal vectors: $\cos^2 2\alpha + \sin^2 2\alpha = 1$, $\cos 2\alpha \sin 2\alpha - \cos 2\alpha \sin 2\alpha = 0$). So, $\mathrm{Ref}^{-1}(\alpha) = \mathrm{Ref}^T(\alpha)$. Its determinant is equal to -1: $-\cos^2 2\alpha - \sin^2 2\alpha = -1$. It is also an involution:

$$\text{Ref}^2(\alpha) = \begin{bmatrix} -\cos 2\alpha & -\sin 2\alpha \\ -\sin 2\alpha & \cos 2\alpha \end{bmatrix} \begin{bmatrix} -\cos 2\alpha & -\sin 2\alpha \\ -\sin 2\alpha & \cos 2\alpha \end{bmatrix} = \begin{bmatrix} 1 & 0 \\ 0 & 1 \end{bmatrix} = I_2$$

or $\text{Ref}(\alpha) = \text{Ref}^{-1}(\alpha)$, implying $\text{Ref}^T(\alpha) = \text{Ref}(\alpha)$, i.e, it is a symmetric matrix, which is obvious from its form (off-diagonal elements are the same).

Since this matrix is symmetric, it can be orthogonally diagonalized, which can serve as a verification that this operation in $\mathbb{R}^2$ represents a reflection $\mathscr{B}$.

We shall first find its eigenvalues by solving the characteristic equation $\det(\mathscr{B} - \lambda I_2) = 0$. Indeed,

$$\begin{vmatrix} -\cos 2\alpha - \lambda & -\sin 2\alpha \\ -\sin 2\alpha & \cos 2\alpha - \lambda \end{vmatrix} = 0 \Rightarrow$$

$$\Rightarrow -(\cos^2 2\alpha - \lambda^2) - \sin^2 2\alpha = 0 \Rightarrow \lambda^2 - \underbrace{(\sin^2 2\alpha + \cos^2 2\alpha)}_{1} = 0 \Rightarrow$$

$$\Rightarrow \lambda^2 = 1 \Rightarrow \lambda_{1,2} = \mp 1.$$

The corresponding eigenvectors must be orthogonal, and they are obtained by the GJM reduction process:

For $\lambda_1 = -1$

$$\begin{bmatrix} -\cos 2\alpha + 1 & -\sin 2\alpha \\ -\sin 2\alpha & \cos 2\alpha + 1 \end{bmatrix} \sim \begin{bmatrix} 1 & \frac{\sin 2\alpha}{\cos 2\alpha - 1} \\ -\sin 2\alpha & \cos 2\alpha + 1 \end{bmatrix} \sim \begin{bmatrix} 1 & \frac{\sin 2\alpha}{\cos 2\alpha - 1} \\ 0 & \frac{\sin^2 2\alpha}{\cos 2\alpha - 1} + \cos 2\alpha + 1 \end{bmatrix} \sim$$

$$\sim \begin{bmatrix} 1 & \frac{\sin 2\alpha}{\cos 2\alpha - \cos 0^\circ} \\ 0 & \frac{\sin^2 2\alpha + \cos^2 2\alpha - 1}{\cos 2\alpha - 1} \end{bmatrix} \sim \begin{bmatrix} 1 & \frac{2\sin \alpha \cos \alpha}{-2\sin^2 \alpha} \\ 0 & 0 \end{bmatrix} \sim \begin{bmatrix} 1 & -\frac{\cos \alpha}{\sin \alpha} \\ 0 & 0 \end{bmatrix} \xrightarrow{\text{GJM}} \begin{bmatrix} 1 & -\frac{\cos \alpha}{\sin \alpha} \\ 0 & -1 \end{bmatrix}.$$

So, the unnormalized eigenvector corresponding to $\lambda_1 = -1$ is $\bar{u}_1 = \begin{bmatrix} \frac{\cos \alpha}{\sin \alpha} \\ 1 \end{bmatrix}$ (the change of sign is irrelevant, but convenient). Its norm is $\|\bar{u}_1\| = \sqrt{\frac{\cos^2 \alpha}{\sin^2 \alpha} + 1} = \sqrt{\frac{\cos^2 \alpha + \sin^2 \alpha}{\sin^2 \alpha}} = \frac{1}{\sin \alpha}$. Dividing by this norm, i.e., multiplying by $\sin \alpha$, we get the normalized eigenvector for $\lambda_1 = -1$ $\quad \bar{c}_1 = \begin{bmatrix} \cos \alpha \\ \sin \alpha \end{bmatrix}$. Similarly, for $\lambda_2 = +1$ we get

$$\begin{bmatrix} -\cos 2\alpha - 1 & -\sin 2\alpha \\ -\sin 2\alpha & \cos 2\alpha - 1 \end{bmatrix} \sim \begin{bmatrix} 1 & \frac{\sin 2\alpha}{\cos 2\alpha + 1} \\ -\sin 2\alpha & \cos 2\alpha - 1 \end{bmatrix} \sim \begin{bmatrix} 1 & \frac{\sin 2\alpha}{\cos 2\alpha + \cos 0^\circ} \\ 0 & \frac{\sin^2 2\alpha}{\cos 2\alpha + 1} + \cos 2\alpha - 1 \end{bmatrix} \sim$$

$$\sim \begin{bmatrix} 1 & \frac{2\sin \alpha \cos \alpha}{2\cos^2 \alpha} \\ 0 & \frac{\sin^2 2\alpha + \cos^2 2\alpha - 1}{\cos 2\alpha + 1} \end{bmatrix} \sim \begin{bmatrix} 1 & \frac{\sin \alpha}{\cos \alpha} \\ 0 & 0 \end{bmatrix} \xrightarrow{\text{GJM}} \begin{bmatrix} 1 & \frac{\sin \alpha}{\cos \alpha} \\ 0 & -1 \end{bmatrix}.$$

The normalized eigenvector for $\lambda_2 = +1$ is $\bar{c}_2 = \begin{bmatrix} -\sin \alpha \\ \cos \alpha \end{bmatrix}$.

The orthogonal transition matrix $\mathscr{T}$ that diagonalizes $\mathscr{B}$ is

$$\mathscr{T} = [\bar{c}_1 \ \bar{c}_2] = \begin{bmatrix} \cos\alpha & -\sin\alpha \\ \sin\alpha & \cos\alpha \end{bmatrix} = \mathrm{Rot}(\alpha).$$

Consequently, $\mathscr{T}^{-1}\mathscr{B}\mathscr{T} = \begin{bmatrix} -1 & 0 \\ 0 & -1 \end{bmatrix}$, implying $\mathscr{B} = \mathscr{T} \begin{bmatrix} -1 & 0 \\ 0 & 1 \end{bmatrix} \mathscr{T}^{-1}$. But, this is exactly the expression we already had for $\mathscr{B} = \mathrm{Ref}(\alpha) = \mathrm{Rot}(\alpha)\mathrm{Ref}(0°)\mathrm{Rot}(-\alpha)$.

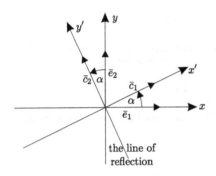

the line of reflection

Geometrically, $\mathscr{T}$ performs the rotation by α, $\bar{c}_1 = \cos\alpha\,\bar{e}_1 + \sin\alpha\,\bar{e}_2$, $\bar{c}_2 = -\sin\alpha\,\bar{e}_1 + \cos\alpha\,\bar{e}_2$. After this rotation, the new y'-axis becomes the line of reflection, so that the reflection matrix in the new coordinate system $x'y'$ becomes just $\begin{bmatrix} -1 & 0 \\ 0 & 1 \end{bmatrix}$.

Now, let us consider the general orthogonal 2×2 matrix $\mathscr{B}$ with determinant equal to -1:

$$\mathscr{B} = \begin{bmatrix} a & b \\ c & d \end{bmatrix}, \quad \mathscr{B}^{-1} = \mathscr{B}^T, \quad \det\mathscr{B} = -1.$$

The inverse $\mathscr{B}^{-1}$ can be obtained by the general rule

$$\mathscr{B}^{-1} = \frac{1}{\det\mathscr{B}}\mathrm{adj}\mathscr{B} = \frac{1}{-1}\begin{bmatrix} d & -b \\ -c & a \end{bmatrix} = \begin{bmatrix} -d & b \\ c & -a \end{bmatrix},$$

and by $\mathscr{B}^{-1} = \mathscr{B}^T = \begin{bmatrix} a & c \\ b & d \end{bmatrix}$. Equality of these two representations of $\mathscr{B}^{-1}$ gives immediately $a = -d$ and $b = c$. This reduces the number of parameters to two: $\mathscr{B} = \begin{bmatrix} a & c \\ c & -a \end{bmatrix}$. The determinant condition $-a^2 - c^2 = -1$ or $a^2 + c^2 = 1$, further reduces the number of parameters to one: $c = \sqrt{1-a^2}$, $|a| \leq 1$ or $\mathscr{B} = \begin{bmatrix} a & \sqrt{1-a^2} \\ \sqrt{1-a^2} & -a \end{bmatrix}$, $|a| \leq 1$.

Alternately, we can use one parameter in trigonometric functions, $a = -\cos\theta$ and $c = -\sin\theta$, $|a| \leq 1$. Now, the matrix $\mathscr{B} = \begin{bmatrix} -\cos\theta & -\sin\theta \\ -\sin\theta & \cos\theta \end{bmatrix}$ is a reflection in the axis which makes an angle $\alpha = \frac{\theta}{2}$ with the y-axis:

$$\mathscr{B} = \text{Ref}(\alpha) = \begin{bmatrix} -\cos 2\alpha & -\sin 2\alpha \\ -\sin 2\alpha & \cos 2\alpha \end{bmatrix}, \quad \alpha = \frac{1}{2}\arccos(-a), \ |a| \leq 1.$$

The reflection $\text{Ref}(\alpha)$ can be expressed in two more different ways, which are important, among other things, in factorization of dihedral groups D_n (the symmetry groups of regular n-polygons, $n = 3, 4, \ldots$).
Namely,

$$\text{Ref}(\alpha) = \begin{bmatrix} -\cos 2\alpha & -\sin 2\alpha \\ -\sin 2\alpha & \cos 2\alpha \end{bmatrix} =$$

$$= \begin{bmatrix} \cos 2\alpha & -\sin 2\alpha \\ \sin 2\alpha & \cos 2\alpha \end{bmatrix} \begin{bmatrix} -1 & 0 \\ 0 & 1 \end{bmatrix} = \text{Rot}(2\alpha)\text{Ref}(0°), \ \text{or}$$

$$\text{Ref}(\alpha) = \begin{bmatrix} -1 & 0 \\ 0 & 1 \end{bmatrix} \begin{bmatrix} \cos 2\alpha & \sin 2\alpha \\ -\sin 2\alpha & \cos 2\alpha \end{bmatrix} = \text{Ref}(0°)\text{Rot}(-2\alpha) = \text{Ref}(0°)\text{Rot}^{-1}(2\alpha).$$

(see next two figures)

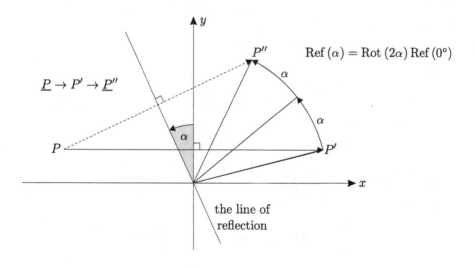

Two Examples

1. Dihedral Group D_4 (the symmetry group of a square).
   There are eight symmetry transformations from $O(2)$ that leave the square in its place:

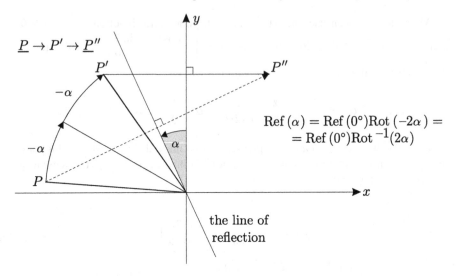

$$P \to P' \to P''$$

$$\text{Ref}\,(\alpha) = \text{Ref}\,(0°)\text{Rot}\,(-2\alpha) =$$
$$= \text{Ref}\,(0°)\text{Rot}^{-1}(2\alpha)$$

the line of
reflection

I

$$A = \text{Rot}(\frac{360°}{4}) =$$
$$= \text{Rot}(90°) =$$
$$= \begin{bmatrix} \cos 90° & -\sin 90° \\ \sin 90° & \cos 90° \end{bmatrix} =$$
$$= \begin{bmatrix} 0 & -1 \\ 1 & 0 \end{bmatrix},$$

II

$$A^2 = \text{Rot}(180°) =$$
$$= \begin{bmatrix} -1 & 0 \\ 0 & -1 \end{bmatrix},$$

III $A^3 = \text{Rot}(270°) = \begin{bmatrix} 0 & 1 \\ -1 & 0 \end{bmatrix},$

IV

$$A^4 = \text{Rot}(360°) = I_2 = \begin{bmatrix} 1 & 0 \\ 0 & 1 \end{bmatrix},$$

V $B = \text{Ref}(0°) = \begin{bmatrix} -1 & 0 \\ 0 & 1 \end{bmatrix},$

VI $\text{Ref}(45°) = \text{Ref}(y = -x) = \text{Rot}(90°)\text{Ref}(0°) = \text{Ref}(0°)\text{Rot}^{-1}(90°),$
VII $\text{Ref}(90°) = \text{Ref}(y = 0) = \text{Rot}(180°)\text{Ref}(0°) = \text{Ref}(0°)\text{Rot}^{-1}(180°),$
VIII $\text{Ref}(135°) = \text{Ref}(y = x) = \text{Rot}(270°)\text{Ref}(0°) = \text{Ref}(0°)\text{Rot}^{-1}(270°).$

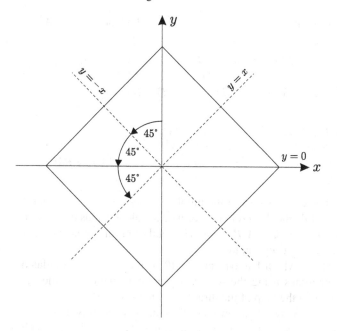

Since $A^{-1} = \text{Rot}^{-1}(90°) = \text{Rot}(270°) = A^3$, we have $\text{Ref}(45°) = AB = BA^3$.
Similarly, $A^{-2} = \text{Rot}^{-1}(180°) = \text{Rot}(180°) = A^2$, so that $\text{Ref}(90°) = A^2B = BA^2$.
Also, $A^{-3} = \text{Rot}^{-1}(270°) = \text{Rot}(90°) = A$, and $\text{Ref}(135°) = A^3B = BA$.
The group of all these eight symmetry transformations is $D_4 = \{I_2, A, A^2, A^3,$
$B, AB, A^2B, A^3B\} = \{I_2, A, A^2, A^3, B, BA^3, BA^2, BA\}$. It has two cyclic subgroups
$C_4 = \{I_2, A, A^2, A^3\}$ and $C_2 = \{I_2, B\}$. But the subgroup C_4 is normal, since it
has only one coset (the coset of reflections), which is left and right at the same
time: $BC_4 = \{B, BA^3, BA^2, BA\} = \{B, AB, A^2B, A^3B\} = C_4B$.
The intersection of these two subgroups is only I_2 : $C_4 \cap C_2 = \{I_2\}$, and their
product is just the group D_4 : $C_4C_2 = D_4$.
We call D_4 a *semi-direct product* $D_4 = C_4 \wedge C_2$ of its two cyclic subgroups C_4
and C_2 since the following three conditions are satisfied:

(1) $D_4 = C_4C_2$,
(2) $C_4 \cap C2 = \{I_2\}$,
(3) C_4 is a normal subgroup of D_4 : $C_4 \triangleleft D_4$.

But the most economical way of defining a group is by giving its generators
and the corresponding generator relations. The group is obtained by multiplying
all different powers of the generators in all possible ways allowed by generator
relations. The group table is then easily constructed by using these relations.
The group D_4 has two generators $A = \text{Rot}(90°)$ and $B = \text{Ref}(0°)$, and three generator relations

$$A^4 = I_2, \quad B^2 = I_2, \quad (AB)^2 = I_2$$

(the last one is equivalent to any of the three: $AB = BA^3$ or $A^2B = BA^2$ or $A^3B = BA$.

Indeed, $AB = BA^3 / \cdot A \Rightarrow ABA = B / \cdot B \Rightarrow ABAB = I_2 \Rightarrow (AB)^2 = I_2$,
$\quad A \cdot /A^3B = BA \Rightarrow B = ABA / \cdot B \Rightarrow ABAB = I_2 \Rightarrow (AB)^2 = I_2$,
$\quad A^2B = BA^2 / \cdot A^3 \Rightarrow A^2(BA^3) = BA \Rightarrow A^2(AB) = BA \Rightarrow A^3B = BA \Rightarrow$
$\quad \Rightarrow (AB)^2 = I_2$.

Also, $ABAB = I_2 / \cdot B \Rightarrow ABA = B / \cdot A^3 \Rightarrow AB = BA^3$,
$\quad A^3 \cdot /ABAB = I_2 \Rightarrow BAB = A^3 / \cdot B \Rightarrow BA = A^3B$,
$\quad ABAB = I_2 / \cdot B \Rightarrow ABA = B / \cdot A^2 \Rightarrow A(BA^3) = BA^2 \Rightarrow A(AB) = BA^2 \Rightarrow$
$\quad \Rightarrow A^2B = BA^2$.)

The first two generator relations restrict the number of possible group elements to 5 : I_2, B, A, A^2, A^3, while the third relation (or its equivalents) adds three more elements: $AB = BA^3$, $A^2B = BA^2$, $A^3B = BA$. The total of group elements in D_4 is therefore 8, as we have already found.

In the group table of D_4 which has 64 entries, 40 of them can be calculated straightforwardly sometimes using the first two generator relations, while the other 24 are obtained with the help of the three equivalents of $(AB)^2 = I_2$.

2. Dihedral Groups D_n, $n \geq 3$ (symmetry groups of regular n-sided polygons)

The most important conclusion is that all dihedral groups D_n, $n = 3,4,5,\ldots$, with $2n$ symmetry transformations from $O(2)$ are semi-direct products of two cyclic subgroups $C_n = \{I_2, A, A^2, \ldots, A^{n-1}\}$ and $C_2 = \{I_2, B\}$, where

$$A = \mathrm{Rot}(\frac{360°}{n}) = \begin{bmatrix} \cos \frac{360°}{n} & -\sin \frac{360°}{n} \\ \sin \frac{360°}{n} & \cos \frac{360°}{n} \end{bmatrix} \quad \text{and } B = \mathrm{Ref}(0°) = \begin{bmatrix} -1 & 0 \\ 0 & 1 \end{bmatrix}.$$

In other words, $D_n = C_n \wedge C_2$, since

(1) $D_n = C_n C_2$,
(2) $C_n \cap C_2 = \{I_2\}$,
(3) $C_n \triangleleft D_n$ or $BC_n = C_n B$ (the coset of reflections)(C_n is a normal subgroup of D_n)

Condition (3) in more detail: $BC_n = \{B, BA, BA^2, \ldots, BA^{n-1}\} = \{B, AB, A^2B, \ldots, A^{n-1}B\} = C_n B$, since $AB = BA^{n-1}$, $A^2B = BA^{n-2}, \ldots, A^{n-1}B = BA$ [$(n-1)$ equalities], which are all equivalent to the third generator relation $(AB)^2 = I_2$. The first two generator relations are $A^n = I_2$ and $B^2 = I_2$, and the two generators are naturally A and B.

The simplest dihedral group is D_3, the symmetry group of an equilateral triangle:

$$A = \mathrm{Rot}(\frac{360°}{3}) = \mathrm{Rot}(120°) = \begin{bmatrix} -\frac{1}{2} & -\frac{\sqrt{3}}{2} \\ \frac{\sqrt{3}}{2} & -\frac{1}{2} \end{bmatrix},$$

$$A^2 = \mathrm{Rot}(240°), \quad A^3 = I_2,$$

$$B = \mathrm{Ref}(0°) = \begin{bmatrix} -1 & 0 \\ 0 & 1 \end{bmatrix},$$

D_4	I_2	A	A^2	A^3	B	AB	A^2B	A^3B
I_2	I_2	A	A^2	A^3	B	AB	A^2B	A^3B
A	A	A^2	A^3	I_2	AB	A^2B	A^3B	B
A^2	A^2	A^3	I_2	A	A^2B	A^3B	B	AB
A^3	A^3	I_2	A	A^2	A^3B	B	AB	A^2B
B	B	A^3B	A^2B	AB	I_2	A^3	A^2	A
AB	AB	B	A^3B	A^2B	A	I_2	A^3	A^2
A^2B	A^2B	AB	B	A^3B	A^2	A	I_2	A^3
A^3B	A^3B	A^2B	AB	B	A^3	A^2	A	I_2

$$(AB)A = A(BA) = A(A^3B) = B$$
$$(AB)A^2 = A(BA^2) = A(A^2B) = A^3B$$
$$(AB)A^3 = A(BA^3) = A(AB) = A^2B$$
$$(A^2B)A = A^2(BA) = A^2(A^3B) = AB$$
$$(A^2B)A^2 = A^2(BA^2) = A^2(A^2B) = B$$
$$(A^2B)A^3 = A^2(BA^3) = A^2(AB) = A^3B$$

$$(A^3B)A = A^3(BA) = A^3(A^3B) = A^2B$$
$$(A^3B)A^2 = A^3(BA^2) = A^3(A^2B) = AB$$
$$(A^3B)A^3 = A^3(BA^3) = A^3(AB) = B.$$

$$B(AB) = (BA)B = (A^3B)B = A^3$$
$$B(A^2B) = (BA^2)B = (A^2B)B = A^2$$
$$B(A^3B) = (BA^3)B = (AB)B = A$$
$$(AB)(AB) = A(BA)B = A(A^3B)B = I_2$$
$$(AB)(A^2B) = A(BA^2)B = A(A^2B)B = A^3$$
$$(AB)(A^3B) = A(BA^3)B = A(AB)B = A^2$$
$$(A^2B)(AB) = A^2(BA)B = A^2(A^3B)B = A$$
$$(A^2B)(A^2B) = A^2(BA^2)B = A^2(A^2B)B = I_2$$
$$(A^2B)(A^3B) = A^2(BA^3)B = A^2(AB)B = A^3$$
$$(A^3B)(AB) = A^3(BA)B = A^3(A^3B)B = A^2$$
$$(A^3B)(A^2B) = A^3(BA^2)B = A^3(A^2B)B = A$$
$$(A^3B)(A^3B) = A^3(BA^3)B = A^3(AB)B = I_2$$

The group table for D_4 obtained with the help of the generator relations.

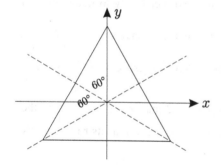

$$\text{Ref}(60°) = \text{Rot}(120°)\text{Ref}(0°) = \text{Ref}(0°)\text{Rot}(240°) =$$
$$AB = BA^2, \text{Ref}(120°) =$$
$$\text{Rot}(240°)\text{Ref}(0°) = \text{Ref}(0°)\text{Rot}(120°) = A^2B = BA.$$

So, $D_3 = \{I_2, A, A^2, B, AB = BA^2, A^2B = BA\} = C_3C_2$, where the two cyclic subgroups are $C_3 = \{I_2, A, A^2\}$ and $C_2 = \{I_2, B\}$, with $C_3 \cap C_2 = \{I_2\}$ and $C_3 \lhd D_3$, since $BC_3 = C_3B$. Therefore, $D_3 = C_3 \wedge C_2$ (a semi-direct product). The two generators are A and B, and three generator relations are

$$A^3 = I_2, \quad B^2 = I_2, \quad (AB)^2 = I_2.$$

5.7.3 The Canonical Forms of Orthogonal Matrices in $\mathbb{R}^3$ (Rotations and Rotations with Inversions)

We start by investigating the eigenvalues of 3×3 orthogonal matrices $\mathscr{A}$ in $\mathbb{R}^3$. We already know that the cubic characteristic equation of $\mathscr{A}$ must have at least one real solution. Now, we show that this eigenvalue is either $+1$ or -1:

$$\mathscr{A}\bar{x} = \lambda\bar{x}, \ \bar{x} \neq \bar{0} \text{ and } (\mathscr{A}\bar{x}, \mathscr{A}\bar{x}) = (\bar{x}, \bar{x}) \Rightarrow \lambda^2(\bar{x}, \bar{x}) = (\bar{x}, \bar{x}) \Rightarrow \lambda^2 = 1 \Rightarrow \lambda_{1,2} = \pm 1.$$

To the eigenvalue $\lambda_1 = 1$, there corresponds an eigenvector $\bar{x}_1$ $(\mathscr{A}\bar{x}_1 = \bar{x}_1)$, and let $E_1 = L(\bar{x}_1)$ be the one-dimensional eigenspace (i.e., the eigenline). We want to show that the two-dimensional orthocomplement $E_1^\perp$ is an invariant plane with respect to $\mathscr{A}$: Let $\bar{y}_1 \in E_1^\perp \Leftrightarrow (\bar{y}_1, \bar{x}_1) = 0$. Then $(\mathscr{A}\bar{y}_1, \bar{x}_1) = (\mathscr{A}\bar{y}_1, \mathscr{A}\bar{x}_1) = (\bar{y}_1, \bar{x}_1) = 0 \Rightarrow \mathscr{A}\bar{y}_1 \in E_1^\perp$.

An analogous proof is in the case of $\lambda_2 = -1$, i.e., $E_{-1}^\perp$ is also an invariant plane with respect to $\mathscr{A}$.

Conclusion: every 3×3 real orthogonal matrix $\mathscr{A}$ has one real eigenvalue which is either $+1$ or -1, and in either case, the orthocomplement of the corresponding eigenline is an invariant plane with respect to $\mathscr{A}$.

(Note that another method to come to this conclusion is to use the analogue of the theorem in Sect. 5.5.1: the orthocomplement of an invariant subspace of an orthogonal matrix is also an invariant subspace. We need only to notice that an eigenline is in fact an invariant subspace.)

In an orthonormal (ON) basis adapted to the orthogonal sum $E_1 \oplus E_1^\perp$ (or $E_{-1} \oplus E_{-1}^\perp$), the matrix $\mathscr{A}$ takes on one of the two forms respectively: $\begin{bmatrix} \pm 1 & 0 & 0 \\ 0 & a & b \\ 0 & c & d \end{bmatrix}$

with the determinant $\pm \det \begin{bmatrix} a & b \\ c & d \end{bmatrix}$. The reduced 2×2 matrix $\mathscr{A}_1 = \begin{bmatrix} a & b \\ c & d \end{bmatrix}$ in the orthocomplement is of course orthogonal (its columns, as well as its rows are ON vectors).

1. If we assume $\det \mathscr{A}_1 = +1$, we can perform the same analysis as we did in Sect. 5.7.2, and conclude

$$\mathscr{A}_1 = \text{Rot}(\alpha) = \begin{bmatrix} \cos\alpha & -\sin\alpha \\ \sin\alpha & \cos\alpha \end{bmatrix}.$$

2. If the determinant of $\mathscr{A}_1$ is -1, we can show that the invariant plane is reducible, i.e., that it necessarily breaks up into the orthogonal sum of two eigenlines which correspond to the eigenvalues -1 and $+1$.

Indeed, the characteristic equation of the matrix $\mathscr{A}_1 = \begin{bmatrix} a & b \\ c & d \end{bmatrix}$ with $\det \mathscr{A}_1 = -1$ is

$\lambda^2 - (a+d)\lambda - 1 = 0$ [since it is $\lambda^2 - (\text{Tr } \mathscr{A}_1)\lambda + \det \mathscr{A}_1 = 0$]. The discriminant of this quadratic equation is $(a+d)^2 + 4$ and it is always positive, so the eigenvalues are real and distinct, in fact -1 and $+1$, since an orthogonal matrix can have only such eigenvalues. The eigenlines that correspond to these distinct eigenvalues are orthogonal, since we have already proved that $\mathscr{A}_1$ is also a symmetric matrix (Sect. 5.6.1). Therefore, the orthogonal matrix $\mathscr{A}_1$ with $\det \mathscr{A}_1 = -1$ has the diagonal form $\begin{bmatrix} -1 & 0 \\ 0 & 1 \end{bmatrix}$ in its ON eigenbasis.

Returning to $\mathscr{A} = \begin{bmatrix} \pm 1 & 0 & 0 \\ 0 & a & b \\ 0 & c & d \end{bmatrix}$, we conclude that $\det \mathscr{A} = +1$ can be achieved in two cases:

$$\begin{bmatrix} +1 & 0 & 0 \\ 0 & \cos\alpha & -\sin\alpha \\ 0 & \sin\alpha & \cos\alpha \end{bmatrix} \begin{array}{l} \text{(an eigenline and} \\ \text{the orthogonal} \\ \text{invariant plane)} \end{array} \text{ and } \begin{bmatrix} -1 & 0 & 0 \\ 0 & -1 & 0 \\ 0 & 0 & +1 \end{bmatrix} \begin{array}{l} \text{(three} \\ \text{orthogonal} \\ \text{eigenlines).} \end{array}$$

The first case represents a rotation in $\mathbb{R}^3$ determined by the only eigenvector $\bar{x}_1$ of $\lambda = +1$ (this vector defines the axis of rotation) performed through the angle α (counterclockwise). We can write it as $\mathrm{Rot}_{\bar{x}_1}(\alpha)$.

The second case is the rotation through the angle π ($\cos\pi = -1$, $\sin\pi = 0$) around the third eigenvector $\bar{x}_3$ (the axis of rotation) corresponding to $\lambda = +1$. A convenient notation for this rotation is $\mathrm{Rot}_{\bar{x}_3}(\pi)$.

In both cases (with $\det \mathscr{A} = +1$), the matrix $\mathscr{A}$ represents a rotation in the space $\mathbb{R}^3$ around the eigenline with $\lambda = +1$ as an axis of rotation. The angles of rotation being α and π, respectively.

If $\det \mathscr{A} = -1$, we also have two possibilities:

$$\begin{bmatrix} -1 & 0 & 0 \\ 0 & \cos\alpha & -\sin\alpha \\ 0 & \sin\alpha & \cos\alpha \end{bmatrix} \text{ or } \begin{bmatrix} +1 & 0 & 0 \\ 0 & -1 & 0 \\ 0 & 0 & +1 \end{bmatrix}.$$

If we denote by $\mathrm{Inv}(\bar{x}_1)$ the inversion of the first coordinate axis (i.e., the 3×3 matrix $\begin{bmatrix} -1 & 0 & 0 \\ 0 & 1 & 0 \\ 0 & 0 & 1 \end{bmatrix}$ representing the reflection with respect to the plane which is orthogonal to the first eigenvector $\bar{x}_1$), then the first case is $\mathrm{Rot}_{\bar{x}_1}(\alpha)\mathrm{Inv}(\bar{x}_1) = \mathrm{Inv}(\bar{x}_1)\mathrm{Rot}_{\bar{x}_1}(\alpha)$. (We have the rotation around the first coordinate axis through the angle α followed or preceded by the inversion of that axis).

In the second case, we have $\mathrm{Inv}(\bar{x}_2) = \begin{bmatrix} 1 & 0 & 0 \\ 0 & -1 & 0 \\ 0 & 0 & 1 \end{bmatrix}$, the inversion of the second coordinate axis which can be followed or preceded by the rotation around that axis through the angle $0°$: $\mathrm{Rot}_{\bar{x}_2}(0°)\mathrm{Inv}(\bar{x}_2) = \mathrm{Inv}(\bar{x}_2)\mathrm{Rot}_{\bar{x}_2}(0°)$, since $\mathrm{Rot}_{\bar{x}_2}(0°) = I_3 = \begin{bmatrix} 1 & 0 & 0 \\ 0 & 1 & 0 \\ 0 & 0 & 1 \end{bmatrix}$.

Theorem (on canonical forms of orthogonal matrices in $\mathbb{R}^3$) Orthogonal matrices $\mathscr{A}$ in $\mathbb{R}^3$ are divided into two sets according to their determinants: if the determinant is $+1$ we have rotations and if $\det \mathscr{A} = -1$ we have rotations with inversion of the axis of rotation. Rotations have one eigenline (the axis of rotation) with the eigenvalue $+1$ and one invariant plane orthogonal to that eigenline. In the basis composed of the ort of the eigenline and any ON basis in the plane the rotation matrix has the canonical form

$$\mathrm{Rot}_{\bar{x}_1}(\alpha) = \begin{bmatrix} 1 & 0 & 0 \\ 0 & \cos\alpha & -\sin\alpha \\ 0 & \sin\alpha & \cos\alpha \end{bmatrix}.$$

But, the ort $\frac{\bar{x}_1}{\|\bar{x}_1\|}$ of the eigenline should be orientated with respect to the direction of the rotation to make a right screw. The matrix of a rotation with inversion has the canonical form

$$\mathrm{Rot}_{\bar{x}_1}(\alpha)\mathrm{Inv}(\bar{x}_1) = \mathrm{Inv}(\bar{x}_1)\mathrm{Rot}_{\bar{x}_1}(\alpha)$$

(where $\mathrm{Inv}(\bar{x}_1)$ is the matrix of the inversion of the axis of rotation, i.e., of the first coordinate axis $\begin{bmatrix} -1 & 0 & 0 \\ 0 & 1 & 0 \\ 0 & 0 & 1 \end{bmatrix}$); more explicitly, this is $\begin{bmatrix} -1 & 0 & 0 \\ 0 & \cos\alpha & -\sin\alpha \\ 0 & \sin\alpha & \cos\alpha \end{bmatrix}$. There are also two special cases of diagonal forms: the rotation for the angle π around the third eigenline with $\lambda = +1$, i.e., $\mathrm{Rot}_{\bar{x}_3}(\pi) = \begin{bmatrix} -1 & 0 & 0 \\ 0 & -1 & 0 \\ 0 & 0 & +1 \end{bmatrix}$, and the trivial rotation through $0°$ around the second eigenline $\mathrm{Rot}_{\bar{x}_2}(0°) = I_3$ with the inversion of that eigenline, i.e., $\mathrm{Inv}(\bar{x}_2) = \begin{bmatrix} 1 & 0 & 0 \\ 0 & -1 & 0 \\ 0 & 0 & 1 \end{bmatrix}$.

Proof Everything has already been proved. $\triangle$

Chapter 6
Tensor Product of Unitary Spaces

6.1 Kronecker Product of Matrices

Definition Let us consider two square matrices (real or complex) $\mathscr{A} = [a_{ij}]_{m \times m}$ and $\mathscr{B} == [b_{\alpha\beta}]_{n \times n}$ (the Greek indices in $\mathscr{B}$ are used to avoid confusion in double index notations). The Kronecker product of the matrices $\mathscr{A}$ and $\mathscr{B}$ we call the square block matrix $\mathscr{A} \, \textcircled{K} \, \mathscr{B}$ which has m rows of blocks and m columns of blocks, such that the (i, j) block is the square matrix $a_{ij}\mathscr{B}$ which is an $n \times n$ ordinary matrix:

$$\mathscr{A} \, \textcircled{K} \, \mathscr{B} = \begin{bmatrix} a_{11}\mathscr{B} & a_{12}\mathscr{B} & \cdots & a_{1m}\mathscr{B} \\ a_{21}\mathscr{B} & a_{22}\mathscr{B} & \cdots & a_{2m}\mathscr{B} \\ \vdots & \vdots & \ddots & \vdots \\ a_{m1}\mathscr{B} & a_{m2}\mathscr{B} & \cdots & a_{mm}\mathscr{B} \end{bmatrix}.$$

More explicitly, we have an ordinary matrix of size $mn \times mn$:

$$\mathscr{A} \, \textcircled{K} \, \mathscr{B} = \begin{bmatrix} a_{11}b_{11} & \cdots & a_{11}b_{1n} & & a_{1m}b_{11} & \cdots & a_{1m}b_{1n} \\ \vdots & \vdots & \vdots & \cdots & \vdots & \vdots & \vdots \\ a_{11}b_{n1} & \cdots & a_{11}b_{nn} & & a_{1m}b_{n1} & \cdots & a_{1m}b_{nn} \\ \vdots & \vdots & \vdots & \vdots & \vdots & \vdots & \vdots \\ a_{m1}b_{11} & \cdots & a_{m1}b_{1n} & & a_{mm}b_{11} & \cdots & a_{mm}b_{1n} \\ \vdots & \vdots & \vdots & \cdots & \vdots & \vdots & \vdots \\ a_{m1}b_{n1} & \cdots & a_{m1}b_{nn} & & a_{mm}b_{n1} & \cdots & a_{mm}b_{nn} \end{bmatrix}.$$

The elements of $\mathscr{A} \, \textcircled{K} \, \mathscr{B}$ are all possible products of elements from $\mathscr{A}$ with those from $\mathscr{B}$.

So, we may use the double index notation for the elements of $\mathscr{A} \, \textcircled{K} \, \mathscr{B}$

$$[\mathscr{A} \, \textcircled{K} \, \mathscr{B}]_{i\alpha, j\beta} = a_{ij}b_{\alpha\beta}, \quad i, j = 1, 2, \ldots, m; \quad \alpha, \beta = 1, 2, \ldots, n.$$

To simplify this notation, we replace the double index notation by the single index one using the leksicography method: $11 \to 1, 12 \to 2, \dots, 1n \to n$, $21 \to n+1, 22 \to n+2, \dots, 2n \to 2 \cdot n$, $31 \to 2 \cdot n + 1, 32 \to 2 \cdot n + 2, \dots$, or generally

$$i\alpha \to (i-1)n + \alpha.$$

With the single index notation, we have $a_{ij} b_{\alpha\beta} = [\mathscr{A} \, Ⓚ \, \mathscr{B}]_{(i-1)n+\alpha,\,(j-1)n+\beta}$, where both indices in $\mathscr{A} \, Ⓚ \, \mathscr{B}$ go from 1 to $m \cdot n$, since $i, j = 1, 2, \dots, m$; $\alpha, \beta = 1, 2, \dots, n$.

Theorem (about the main properties of the Kronecker product of matrices)

(1) $\mathrm{Tr}(\mathscr{A} \, Ⓚ \, \mathscr{B}) = \mathrm{Tr}\,\mathscr{A} \cdot \mathrm{Tr}\,\mathscr{B}$ (this property is used very much in the theory of linear representation of groups);

(2) $(\mathscr{A} \, Ⓚ \, \mathscr{B}) \, Ⓚ \, \mathscr{C} = \mathscr{A} \, Ⓚ \, (\mathscr{B} \, Ⓚ \, \mathscr{C}) = \mathscr{A} \, Ⓚ \, \mathscr{B} \, Ⓚ \, \mathscr{C}$ (this is the property of associativity of the Kronecker product of matrices and it enables the definition of the Kronecker product of three and more matrices);

(3) $\mathscr{A} \, Ⓚ \, (\mathscr{B} + \mathscr{C}) = \mathscr{A} \, Ⓚ \, \mathscr{B} + \mathscr{A} \, Ⓚ \, \mathscr{C}$, $(\mathscr{A} + \mathscr{B}) \, Ⓚ \, \mathscr{C} = \mathscr{A} \, Ⓚ \, \mathscr{C} + \mathscr{B} \, Ⓚ \, \mathscr{C}$, $(a\mathscr{A}) \, Ⓚ \, \mathscr{B} = = \mathscr{A} \, Ⓚ \, (a\mathscr{B}) = a(\mathscr{A} \, Ⓚ \, \mathscr{B})$ (these three properties can be generalized into a single one—the Kronecker product is bilinear:

$$\left(\sum_i a_i \mathscr{A}_i\right) Ⓚ \left(\sum_j b_j \mathscr{B}_j\right) = \sum_{i,j} a_i b_j (\mathscr{A}_i \, Ⓚ \, \mathscr{B}_j), \quad a_i, b_j \text{ any scalars;}$$

(4) $(\mathscr{A} \, Ⓚ \, \mathscr{B})(\mathscr{C} \, Ⓚ \, \mathscr{D}) = \mathscr{A}\mathscr{C} \, Ⓚ \, \mathscr{B}\mathscr{D}$, where $\mathscr{A}$ and $\mathscr{C}$ are of size $m \times m$, whereas $\mathscr{B}$ and $\mathscr{D}$ are of size $n \times n$ (this property connects Kronecker multiplication with ordinary matrix multiplication, and it is important for the representation of the tensor product of linear operators).

Proof

1.
$$\mathrm{Tr}(\mathscr{A} \, Ⓚ \, \mathscr{B}) = \sum_{i=1}^{m} \sum_{\alpha=1}^{n} [\mathscr{A} \, Ⓚ \, \mathscr{B}]_{i\alpha, i\alpha} = \sum_{i=1}^{m} \sum_{\alpha=1}^{n} a_{ii} b_{\alpha\alpha} =$$
$$= \left(\sum_{i=1}^{m} a_{ii}\right)\left(\sum_{\alpha=1}^{n} b_{\alpha\alpha}\right) = \mathrm{Tr}\,\mathscr{A} \cdot \mathrm{Tr}\,\mathscr{B};$$

2.
$$(\mathscr{A} \, Ⓚ \, \mathscr{B}) \, Ⓚ \, \mathscr{C} = \begin{bmatrix} a_{11}\mathscr{B} & \cdots & a_{1m}\mathscr{B} \\ \vdots & \cdots & \vdots \\ a_{m1}\mathscr{B} & \cdots & a_{mm}\mathscr{B} \end{bmatrix} Ⓚ \, \mathscr{C} =$$
$$= \begin{bmatrix} a_{11}(\mathscr{B} \, Ⓚ \, \mathscr{C}) & \cdots & a_{1m}(\mathscr{B} \, Ⓚ \, \mathscr{C}) \\ \vdots & \cdots & \vdots \\ a_{m1}(\mathscr{B} \, Ⓚ \, \mathscr{C}) & \cdots & a_{mm}(\mathscr{B} \, Ⓚ \, \mathscr{C}) \end{bmatrix}$$
$$= \mathscr{A} \, Ⓚ \, (\mathscr{B} \, Ⓚ \, \mathscr{C});$$

3. $\mathscr{A} \otimes (\mathscr{B}+\mathscr{C}) = \begin{bmatrix} a_{11}(\mathscr{B}+\mathscr{C}) & \cdots & a_{1m}(\mathscr{B}+\mathscr{C}) \\ \vdots & \cdots & \vdots \\ a_{m1}(\mathscr{B}+\mathscr{C}) & \cdots & a_{mm}(\mathscr{B}+\mathscr{C}) \end{bmatrix} =$

$$= \begin{bmatrix} a_{11}\mathscr{B}+a_{11}\mathscr{C} & \cdots & a_{1m}\mathscr{B}+a_{1m}\mathscr{C} \\ \vdots & \cdots & \vdots \\ a_{m1}\mathscr{B}+a_{m1}\mathscr{C} & \cdots & a_{mm}\mathscr{B}+a_{mm}\mathscr{C} \end{bmatrix} =$$

$$= \begin{bmatrix} a_{11}\mathscr{B} & \cdots & a_{1m}\mathscr{B} \\ \vdots & \cdots & \vdots \\ a_{m1}\mathscr{B} & \cdots & a_{mm}\mathscr{B} \end{bmatrix} + \begin{bmatrix} a_{11}\mathscr{C} & \cdots & a_{1m}\mathscr{C} \\ \vdots & \cdots & \vdots \\ a_{m1}\mathscr{C} & \cdots & a_{mm}\mathscr{C} \end{bmatrix} = \mathscr{A} \otimes \mathscr{B} + \mathscr{A} \otimes \mathscr{C},$$

$$(\mathscr{A}+\mathscr{B}) \otimes \mathscr{C} = \begin{bmatrix} (a_{11}+b_{11})\mathscr{C} & \cdots & (a_{1m}+b_{1m})\mathscr{C} \\ \vdots & \cdots & \vdots \\ (a_{m1}+b_{m1})\mathscr{C} & \cdots & (a_{mm}+b_{mm})\mathscr{C} \end{bmatrix} =$$

$$= \begin{bmatrix} a_{11}\mathscr{C} & \cdots & a_{1m}\mathscr{C} \\ \vdots & \cdots & \vdots \\ a_{m1}\mathscr{C} & \cdots & a_{mm}\mathscr{C} \end{bmatrix} + \begin{bmatrix} b_{11}\mathscr{C} & \cdots & b_{1m}\mathscr{C} \\ \vdots & \cdots & \vdots \\ b_{m1}\mathscr{C} & \cdots & b_{mm}\mathscr{C} \end{bmatrix} =$$

$$= \mathscr{A} \otimes \mathscr{C} + \mathscr{B} \otimes \mathscr{C},$$

$$a(\mathscr{A} \otimes \mathscr{B}) = a \begin{bmatrix} a_{11}\mathscr{B} & \cdots & a_{1m}\mathscr{B} \\ \vdots & \cdots & \vdots \\ a_{m1}\mathscr{B} & \cdots & a_{mm}\mathscr{B} \end{bmatrix} = \begin{bmatrix} (aa_{11})\mathscr{B} & \cdots & (aa_{1m})\mathscr{B} \\ \vdots & \cdots & \vdots \\ (aa_{m1})\mathscr{B} & \cdots & (aa_{mm})\mathscr{B}) \end{bmatrix} =$$

$$= \begin{bmatrix} a_{11}(a\mathscr{B}) & \cdots & a_{1m}(a\mathscr{B}) \\ \vdots & \cdots & \vdots \\ a_{m1}(a\mathscr{B}) & \cdots & a_{mm}(a\mathscr{B}) \end{bmatrix} = (a\mathscr{A}) \otimes \mathscr{B} = \mathscr{A} \otimes (a\mathscr{B});$$

4. We shall calculate the general element $(i\alpha, j\beta)$ of both sides and find that we get the same expression:

$$[(\mathscr{A} \otimes \mathscr{B})(\mathscr{C} \otimes \mathscr{D})]_{i\alpha,j\beta} = \sum_{k=1}^{m} \sum_{\gamma=1}^{n} [\mathscr{A} \otimes \mathscr{B}]_{i\alpha,k\gamma}[\mathscr{C} \otimes \mathscr{D}]_{k\gamma,j\beta} =$$

$$= \sum_{k=1}^{m} \sum_{\gamma=1}^{n} a_{ik}b_{\alpha\gamma}c_{kj}d_{\gamma\beta},$$

$$[\mathscr{A}\mathscr{C} \otimes \mathscr{B}\mathscr{D}]_{i\alpha,j\beta} = [\mathscr{A}\mathscr{C}]_{ij}[\mathscr{B}\mathscr{D}]_{\alpha\beta} = \sum_{k=1}^{m} a_{ik}c_{kj} \sum_{\gamma=1}^{n} b_{\alpha\gamma}d_{\gamma\beta} =$$

$$= \sum_{k=1}^{m} \sum_{\gamma=1}^{n} a_{ik}c_{kj}b_{\alpha\gamma}d_{\gamma\beta}. \ \Delta$$

The Kronecker product of square matrices can be immediately generalized to the Kronecker product of rectangular matrices using the same definition. It should be emphasized that for Kronecker multiplication there is NO restriction on the sizes of factor matrices, unlike the case for ordinary multiplication of matrices where we can multiply only matrices $\mathscr{A} = [a]_{m \times p}$ and $\mathscr{B} = [b]_{p \times n}$.

Two Examples of Kronecker multiplication of rectangular matrices which we shall need further.

A. Matrix-columns of length m Kronecker-multiplied with matrix-columns of length n will give again matrix-columns, but of length $m \cdot n$:

$$\xi \otimes \eta = \begin{bmatrix} \xi_1 \\ \xi_2 \\ \vdots \\ \xi_m \end{bmatrix} \otimes \begin{bmatrix} \eta_1 \\ \vdots \\ \eta_n \end{bmatrix} = \underbrace{[\xi_1\eta_1 \ \xi_1\eta_2 \ \cdots \ \xi_1\eta_n \ \xi_2\eta_1 \ \cdots \ \xi_2\eta_n \ \cdots \ \xi_m\eta_1 \ \cdots \ \xi_m\eta_n]^T}_{m \cdot n}$$

We will use this product (Sect. 6.3) to represent elementary vectors in the tensor product $U_m \otimes U_n$.

B. Kronecker product of matrix-columns of length m with matrix-rows of length n will give matrices of size $m \times n$:

$$\text{Let } \xi = \begin{bmatrix} \xi_1 \\ \xi_2 \\ \vdots \\ \xi_m \end{bmatrix}, \ \eta = \begin{bmatrix} \eta_1 \\ \eta_2 \\ \vdots \\ \eta_n \end{bmatrix},$$

$$\xi \otimes \eta^T = [\xi_1 \ \xi_2 \ \cdots \ \xi_m]^T \otimes [\eta_1 \ \eta_2 \ \cdots \ \eta_n] = \begin{bmatrix} \xi_1\eta^T \\ \xi_2\eta^T \\ \cdots \\ \xi_m\eta^T \end{bmatrix} =$$

$$= \begin{bmatrix} \xi_1\eta_1 & \xi_1\eta_2 & \cdots & \xi_1\eta_n \\ \xi_2\eta_1 & \xi_2\eta_2 & \cdots & \xi_2\eta_n \\ \vdots & \vdots & \ddots & \vdots \\ \xi_m\eta_1 & \xi_m\eta_2 & \cdots & \xi_m\eta_n \end{bmatrix} = [\xi_i\eta_j]_{m \times n}$$

It should be noticed that the above Kronecker product gives the same result as the ordinary matrix product of these factors

$$\xi\eta^T = \begin{bmatrix} \xi_1 \\ \xi_2 \\ \vdots \\ \xi_m \end{bmatrix} [\eta_1 \ \eta_2 \ \cdots \ \eta_n] = \begin{bmatrix} \xi_1\eta_1 & \xi_1\eta_2 & \cdots & \xi_1\eta_n \\ \xi_2\eta_1 & \xi_2\eta_2 & \cdots & \xi_2\eta_n \\ \vdots & \vdots & \ddots & \vdots \\ \xi_m\eta_1 & \xi_m\eta_2 & \cdots & \xi_m\eta_n \end{bmatrix} = [\xi_i \ \eta_j]_{m \times n}.$$

Thus, $\xi \otimes \eta^T = \xi\eta^T$.

We will use this product (Sect. 7.3.2) to represent elementary vectors (diads) in the tensor product $U_n \otimes U_n^*$ of a unitary space U_n and its U_n^* dual in an ON basis of U_n.

6.2 Axioms for the Tensor Product of Unitary Spaces

The tensor product of unitary spaces is of great importance for the Quantum Mechanics of many particles. Also, the multiple tensor product of U_n with the multiple tensor product of U_n^* is the principal example for the notion of unitary tensors. To avoid an abstract exposition of axioms without giving one of the most typical examples, we shall start with the case that we have already described.

6.2.1 The Tensor product of Unitary Spaces $\mathbb{C}^m$ and $\mathbb{C}^n$

Let us consider the Cartesian product $\mathbb{C}^m \times \mathbb{C}^n$ which consists of all ordered pairs $[\xi, \eta]$ from $\mathbb{C}^m$ and $\mathbb{C}^n$, i.e., $\xi \in \mathbb{C}^m$, $\eta \in \mathbb{C}^n$. (We use square brackets for ordered pairs to avoid confusion with the inner-product notation.)

Now, we shall consider the $m \cdot n$-dimensional unitary space $\mathbb{C}^{m \cdot n}$ which consists of matrix-columns of complex numbers of length $m \cdot n$, and concentrate our attention on those elements from $\mathbb{C}^{m \cdot n}$ which can be expressed as the Kronecker product of vectors from $\mathbb{C}^m$ and $\mathbb{C}^n$. We call them *elementary vectors* in $\mathbb{C}^{m \cdot n}$. We see that there is a natural map φ_0 which takes every ordered pair $[\xi, \eta]$ from $\mathbb{C}^m \times \mathbb{C}^n$ onto the unique Kronecker product $\xi \, \text{\textcircled{K}} \, \eta$ in $\mathbb{C}^{m \cdot n}$:

$$\varphi_0[\xi, \eta] = \xi \, \text{\textcircled{K}} \, \eta.$$

This map φ_0 is obviously bilinear, since the Kronecker product is such:

$$\varphi_0[\sum_i a_i \xi_i, \sum_j b_j \eta_j] = (\sum_i a_i \xi_i) \, \text{\textcircled{K}} \, (\sum_j b_j \eta_j) = \sum_{i,j} a_i b_j (\xi_i \, \text{\textcircled{K}} \, \eta_j).$$

The ordered pairs of vectors from the ON standard bases $\{e_1, e_2, \ldots, e_m\}$ and $\{e_1', e_2', \ldots, e_n'\}$ in $\mathbb{C}^m$ and $\mathbb{C}^n$, respectively, are obviously mapped onto the ON standard basis in $\mathbb{C}^{m \cdot n}$. As the general rule, the column e_i from $\mathbb{C}^m$ which has 1 at the i-th row when Kronecker multiplied with the column e_α from $\mathbb{C}^n$ which has 1 at the α-th row will give the column in $\mathbb{C}^{m \cdot n}$ which has 1 at the $[(i-1)n + \alpha]$th row. The rest of the elements in all three columns are 0.

These two properties (bilinearity of φ_0 and that ON standard bases in factor spaces give the ON standard basis in the product space) of the three unitary spaces (the dimension of the third one is the product of the dimensions of the other two) with the map $\varphi_0 : \mathbb{C}^m \times \mathbb{C}^n \to \mathbb{C}^{m \cdot n}$ are sufficient that we can consider $\mathbb{C}^{m \cdot n}$ as the *tensor product* of $\mathbb{C}^m$ and $\mathbb{C}^n$. We denote this as

$$\mathbb{C}^{m\cdot n} = \mathbb{C}^m \otimes \mathbb{C}^n,$$

while for elementary vectors $\xi \;\text{\textcircled{K}}\; \eta$, we use the general notation $\xi \otimes \eta$ knowing how we define them.

There are three important consequences of this definition of the tensor product between $\mathbb{C}^m$ and $\mathbb{C}^n$.

1. The elementary vectors in $\mathbb{C}^{m\cdot n}$, which make the range of φ_0, span the whole $\mathbb{C}^{m\cdot n}$. As a matter of fact, only a subset of elementary vectors $\{e_i \otimes e'_j \,|\, i = 1,2,\ldots,m; \; j = 1,2,\ldots,n\}$ span the whole $\mathbb{C}^{m\cdot n}$ since this subset is the standard ON basis in $\mathbb{C}^{m\cdot n}$. Thus, $\text{LIN}[\varphi_0(\mathbb{C}^m \times \mathbb{C}^n)] = \mathbb{C}^m \otimes \mathbb{C}^n = \mathbb{C}^{m\cdot n}$. Δ

2. The inner product between elementary vectors in $\mathbb{C}^{m\cdot n}$ is directly expressible through the inner products in $\mathbb{C}^m$ and $\mathbb{C}^n$:

$$(\xi \otimes \eta, \xi' \otimes \eta') = (\xi, \xi')(\eta, \eta'),$$

which is easy to verify:

$$(\xi \otimes \eta, \xi' \otimes \eta') = \sum_{i=1}^{m}\sum_{j=1}^{n} \xi_i^* \eta_j^* \xi_i' \eta_j' = \Big(\sum_{i=1}^{m} \xi_i^* \xi_i'\Big)\Big(\sum_{j=1}^{n} \eta_j^* \eta_j'\Big) =$$
$$= (\xi, \xi')(\eta, \eta'). \; \Delta$$

3. The consequences 1 and 2, together with the requirement of bilinearity of φ_0, can be used as three equivalent definitions of the tensor product, so that the property that ON bases in the factor spaces give an ON basis in the product space follows. We shall not give the easy proof for this statement.

   Instead, we shall show that if any two ON bases in the factor spaces give an ON basis in the product space (like e.g. standard bases), then this will be valid for any other pair of ON bases from the factor spaces.

Proof Let $\{u_1,\ldots,u_m\}$ and $\{v_1,\ldots,v_n\}$ be arbitrary ON bases in the factor spaces, so that $(u_k \otimes v_l, u_p \otimes v_q) = (u_k,u_p)(v_l,v_q) = \delta_{kp}\delta_{lq}, \;\; k,p = 1,2,\ldots,m, \;\; l,q = 1,2,\ldots,n$, (an ON basis in the product space).

If we now choose another pair of ON bases $\{u'_1,\ldots,u'_m\}$ and $\{v'_1,\ldots,v'_n\}$ in the factor spaces obtained by the unitary replacement matrices $\mathcal{R}$ and $\mathcal{R}'$, then

$$(u'_i \otimes v'_j, u'_s \otimes v'_t) = \Big(\big(\sum_k r_{ik}u_k\big) \otimes \big(\sum_l r'_{jl}v_l\big), \big(\sum_p r_{sp}u_p\big) \otimes \big(\sum_q r'_{tq}v_q\big)\Big) =$$
$$= \sum_{k,l,p,q} r_{ik}^* r_{jl}'^* r_{sp} r'_{tq} \underbrace{(u_k \otimes v_l, u_p \otimes v_q)}_{\delta_{kp}\delta_{lq}} =$$
$$= \sum_k r_{ik}^* r_{sk} \sum_l r_{jl}'^* r'_{tl} = \delta_{is}\delta_{jt},$$
$$i,s = 1,2,\ldots,m, \;\; j,t = 1,2,\ldots,n,$$

since $(\mathcal{R}\mathcal{R}^{\dagger} = I)^* \Leftrightarrow \mathcal{R}^* \mathcal{R}^T = I$. Δ

Now, we shall give the general definition of the tensor product of unitary spaces.

6.2.2 Definition of the Tensor Product of Unitary Spaces, in Analogy with the Previous Example

Let us take into consideration three unitary spaces U_m, U_n, and $U_{m \cdot n}$. We consider $U_{m \cdot n}$ as the tensor product of U_m and U_n, and write that as

$$U_{m \cdot n} = U_m \otimes U_n.$$

If there exists a map φ of $U_m \times U_n$ (the Cartesian product made of ordered pairs from U_m and U_n) into $U_{m \cdot n}$, i.e., $\varphi[\bar{x}, \bar{y}] = \bar{x} \otimes \bar{y}$, $\bar{x} \in U_m$, $\bar{y} \in U_n$, $\bar{x} \otimes \bar{y} \in U_{m \cdot n}$ (the image $\bar{x} \otimes \bar{y}$ is called an *elementary vector* in $U_{m \cdot n}$), which satisfies only two properties:

1. The map φ is bilinear

$$\varphi[\sum_i a_i \bar{x}_i, \sum_j b_j \bar{y}_j] = (\sum_i a_i \bar{x}_i) \otimes (\sum_j b_j \bar{y}_j) = \sum_{i,j} a_i b_j (\bar{x}_i \otimes \bar{y}_j);$$

2. If $\{\bar{u}_1, \ldots, \bar{u}_m\}$ and $\{\bar{v}_1, \ldots, \bar{v}_n\}$ are any two ON bases in U_m and U_n, respectively, then $\{\varphi[\bar{u}_i, \bar{v}_j] \mid i = 1, \ldots, m; \ j = 1, \ldots, n\} = \{\bar{u}_i \otimes \bar{v}_j \mid i = 1, \ldots, m; \ j = 1, \ldots, n\}$ is an ON basis in $U_{m \cdot n}$.

   For linear operators A from U_m and B from U_n, we can define their tensor product $A \otimes B$ as the linear operator in $U_{m \cdot n}$ which is given by its action on the vectors from an ON basis in $U_{m \cdot n}$

$$(A \otimes B)(\bar{u}_i \otimes \bar{v}_j) = (A\bar{u}_i) \otimes (B\bar{v}_j).$$

The action of $A \otimes B$ on other vectors from $U_{m \cdot n}$ is immediately obtained when we expand them in the above basis and then apply the linear operator on these expansions.

Note that not all linear operators in $U_{m \cdot n}$ are in the form of the tensor product of linear operators from U_m and U_n.

For the general case of the tensor product of unitary spaces $U_{m \cdot n} = U_m \otimes U_n$, we could easily demonstrate the two important consequences 1 and 2, which we proved in the case of $\mathbb{C}^{m \cdot n} = \mathbb{C}^m \otimes \mathbb{C}^n$:

1. The elementary vectors $\bar{x} \otimes \bar{y}$ in $U_{m \cdot n}$ span the whole $U_{m \cdot n}$:

$$\text{LIN}[\varphi(U_m \times U_n)] = U_{m \cdot n} = U_m \otimes U_n;$$

2. The inner product between elementary vectors in $U_{m \cdot n}$ is directly expressible through the inner products in U_m and U_n:

$$(\bar{x} \otimes \bar{y}, \bar{x}' \otimes \bar{y}') = (\bar{x}, \bar{x}')(\bar{y}, \bar{y}').$$

The tensor product of unitary spaces is associative, so it can be defined for more than two factors: e.g., $\underbrace{U_n \otimes U_n \otimes \cdots \otimes U_n}_{m} = \overset{m}{\otimes} U_n$—the m-th tensorial power of U_n.

6.3 Matrix Representation of the Tensor Product
of Unitary Spaces

Let the two unitary spaces U_m and U_n be given, together with ON bases $\{\bar{u}_1,\ldots,\bar{u}_m\}$ and $\{\bar{v}_1,\ldots,\bar{v}_n\}$ in them, respectively. As a consequence, we have two isomorphisms i_1 and i_2 between these spaces and unitary spaces $\mathbb{C}^m$ and $\mathbb{C}^n$, respectively. So, for $\bar{x} \in U_m$, we have the image $i_1(\bar{x}) = \xi = [\xi_1,\ldots,\xi_m]^T$, where $\bar{x} = \sum_{i=1}^{m}\xi_i\bar{u}_i$, and analogously for $\bar{y} \in U_n$, $i_2(\bar{y}) = \eta = [\eta_1,\ldots,\eta_n]^T$.

If the third unitary space $U_{m\cdot n}$ is also given, and a map $\varphi : U_m \times U_n \to U_{m\cdot n}$, which satisfies the two axioms of the tensor product, then due to bilinearity of φ

$$\varphi[\bar{x},\bar{y}] = \bar{x} \otimes \bar{y} = \sum_{i=1}^{m}\sum_{j=1}^{n}\xi_i\eta_j(\bar{u}_i \otimes \bar{v}_j) = \sum_{i=1}^{m}\sum_{j=1}^{n}(\xi \,\text{Ⓚ}\, \eta)_{ij}(\bar{u}_i \otimes \bar{v}_j).$$

Since $\{\bar{u}_i \otimes \bar{v}_j \,|\, i = 1,\ldots,m;\ j = 1,\ldots,n\}$ is an ON basis in $U_{m\cdot n}$, we have the isomorphism i_3 between unitary spaces $U_{m\cdot n}$ and $\mathbb{C}^{m\cdot n}$. So, by choosing ON bases in U_m and U_n, we realized a map $\varphi' = i_3 \circ \varphi$ of $U_m \times U_n$ on the set of elementary vectors in $\mathbb{C}^{m\cdot n}$

$$\varphi'[\bar{x},\bar{y}] = \xi \,\text{Ⓚ}\, \eta = \xi \otimes \eta.$$

It is easy to verify that the map φ' is bilinear and that the tensor product of two ON bases in U_m and U_n is mapped onto the standard ON basis in $\mathbb{C}^{m\cdot n}$.

Thus, the unitary space $\mathbb{C}^{m\cdot n}$ represents the tensor product $U_{m\cdot n} = U_m \otimes U_n$ through the isomorphism i_3, and this representation is achieved by the choice of ON bases in U_m and U_n.

More directly, the map $\varphi' = i_3 \circ \varphi$ enables us to write $\mathbb{C}^{m\cdot n} = U_m \otimes U_n$. This is a typical example of a basis-dependent definition of the tensor product.

In a very similar way, one can prove that the operator $A \otimes B$ in $U_{m\cdot n}$ is represented by the Kronecker product $\mathscr{A} \,\text{Ⓚ}\, \mathscr{B}$ of matrices $\mathscr{A}$ and $\mathscr{B}$ that represent A and B in the chosen bases:

$$(A \otimes B)(\bar{u}_i \otimes \bar{v}_p) = (A\bar{u}_i) \otimes (B\bar{v}_p) = \Big(\sum_{j=1}^{m}a_{ji}\bar{u}_j\Big) \otimes \Big(\sum_{q=1}^{n}b_{qp}\bar{v}_q\Big) =$$

$$= \sum_{j=1}^{m}\sum_{q=1}^{n}a_{ji}b_{qp}(\bar{u}_j \otimes \bar{v}_q) =$$

$$= \sum_{j=1}^{m}\sum_{q=1}^{n}[\mathscr{A} \,\text{Ⓚ}\, \mathscr{B}]_{jq,ip}(\bar{u}_j \otimes \bar{v}_q).$$

Two Remarks

(1) We notice that there is a second route by which we can reach $\xi \,\text{Ⓚ}\, \eta$ (see the figure below), once the ON bases in U_m and U_n have been chosen, without taking into consideration the map φ and the particular tensor product $U_m \otimes U_n = U_{m\cdot n}$

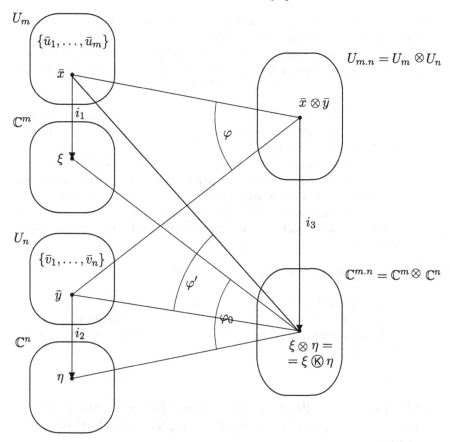

that this map implies. Namely, the ordered pair $[\xi, \eta]$ of $i_1(\bar{x}) = \xi$ and $i_2(\bar{y}) = \eta$ is taken by the natural map φ_0 onto the Kronecker product $\xi \, \text{\textcircled{K}} \, \eta$ in $\mathbb{C}^m \otimes \mathbb{C}^n = \mathbb{C}^{m \cdot n}$

$$\varphi_0[i_1(\bar{x}), i_2(\bar{y})] = \xi \, \text{\textcircled{K}} \, \eta \quad \text{(see Sect. 6.2.1).}$$

So, we can conclude that every other realization of the tensor product of U_m and U_n is isomorphic to this one

$$\mathbb{C}^{m \cdot n} = U_m \otimes U_n.$$

In other words, if $U_m \otimes U_n$ and $U_m \otimes' U_n$ are two distinct tensor products defined by bilinear maps φ and φ', respectively, then there exists a unique isomorphism T whose domain is $U_m \otimes U_n$ and the range is $U_m \otimes' U_n$, so that

$$T\varphi[\bar{x}, \bar{y}] = \varphi'[\bar{x}, \bar{y}], \quad \text{for all } \bar{x} \in U_m \text{ and } \bar{y} \in U_n.$$

The tensor product $U_n \otimes U_n^*$ of a unitary space U_n and its dual space U_n^* is especially important for its application in Quantum Mechanics (the Dirac notation). In Sect. 7.3, we shall study two isomorphic realizations of this tensor product, namely the superspaces $\hat{L}(U_n, U_n)$ and $\hat{L}(U_n^*, U_n^*)$, which are connected by an invariant isomorphism-isodualism, which even enables their identification (Sects. 4.7 and 7.2.2).

(2) The matrix representation $\mathbb{C}^{m \cdot n} = U_m \otimes U_n$ of the tensor product is basis dependent. In another ON basis in U_m, the same vector $\bar{x}$ is represented by the matrix-column $\xi' = \mathscr{R}^* \xi$ (Sect. 4.4), where $\mathscr{R}$ is the unitary replacement matrix from the old to the new ON basis, and $\mathscr{R}^* = (\mathscr{R}^\dagger)^T = (\mathscr{R}^{-1})^T = \mathscr{S}$ is the unitary contragredient matrix of $\mathscr{R}$. The analogous situation is in U_n, the vector $\bar{y}$ is now represented by the matrix-column $\eta' = \bar{\mathscr{R}}^* \eta$, where the unitary matrix $\bar{\mathscr{R}}^* = \bar{\mathscr{S}}$ is contragredient to the unitary matrix $\bar{\mathscr{R}}$ that connects ON bases in U_n. The elementary vector $\bar{x} \otimes \bar{y}$ is now represented by another elementary vector in $\mathbb{C}^{m \cdot n}$

$$\xi' \otimes\!\!\!\!\!K\;\; \eta' = \mathscr{R}^* \xi \otimes\!\!\!\!\!K\;\; \bar{\mathscr{R}}^* \eta = (\mathscr{R}^* \otimes\!\!\!\!\!K\;\; \bar{\mathscr{R}}^*)(\xi \otimes\!\!\!\!\!K\;\; \eta),$$

by the fourth property of the Kronecker product.

Thus, even though the matrix representation of the tensor product $U_m \otimes U_n$ is basis dependent, we know the precise procedure for representing $\bar{x} \otimes \bar{y}$ in other ON bases in U_m and U_n.

6.4 Multiple Tensor Products of a Unitary Space U_n and of its Dual Space U_n^* as the Principal Examples of the Notion of Unitary Tensors

When we multiply the unitary space U_n tensorially with itself, i.e., $U_n \otimes U_n$ (this is called the tensorial square of U_n and denoted as $\overset{2}{\otimes} U_n$), this means that there is a map φ which takes the Cartesian product $U_n \times U_n$ into another unitary space U_{n^2}, so that the two axioms of the tensor product are satisfied.

If we represent that product in an ON basis $\{\bar{v}_1, \ldots, \bar{v}_n\}$ in U_n, then every elementary vector $\bar{x} \otimes \bar{y}$, $\bar{x}, \bar{y} \in U_n$, is taken to the Kronecker product $\xi \otimes\!\!\!\!\!K\;\; \eta$, where $\xi = i(\bar{x})$ and $\eta = i(\bar{y})$ are the representative columns of $\bar{x}$ and $\bar{y}$ realized by the isomorphism $i : U_n \to \mathbb{C}^n$, induced by the choice of the above ON basis in U_n.

When we go over to another ON basis $\{\bar{v}_1', \ldots, \bar{v}_n'\}$ in U_n by the unitary replacement matrix $\mathscr{R} = [r_{ij}]_{n \times n}$, then $\bar{x} \otimes \bar{y}$ will be represented by the Kronecker product of columns $\xi' \otimes\!\!\!\!\!K\;\; \eta' = \mathscr{R}^* \xi \otimes\!\!\!\!\!K\;\; \mathscr{R}^* \eta = (\mathscr{R}^* \otimes\!\!\!\!\!K\;\; \mathscr{R}^*)(\xi \otimes\!\!\!\!\!K\;\; \eta)$, (Sects. 4.4 and 6.1) where $\mathscr{R}^* = (\mathscr{R}^\dagger)^T = (\mathscr{R}^{-1})^T = \mathscr{S}$ is the contragredient matrix to $\mathscr{R}$. In terms of components, this reads as

$$\xi_\alpha' \eta_\beta' = \sum_{i,j=1}^{n} r_{\alpha i}^* r_{\beta j}^* \xi_i \eta_j, \quad \alpha, \beta = 1, \ldots, n. \quad (*)$$

In this way, we get a two index system of n^2 complex numbers $\{\xi_i\eta_j \mid i,j = 1,\ldots,n\}$ which is attached to the first ON basis in U_n, and which changes into the new system [see above expression $(*)$] when the ON basis is changed to the new one by the replacement matrix $\mathscr{R} \in U(n): \bar{v}_i' = \sum_{j=1}^n r_{ij}\bar{v}_j$. This is the principal example of the unitary tensor which is twice contravariant since both indices change by the contragredient matrix $\mathscr{R}^* = \mathscr{S} = (\mathscr{R}^{-1})^T$. So, the tensorial square of the unitary space U_n generates such a tensor when represented by matrices in ON bases in U_n.

We can generalize this idea to the tensor product of U_n with itself m times $\overset{m}{\otimes} U_m$ (the m-th tensorial power of U_n). When we represent that product in an ON basis in U_n, then every elementary vector $\bar{x}_1 \otimes \cdots \otimes \bar{x}_m$ is mapped on the Kronecker product $\xi_1 \mathbin{\text{Ⓚ}} \cdots \mathbin{\text{Ⓚ}} \xi_m$ which is an m-index system of n^m complex numbers. In another ON basis in U_n, which is obtained by the unitary replacement matrix $\mathscr{R}$, this system is multiplied by the Kronecker product $\underbrace{\mathscr{R}^* \mathbin{\text{Ⓚ}} \cdots \mathbin{\text{Ⓚ}} \mathscr{R}^*}_{m}$ of m contragredient ma-

trices $\mathscr{R}^* = \mathscr{S}$. This is the principal example of the unitary tensor which is m times contravariant.

In Sect. 4.1, we analyzed the change of the representation of the dual vector $D\bar{x} = f \in U_n^*$ which is represented by the row $\psi = [f(\bar{u}_1) \ldots f(\bar{u}_n)]$ in an ON basis $\{\bar{u}_1,\ldots,\bar{u}_n\}$ in U_n. When the first basis is replaced by the second ON basis $\{\bar{u}_1',\ldots,\bar{u}_n'\}$ by the unitary replacement matrix $\mathscr{R} = [r_{ij}]_{n\times n}$, $\mathscr{R}^{-1} = \mathscr{R}^\dagger$: $\bar{u}_i' = \sum_{j=1}^n r_{ij}\bar{u}_j$, then the representing row of f is changed into $\psi' = \psi\mathscr{R}^T$ or in terms of components

$$f(\bar{u}_i') = \sum_{j=1}^n r_{ij}f(\bar{u}_j), \quad i = 1,\ldots,n.$$

[We may repeat the proof: $f(\bar{u}_i') = f(\sum_{j=1}^n r_{ij}\bar{u}_j) = \sum_{j=1}^n r_{ij}f(\bar{u}_j) = \sum_{j=1}^n f(\bar{u}_j)\{\mathscr{R}^T\}_{ji}$, which is expressed in a concise matrix form as $\psi' = \psi\mathscr{R}^T$].

Here, we have a one index system of n complex numbers which is connected with an ON basis in U_n, and which system changes (varies) in the same way as the ON basis (we say that this system is a *covariant n-vector*).

The tensor product of U_n^* with itself (the tensorial square $\overset{2}{\otimes} U_n^*$) is represented so that each elementary vector $f_1 \otimes f_2$ is represented in an ON basis in U_n by the Kronecker product of representing rows $\psi_1 \mathbin{\text{Ⓚ}} \psi_2$. In another ON basis in U_n, the same elementary vector is represented by $\psi_1' \mathbin{\text{Ⓚ}} \psi_2' = \psi_1\mathscr{R}^T \mathbin{\text{Ⓚ}} \psi_2\mathscr{R}^T = (\psi_1 \mathbin{\text{Ⓚ}} \psi_2)(\mathscr{R}^T \mathbin{\text{Ⓚ}} \mathscr{R}^T)$. In terms of components

$$f_1(\bar{u}_\alpha')f_2(\bar{u}_\beta') = \sum_{i,j=1}^n r_{\alpha i}r_{\beta j}f(\bar{u}_i)f(\bar{u}_j). \quad (**)$$

We get a two index system of n^2 complex numbers $\{f_1(\bar{u}_i)f_2(\bar{u}_j) \mid i,j = 1,\ldots,n\}$ which is attached to an ON basis in U_n, and which changes by the above expression $(**)$ when the ON basis is changed into the new one by the replacement matrix $\mathscr{R} \in U(n)$. This is the principal example of a unitary tensor which is twice covariant, since both indices vary by the matrix $\mathscr{R}$ itself.

This example is easily generalized to $\overset{p}{\otimes} U_n^*$, so that we get unitary tensors which are p-times covariant.

Now, we shall discuss a further generalization: we consider $\overset{p}{\otimes} U_n^*$, the tensor product of U_n^* with itself p times, which is multiplied tensorially with $\overset{m}{\otimes} U_n$, the tensor product of U_n with itself m times. We represent an elementary vector from

$$(\overset{p}{\otimes} U_n^*) \otimes (\overset{m}{\otimes} U_n)$$

in an ON basis in U_n and get a system of n^{p+m} complex numbers ($p + m$ index system) in the Kronecker product of p rows and m columns (which are all n-vectors). When we go over to a new ON basis in U_n by the unitary replacement matrix $\mathscr{R}$, this system of complex numbers goes to the new one, so that p indices are changed by the matrix $\mathscr{R}$, while the other m indices are changed by the contragredient matrix $\mathscr{R}^* = (\mathscr{R}^\dagger)^T = (\mathscr{R}^{-1})^T = \mathscr{S}$.

This is the principal example of a unitary tensor which is p times covariant and m times contravariant.

6.5 Unitary Space of Antilinear Operators $\hat{L}_a(U_m, U_n)$ as the Main Realization of $U_m \otimes U_n$

We denote by $\hat{L}_a(U_m, U_n)$ the set of all antilinear operators which map the unitary space U_m into the unitary space U_n, where the index "a" denotes the antilinear nature of the members of the set.

The members of this set are all antilinear operators A_a which map the whole U_m onto a subspace (proper or improper) of the space U_n. But, they map a linear combination of vectors from U_m onto the linear combination of unique images in U_n with the complex conjugate coefficients:

$$A_a(a\bar{x}_1 + b\bar{x}_2) = a^* A_a \bar{x}_1 + b^* A_a \bar{x}_2, \quad \bar{x}_1, \bar{x}_2 \in U_m; \ a, b \in \mathbb{C}.$$

This set of antilinear operators is obviously a vector space, because the sum of any two antilinear operators A_a and B_a is defined as the antilinear operator $C_a = A_a + B_a$:

$$C_a \bar{x} = (A_a + B_a)\bar{x} = A_a \bar{x} + B_a \bar{x}, \quad \text{for every } \bar{x} \in U_m.$$

The second "+" sign is the addition in U_n, which makes U_n an Abelian group, so that this addition in $\hat{L}_a(U_m, U_n)$ makes this set an Abelian group as well. It is closed, commutative, associative; there exists the additive inverse $-A_a$ for every A_a; and there exists the additive identity (neutral)—the zero operator $\hat{0}\bar{x} = \bar{0}$, $\forall \bar{x} \in U_m (A_a + \hat{0} = A_a)$. We also define the product of the antilinear operator A_a with a complex number c (scalar) as the antilinear operator $D_a = cA_a : D_a \bar{x} = (cA_a)\bar{x} = c(A_a \bar{x})$, $\forall \bar{x} \in U_m$, $c \in \mathbb{C}$. This product satisfies all four properties of scalar multiplication.

Every operator $A_a \in \hat{L}_a(U_m, U_n)$ generates by means of the expression $(A_a \bar{x}, \bar{y})$, $\bar{x} \in U_m$, $\bar{y} \in U_n$ (this inner product is obviously in U_n), for a fixed $\bar{y}$, a linear map between U_n and U_m^* (the unitary space of linear functionals in U_m): it adjoins the complex number $(A_a \bar{x}, \bar{y})$ to every vector $\bar{x} \in U_m$ in a linear fashion:

$$(A_a(a\bar{x}_1 + b\bar{x}_2), \bar{y}) = (a^* A_a \bar{x}_1 + b^* A_a \bar{x}_2, \bar{y}) = a(A_a \bar{x}_1, \bar{y}) + b(A_2 \bar{x}_2, \bar{y})$$

(remember that the inner product is antilinear in the first factor). We can denote this linear functional in U_m by $f_{A_a,\bar{y}}$, and the map itself as $\bar{y} \mapsto f_{A_a,\bar{y}}$. A linear combination of $\bar{y}$s is mapped on the same linear combination of functionals:

$$f_{A_a, c\bar{y}_1 + d\bar{y}_2} = (A_a \bar{x}, c\bar{y}_1 + d\bar{y}_2) = c(A_a \bar{x}, \bar{y}_1) + d(A_a \bar{x}, \bar{y}_2) =$$
$$= c f_{A_a, \bar{y}_1} + d f_{A_a, \bar{y}_2},$$

so this map $U_n \to U_m^*$ is linear.

Furthermore, every linear functional in U_m (vector in U_m^*) is mapped on the corresponding dual vector $\bar{x}'$ in U_m by the inverse antilinear dualism D^{-1} (Sect. 4.1):

$$\bar{x}' = D^{-1} f_{A_a, \bar{y}}.$$

So, we now have a map $\bar{y} \mapsto \bar{x}'$ which is antilinear

$$D^{-1}(A_a \bar{x}, c\bar{y}_1 + d\bar{y}_2) = D^{-1}(c f_{A_a, \bar{y}_1} + d f_{A_a, \bar{y}_2}) =$$
$$= c^* D^{-1} f_{A_a, \bar{y}_1} + d^* D^{-1} f_{A_a, \bar{y}_2} = c^* \bar{x}_1' + d^* \bar{x}_2'.$$

We denote this antilinear map $U_n \to U_m$ (which depends on A_a) by $A_a^\dagger$:

$$A_a^\dagger \bar{y} = \bar{x}', \quad \bar{y} \in U_n, \ \bar{x}' \in U_m.$$

Remembering *The Fundamental Formula of Dualism* (FFD) (Sect. 4.1) $f(\bar{x}) = (\bar{x}', \bar{x}) = = (\bar{x}, \bar{x}')^*$, $\bar{x}, \bar{x}' \in U_m$; $f \in U_m^*$ where $\bar{x}' = D^{-1} f$, we have $(A_a \bar{x}, \bar{y}) = f_{A_a, \bar{y}}(\bar{x}) = (\bar{x}', \bar{x}) = = (\bar{x}, \bar{x}')^* = (\bar{x}, A_a^\dagger \bar{y})^*$, the last inner product is in U_m.

So, $A_a^\dagger \in \hat{L}_a(U_n, U_m)$, and it is uniquely determined by A_a

$$(A_a \bar{x}, \bar{y}) = (\bar{x}, A_a^\dagger \bar{y})^*, \quad \bar{x} \in U_m, \ \bar{y} \in U_n.$$

We call this antilinear map $A_a^\dagger$ from U_n into U_m the *adjoint* of A_a.

The inner product in complex vector space $\hat{L}_a(U_m, U_n)$ is defined as

$$(A_a, B_a) = \operatorname{Tr} B_a^\dagger A_a,$$

where $B_a^\dagger A_a$ is a linear operator in U_m, and its trace $\operatorname{Tr} B_a^\dagger A_a = \sum_{i=1}^{m}(\bar{u}_i, B_a^\dagger A_a \bar{u}_i)$ we can calculate in any ON basis $\{\bar{u}_1, \ldots, \bar{u}_m\}$ in U_m.

Proof of the three basic properties of the inner product in any complex vector space.

(1) Skew-symmetry: $(A_a, B_a) = (B_a, A_a)^*$.

$$(B_a, A_a)^* = (\operatorname{Tr} A_a^\dagger B_a)^* = \sum_{i=1}^{m}(\bar{u}_i, A_a^\dagger B_a \bar{u}_i)^* = \sum_{i=1}^{m}(A_a^\dagger(B_a \bar{u}_i), \bar{u}_i) =$$
$$= \sum_{i=1}^{m}(B_a \bar{u}_i, A_a \bar{u}_i)^* = \sum_{i=1}^{m}(\bar{u}_i, B_a^\dagger A_a \bar{u}_i) = \operatorname{Tr} B_a^\dagger A_a = (A_a, B_a).$$

(2) Linearity in the second factor: $(A_a, bB_a + cC_a) = b(A_a, B_a) + c(A_a, C_a)$.

$$(A_a, bB_a + cC_a) = \text{Tr}\,(bB_a + cC_a)^\dagger A_a = \sum_{i=1}^m (\bar{u}_i, (bB_a + cC_a)^\dagger A_a \bar{u}_i) =$$

$$= \sum_{i=1}^m (\bar{u}_i, (bB_a)^\dagger A_a \bar{u}_i) + \sum_{i=1}^m (\bar{u}_i, (cC_a)^\dagger A_a \bar{u}_i) =$$

$$= \sum_{i=1}^m b(\bar{u}_i, B_a^\dagger A_a \bar{u}_i) + c \sum_{i=1}^m (\bar{u}_i, C_a^\dagger A_a \bar{u}_i) =$$

$$= b(A_a, B_a) + c(A_a C_a),$$

$$\text{since, } (bB_a)^\dagger = bB_a^\dagger \text{ and } (cC_a)^\dagger = cC_a^\dagger.$$

(Note: $(A_a \bar{x}, \bar{y}) = (\bar{x}, A_a^\dagger \bar{y})^* \Rightarrow a^*(A_a \bar{x}, \bar{y}) = a^*(\bar{x}, A_a^\dagger \bar{y})^* \Rightarrow ((aA_a)\bar{x}, \bar{y}) = (\bar{x}, aA_a^\dagger \bar{y})^*$, so $(aA_a)^\dagger = aA_a^\dagger$).

(3) Positive definiteness: $(A_a, A_a) > 0$ for $A_a \neq \hat{0}$.

$$(A_a, A_a) = \text{Tr}\,A_a^\dagger A_a = \sum_{i=1}^m (\bar{u}_i, A_a^\dagger A_a \bar{u}_i) = \sum_{i=1}^m (A_a \bar{u}_i, A_a \bar{u}_i)^* =$$

$$= \sum_{i=1}^m ||A_a \bar{u}_i||^2 > 0.\ \Delta$$

Thus $\hat{L}_a(U_m, U_n)$ is a unitary space. Its dimension is $m \cdot n$, which can be shown analogously as in Sect. 6.1. for $\hat{L}(V_n, W_m)$.

This $m \cdot n$ dimensional unitary space will become the tensor product of U_m and U_n, i.e., $\hat{L}_a(U_m, U_n) = U_m \otimes U_n$, if we define the map φ which takes the Cartesian product $U_m \times U_n$ into $\hat{L}_a(U_m, U_n)$ so that the two axioms from Sect. 6.2.2 are satisfied.

The map φ is defined so that to each ordered pair of vectors $[\bar{x}, \bar{y}]$ $\bar{x} \in U_m$ and $\bar{y} \in U_n$, we adjoin an antilinear operator [elementary vector $(\bar{x} \otimes \bar{y})_a$ in $\hat{L}_a(U_m, U_n)$] which is given by its action on an arbitrary vector $\bar{z} \in U_m$:

$$(\bar{x} \otimes \bar{y})_a \bar{z} = (\bar{z}, \bar{x})\,\bar{y}.$$

This operator maps the whole space U_m onto the one-dimensional subspace $L(\bar{y})$ in U_n which is spanned by the vector $\bar{y}$. It is antilinear, since z is in the antilinear factor of the inner product in U_m:

$$(\bar{x} \otimes \bar{y})_a(a\bar{z}_1 + b\bar{z}_2) = (a\bar{z}_1 + b\bar{z}_2, \bar{x})\bar{y} = a^*(\bar{z}_1, \bar{x})\bar{y} + b^*(\bar{z}_2, \bar{x})\bar{y} =$$

$$= a^*(\bar{x} \otimes \bar{y})_a \bar{z}_1 + b^*(\bar{x} \otimes \bar{y})_a \bar{z}_2.$$

We shall omit "a" in $(\bar{x} \otimes \bar{y})_a$ for the sake of simplicity in notation.

The adjoint operator of $\bar{x} \otimes \bar{y}$, i.e., $(\bar{x} \otimes \bar{y})^\dagger$, is the elementary vector $\bar{y} \otimes \bar{x}$ from $\hat{L}_a(U_n, U_m)$, which acts analogously on an arbitrary vector $\bar{u} \in U_n$ as $\bar{y} \in U_n$, $\bar{x} \in U_m$, $(\bar{y} \otimes \bar{x})\bar{u} = (\bar{u}, \bar{y})\bar{x}$. [It maps U_n onto $L(\bar{x})$].

Proof

$$((\bar{x} \otimes \bar{y})\bar{u}, \bar{v}) = (\bar{u}, \bar{x})^*(\bar{y}, \bar{v}) = (\bar{x}, \bar{u})(\bar{y}, \bar{v}) \text{ and}$$
$$(\bar{u}, (\bar{y} \otimes \bar{x})\bar{v})^* = (\bar{v}, \bar{y})^*(\bar{u}, \bar{x})^* = (\bar{x}, \bar{u})(\bar{y}, \bar{v}), \text{ so that}$$
$$((\bar{x} \otimes \bar{y})\bar{u}, \bar{v}) = (\bar{u}, (\bar{y} \otimes \bar{x})\bar{v})^*, \text{ and } (\bar{x} \otimes \bar{y})^\dagger = \bar{y} \otimes \bar{x}. \ \Delta$$

We shall now demonstrate that the map φ satisfies the two axioms from Sect. 6.2.2.

(1) The map φ is bilinear:

$$((\bar{x}_1 + \bar{x}_2) \otimes \bar{y})\bar{z} = (\bar{z}, \bar{x}_1 + \bar{x}_2)\bar{y} = (\bar{z}, \bar{x}_1)\bar{y} + (\bar{z}, \bar{x}_2)\bar{y}$$
$$= (\bar{x}_1 \otimes \bar{y})\bar{z} + (\bar{x}_2 \otimes \bar{y})\bar{z};$$
$$(\bar{x} \otimes (\bar{y}_1 + \bar{y}_2))\bar{z} = (\bar{z}, \bar{x})(\bar{y}_1 + \bar{y}_2) = (\bar{z}, \bar{x})\bar{y}_1 + (\bar{z}, \bar{x})\bar{y}_2 =$$
$$= (\bar{x} \otimes \bar{y}_1)\bar{z} + (\bar{x} \otimes \bar{y}_2)\bar{z};$$
$$a(\bar{x} \otimes \bar{y})\bar{z} = a(\bar{z}, \bar{x})\bar{y} = (\bar{z}, \bar{x})(a\bar{y}) = (\bar{z}, a\bar{x})\bar{y} =$$
$$= (\bar{x} \otimes a\bar{y})\bar{z} = (a\bar{x} \otimes \bar{y})\bar{z}.$$

(2) If $\{\bar{u}_1, \ldots, \bar{u}_m\}$ and $\{\bar{v}_1, \ldots, \bar{v}_n\}$ are two ON bases in U_m and U_n, respectively, then $\{\bar{u}_i \otimes \bar{v}_j \mid i = 1, \ldots, m; \ j = 1, \ldots, n\}$ is an ON basis in $\hat{L}_a(U_m, U_n)$:

$$(\bar{u}_k \otimes \bar{v}_l, \bar{u}_p \otimes \bar{v}_q) = \text{Tr} \, (\bar{u}_p \otimes \bar{v}_q)^\dagger (\bar{u}_k \otimes \bar{v}_l) = \sum_{i=1}^{m} (\bar{u}_i, (\bar{v}_q \otimes \bar{u}_p)(\bar{u}_k \otimes \bar{v}_l)\,\bar{u}_i) =$$
$$= \sum_{i=1}^{m} (\bar{u}_i, (\bar{v}_q \otimes \bar{u}_p)(\underbrace{\bar{u}_i, \bar{u}_k}_{\delta_{ik}})\,\bar{v}_l) = (\bar{u}_k, (\underbrace{\bar{v}_l, \bar{v}_q}_{\delta_{lq}})\,\bar{u}_p) = \delta_{kp}\delta_{lq}. \ \Delta$$

Therefore, $U_m \otimes U_n = \hat{L}_a(U_m, U_n)$.

6.6 Comparative Treatment of Matrix Representations of Linear Operators from $\hat{L}(U_m, U_n)$ and Antimatrix Representations of Antilinear Operators from $\hat{L}_a(U_m, U_n) = U_m \otimes U_n$

(A) We shall first consider $\hat{L}(U_m, U_n)$ and find matrix representations of linear operators that map U_m into U_n in a pair of ON bases

$$u = \{\bar{u}_1, \ldots, \bar{u}_m\} \text{ in } U_m \text{ and } v = \{\bar{v}_1, \ldots, \bar{v}_n\} \text{ in } U_n.$$

The general formula for matrix representation was discussed in Sect. 2.6.4: Given $L \in \hat{L}(U_m, U_n)$, we expand the images of vectors from basis u $\{L\bar{u}_1, \ldots, L\bar{u}_m\}$ in basis v, so that the expansion coefficients form the columns of the representation matrix M:

$$L\bar{u}_j = \sum_{k=1}^{n} a_{kj}\bar{v}_k, \ j = 1, \ldots, m. \ (*)$$

More explicitly,

$$[L\bar{u}_j]_v = \begin{bmatrix} a_{1j} \\ a_{2j} \\ \vdots \\ a_{nj} \end{bmatrix}, \ j = 1,\ldots,m \text{ (the } j\text{-th column of M).}$$

Now, multiplying $(*)$ from the left with $\bar{v}_i$, $i = 1,\ldots,n$, we get

$$(\bar{v}_i, L\bar{u}_j) = \sum_{k=1}^{n} a_{kj} \underbrace{(\bar{v}_i, \bar{v}_k)}_{\delta_{ik}} = a_{ij}, \ i = 1,\ldots,n; \ j = 1,\ldots,m.$$

So, we have the $n \times m$ matrix $M = [(\bar{v}_i, L\bar{u}_j)]_{n \times m}$, which represents $L \in \hat{L}(U_m, U_n)$ on the pair of ON bases u and v.

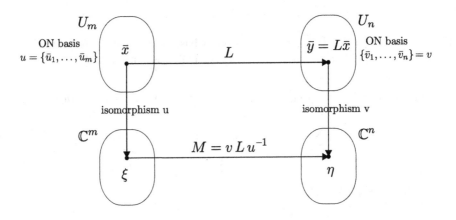

$$[\bar{x}]_u = \xi = \begin{bmatrix} \xi_1 \\ \xi_2 \\ \vdots \\ \xi_m \end{bmatrix} = \begin{bmatrix} (\bar{u}_1, \bar{x}) \\ (\bar{u}_2, \bar{x}) \\ \vdots \\ (\bar{u}_m, \bar{x}) \end{bmatrix}, [\bar{y}]_v = \eta = \begin{bmatrix} \eta_1 \\ \eta_2 \\ \vdots \\ \eta_n \end{bmatrix} = \begin{bmatrix} (\bar{v}_1, L\bar{x}) \\ (\bar{v}_2, L\bar{x}) \\ \vdots \\ (\bar{v}_n, L\bar{x}) \end{bmatrix} = [L\bar{x}]_v.$$

We have to prove $M\xi = \eta$ or $\boxed{M[\bar{x}]_u = [L\bar{x}]_v}$.

Proof

$$\boxed{M\xi = \eta} \text{ or } [(\bar{v}_i, L\bar{u}_j)]_{n \times m} \begin{bmatrix} (\bar{u}_1, \bar{x}) \\ (\bar{u}_2, \bar{x}) \\ \vdots \\ (\bar{u}_m, \bar{x}) \end{bmatrix}_{m \times 1} = \begin{bmatrix} (\bar{v}_1, L\bar{x}) \\ (\bar{v}_2, L\bar{x}) \\ \vdots \\ (\bar{v}_n, L\bar{x}) \end{bmatrix}_{n \times 1}$$

The product of the first row of M with ξ:

$$\sum_{j=1}^{m} (\bar{v}_1, L\bar{u}_j)(\bar{u}_j, \bar{x}) = \sum_{j=1}^{m} (L^\dagger \bar{v}_1, \bar{u}_j)(\bar{u}_j, \bar{x}) = \quad \text{(on using Parseval's identity, Sect. 3.3)}$$

$$= (L^\dagger \bar{v}_1, \bar{x}) = (\bar{v}_1, L\bar{x}).$$

For other rows of M we do the same. $\quad \Delta$

Remark Every operator $L \in \hat{L}(U_m, U_n)$ defines the unique adjoint operator $L^\dagger \in \hat{L}(U_n, U_m)$, analogously as we defined the adjoint operator in $\hat{L}(U_n, U_n)$ (Sect. 4.2),

$$(\bar{x}, L\bar{y}) = (L^\dagger \bar{x}, \bar{y}), \; \bar{x} \in U_m, \; \bar{y} \in U_n.]$$

(B) To handle antilinear operators from $\hat{L}_a(U_m, U_n)$ properly, we have to introduce the standard factorization of A_a into the linear factor and the conjugation: we define a special antilinear operator K_u which leaves invariant all vectors from the chosen ON basis $u = \{\bar{u}_1, \ldots, \bar{u}_m\}$ in U_m

$$K_u \bar{u}_i = \bar{u}_i, \; i = 1, \ldots, m.$$

This operator is obviously an involution $K_u^2 = I_{U_m}$. Next, we consider the standard factorization of A_a

$$A_a = (A_a K_u) K_u,$$

which is obviously basis dependant. So, $A_a K_u$ is a linear operator, and its representation $n \times m$ matrix in the two ON bases u and $v = \{\bar{v}_1, \ldots, \bar{v}_n\}$ in U_m and U_n is obtained as the matrix representing any linear operator

$$(\bar{v}_i, (A_a K_u)\bar{u}_j) = (\bar{v}_i, A_a(K_u \bar{u}_j)) = (\bar{v}_i, A_a \bar{u}_j) = b_{ij}, \; j = 1, \ldots, m; \; i = 1, \ldots, n;$$

so $N = [b_{ij}]_{n \times m} = [(\bar{v}_i, A_a \bar{u}_j)]_{n \times m}$.

The antilinear operator K_u is represented by the operation K which is obtained by taking the complex conjugate of all matrices to the right.

The antilinear operator A_a is represented in u and v by

$$NK = [b_{ij}]_{n \times m} K = [(\bar{v}_i, A_a \bar{u}_j)]_{n \times m} K,$$

which is more complicated than an ordinary matrix, because K, though a very common operator in the space of complex column vectors, is not expressible as a matrix.

We refer to such a product of a matrix factor N and the complex conjugation K as an *antimatrix*.

Since K is the same for all antilinear operators, every antilinear operator is essentially represented by its matrix factor. There is no need to investigate the transformation properties of K because it is the same operation in the space of complex column vectors for any choice of bases.

Let us now consider the following figure

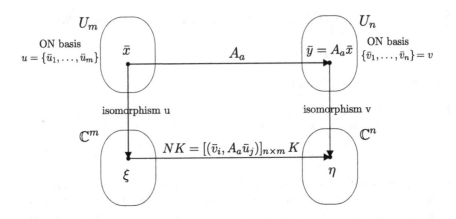

$$[\bar{x}]_u = \xi = \begin{bmatrix} \xi_1 \\ \xi_2 \\ \vdots \\ \xi_m \end{bmatrix} \begin{bmatrix} (\bar{u}_1, \bar{x}) \\ (\bar{u}_2, \bar{x}) \\ \vdots \\ (\bar{u}_m, \bar{x}) \end{bmatrix} \qquad [\bar{y}]_v = \eta = \begin{bmatrix} \eta_1 \\ \eta_2 \\ \vdots \\ \eta_n \end{bmatrix} = \begin{bmatrix} (\bar{v}_1, A_a\bar{x}) \\ (\bar{v}_2, A_a\bar{x}) \\ \vdots \\ (\bar{v}_n, A_a\bar{x}) \end{bmatrix} = [A_a\bar{x}]_v$$

We have to prove $(NK)\xi = \eta$ or $\boxed{NK[\bar{x}]_u = [A_a\bar{x}]_v}$ or

$$[(\bar{v}_i, A_a\bar{u}_j)]_{n\times m} K \begin{bmatrix} (\bar{u}_1, \bar{x}) \\ \vdots \\ (\bar{u}_m, \bar{x}) \end{bmatrix}_{m\times 1} = \begin{bmatrix} (\bar{v}_1, A_a\bar{x}) \\ \vdots \\ (\bar{v}_n, A_a\bar{x}) \end{bmatrix}_{n\times 1}$$

Proof The first row of NK multiplied with ξ:

$$\sum_{j=1}^m (\bar{v}_1, A_a\bar{u}_j) K (\bar{u}_j, \bar{x}) = \sum_{j=1}^m (A_a^\dagger\bar{v}_1, \bar{u}_j)^* (\bar{u}_j, \bar{x})^*$$
$$= (A_a^\dagger\bar{v}_1, \bar{x})^* = (\bar{v}_1, A_a\bar{x}),$$
on using Parseval's identity, Sect. 3.3.

For other rows of NK we do the same. $\triangle$

It is an interesting fact that the matrix factor

$$N = [(\bar{v}_i, A_a\bar{u}_j)]_{n\times m}$$

consists of the expansion coefficients of the antilinear operator A_a in the orthonormal basis $\{\bar{u}_j \otimes \bar{v}_i \,|\, i = 1,\dots,n,\ j = 1,\dots,m\}$ in $\hat{L}_a(U_m, U_n) = U_m \otimes U_n$:

$$A_a = \sum_{j=1}^{m} \sum_{i=1}^{n} (\bar{u}_j \otimes \bar{v}_i, A_a)(\bar{u}_j \otimes \bar{v}_i) = \sum_{j=1}^{m} \sum_{i=1}^{n} (\bar{v}_i, A_a \bar{u}_j)(\bar{u}_j \otimes \bar{v}_i).$$

Proof To prove this, we have only to calculate the Fourier coefficients in the expansion:

$$(\bar{u}_j \otimes \bar{v}_i, A_a) = \mathrm{Tr} A_a^{\dagger}(\bar{u}_j \otimes \bar{v}_i) = \sum_{k=1}^{m} (\bar{u}_k, A_a^{\dagger}(\bar{u}_j \otimes \bar{v}_i)\bar{u}_k) =$$

$$= \sum_{k=1}^{m} (\bar{u}_k, A_a^{\dagger} \underbrace{(\bar{u}_k, \bar{u}_j)}_{\delta_{kj}} \bar{v}_i) = (\bar{u}_j, A_a^{\dagger} \bar{v}_i) = (A_a \bar{u}_j, \bar{v}_i)^* = (\bar{v}_i, A_a \bar{u}_j). \quad \Delta$$

A Bibliographical Note

The first study of antilinear unitary operators was done by E.P.Wigner in J. Math. Phys. <u>1</u>, 409(1960).

Inspired by Wigner's paper, we investigated a few basic problems in *The Algebra of Antilinear Operators* in the paper *"Basic Algebra of Antilinear Operators and Some Applications"* by F.Herbut and M.Vujičić, J. Math. Phys. <u>8</u>, 1345(1967). Later on, we discovered three very important applications of antilinear operators and antimatrices:

(1) In the theory of *Superconductivity* (*"Antilinear Operators in Hartree-Bogolybov Theory"* by F. Herbut and M. Vujičić, Phys. Rev. <u>172</u>, 1031(1968));
(2) In the theory of *Quantum Entanglement* (*"A Quantum-Mechanical Theory of Distant Correlations"* by M. Vujičić and F. Herbut, J. Math. Phys. <u>25</u>, 2253(1984));
(3) In the theory of *Magnetic Groups* (*"Magnetic Line Groups"* by M. Damnjanović and M. Vujičić, Phys. Rev. <u>B25</u>, 6987(1982)).

Chapter 7

The Dirac Notation in Quantum Mechanics: Dualism between Unitary Spaces (Sect. 4.1) and Isodualism between Their Superspaces (Sect. 4.7)

7.1 Repeating the Statements about the Dualism D

Consider an n-dimensional unitary space U_n and choose an ON basis $u = \{\bar{u}_1 \ldots, \bar{u}_n\}$ in it. Then, every vector $\bar{x} \in U_n$ is represented in that basis by a matrix-column of Fourier coefficients

$$\xi = \begin{bmatrix} \xi_1 \\ \vdots \\ \xi_n \end{bmatrix} = \begin{bmatrix} (\bar{u}_1, \bar{x}) \\ \vdots \\ (\bar{u}_n, \bar{x}) \end{bmatrix}.$$

If the basis changes by a unitary replacement matrix $\mathscr{R}$ ($\mathscr{R}^{-1} = \mathscr{R}^\dagger$), then the representing column ξ changes into $\mathscr{R}^* \xi$ ($\mathscr{S} = (\mathscr{R}^{-1})^T = (\mathscr{R}^\dagger)^T = \mathscr{R}^*$) (Sect. 4.4), which means that $\bar{x}$ generates a *contravariant unitary vector* $\{\mathscr{R}^* \xi \mid \mathscr{R} \in U(n)\}$, i.e., a unitary tensor of the first order where a 1-index complex number system changes by the contragredient matrix $\mathscr{S} = \mathscr{R}^*$ with the change of ON basis.

Every linear functional $f \in \hat{L}(U_n, \mathbb{C})$ is represented in the same ON basis u by a matrix-row of the images of this basis $\varphi = [f(\bar{u}_1) \ldots f(\bar{u}_n)]$.

If the basis changes by $\mathscr{R}$ ($\bar{u}_i' = \sum_{j=1}^n r_{ij}\bar{u}_j$), then the representing row φ changes (like the basis) in $\varphi \mathscr{R}^T$ ($f(\bar{u}_i') = \sum_{j=1}^n r_{ij}f(\bar{u}_j)$) (Sect. 4.1), which means that the functional f generates a *covariant unitary vector* $\{\varphi \mathscr{R}^T \mid \mathscr{R} \in U(n)\}$.

All linear functionals in U_n form an n-dimensional unitary space $U_n^* = \hat{L}(U_n, \mathbb{C})$, called the *dual* space.

There is a basis-independent antilinear bijection (called *dualism D*) between U_n and U_n^* established by the fundamental formula of dualism (FFD):

$$\boxed{f(\bar{y}) = (\bar{x}, \bar{y})}, \ \forall \bar{y} \in U_n.$$

This formula determines the dual functional f of $\bar{x}$:

$$D\bar{x} = f,$$

which is consequently represented in u by the *adjoint* of ξ, i.e.,

$$\varphi = \xi^\dagger = [(\bar{u}_1, \bar{x})^* \dots (\bar{u}_n, \bar{x})^*].$$

The same FFD determines the dual vector $\bar{x}$ of f:

$$D^{-1} f = \bar{x},$$

which is now represented in u by the *adjoint* of φ, i.e.,

$$\xi = \varphi^\dagger = [f(\bar{u}_1)^* \dots f(u_n)^*]^T.$$

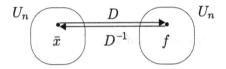

That this bijection is basis independent and antilinear has been shown in detail in Sect. 4.1.

The inner-product in U_n^* is defined by that in U_n:

$$(D\bar{x}_1, D\bar{x}_2) = (\bar{x}_1, \bar{x}_2)^*.$$

Evidently, we can introduce the dual basis $\{D\bar{u}_1, \dots, D\bar{u}_n\}$ of u in U_n^*, which is also orthonormal (ON):

$$(D\bar{u}_i, D\bar{u}_j) = (\bar{u}_i, \bar{u}_j)^* = \delta_{ij}.$$

In all pairs of dual ON bases, the dual vectors $\bar{x}$ and $\bar{f}$ (now we treat f as the vector $\bar{f}$ in U_n^*) are represented by complex conjugate columns ξ and ξ^*, respectively.

Nevertheless, we shall continue to regard all the relations between U_n and U_n^* from the point of view of an ON basis u in U_n: a linear functional f is represented in u as the matrix-row φ, which changes covariantly as $\varphi \mathscr{R}^T$ with the change of basis, while a vector $\bar{x}$ is represented in u as the matrix-column ξ, which changes contravariantly as $\mathscr{R}^* \xi$. If $\bar{x}$ and f are dual to each other, then their representing forms are adjoint to each other

$$\xi = \varphi^\dagger$$

in all ON bases in U_n.

For these reasons, in the Dirac notation (Sect. 4.7), the vector $\bar{x}$ is written as a *ket* $|x>$, where the left vertical segment | implicitly contains two hidden properties (which are rarely explained in the literature):

1. The contravariant change $\mathscr{R}^* | \bar{x} >$ of the representing column [here we use this short simplified form instead of the exact one $\{\mathscr{R}^* \xi \,|\, \mathscr{R} \in U(n)\}$], and

2. All operators are applied to $|x>$ from that side: $A \rightarrow |x>$, because $|x>$ is always represented by matrix-columns.

The linear functional f is written as a *bra* $<f|$, where the right vertical segment | also contains two hidden points of information:

1. The covariant change $<f|\mathcal{R}^T$ of the representing row, and
2. All operators are applied to $<f|$ from the right hand side: $<f|\leftarrow A$, because $<f|$ is represented by matrix-rows.

Remark In Quantum Mechanics, one usually writes between the graphic signs $|>$ and $<|$ the quantum numbers which uniquely determine the state vectors.

The action of the dualism D on the ket $|x>$ gives its dual bra, which we shall write from now on as $<x|$:

$$D|x>=<x|, \text{ and also } <x|D^{-1} = |x>.$$

In the Dirac notation, the fundamental formula of dualism FFD becomes very simple, which is one of the main features of Dirac notation. Namely, FFD is $(\bar{x}, \bar{y}) = [D\bar{x}](\bar{y})$, and in Dirac notation it can be written as

$$(|x>,|y>) = [D|x>](|y>) = [<x|](|y>) =<x||y>=<x|y>.$$

(The terms bra and ket come from this form of the FFD, because the graphic symbol $<|>$ is interpreted as bra¢ket: $<|>=$<bra | ket >).

This simple Dirac form of the inner product of kets by use of the FFD has this symmetric and elegant form, because it has built in the contraction rule. This rule can be formulated as follows: when two vertical segments, one left and one right, meet (as in the case $<x||y>$, where the functional $<x|$ acts on the vector $|y>$), then that can be united as $<x|y>$. The reason is that here, after tensor multiplication of the covariant vector $<x|\mathcal{R}^T$ with the contravariant vector $\mathcal{R}^*|y>$, which gives a mixed second order tensor, we have an annihilation of their tensorial properties: $<x|\mathcal{R}^T\mathcal{R}^*|y>=<x|y>$, because of

$$\mathcal{R}^T\mathcal{R}^* = (\mathcal{R}^\dagger\mathcal{R})^* = I_n \text{ (R is always a unitary matrix).}$$

Because of this contraction, the expression $<x|y>$ becomes a scalar, i.e., the tensor of zeroth order. In any ON basis in U_n, this is the unique number $\xi^\dagger\eta$.
Note The term inner product in tensor algebra is used for the tensor product with a contraction. Every contraction reduces the order of tensor product by two.

We shall have the same rule of merging vertical segments in many applications of Dirac notation, which appears as one of the main advantages of this notation (Sects. 7.2.2 and 7.3.2).

The inner product of bras (vectors in U_n^*) is obviously expressed as follows:

$$(<x|,<y|) = (D|x>,D|y>) = (|x>,|y>)^* = (|y>,|x>) =<y|x>.$$

We are already convinced about the usefulness of Dirac notation, but more is to come. This notation contracts many complicated mathematical properties of vectors and operators in U_n and U_n^* into simple graphical foolproof rules. This notation in an ingenious manner takes care of the tensor nature of these objects, especially when the tensor multiplication is followed by contractions.

7.2 Invariant Linear and Antilinear Bijections between the Superspaces $\hat{L}(U_n, U_n)$ and $\hat{L}(U_n^*, U_n^*)$

7.2.1 Dualism between the Superspaces

In Sect. 7.1, we have analyzed the invariant antilinear bijection D between U_n and U_n^* by means of the FFD. As is the case with every bijection between the spaces, the dualism D induces (by similarity transformation) a bijection $\hat{D}$ between the superspaces $\hat{L}(U_n, U_n)$ and $\hat{L}(U_n^*, U_n^*)$:

$$\hat{D}A = DAD^{-1} = A^* \text{ for every } A \in \hat{L}(U_n, U_n).$$

Obviously, $\hat{D}^{-1}A^* = D^{-1}A^*D = A$.

We call this map $\hat{D}$, the *dualism between the superspaces*, and A and A^* the *dual operators*. This is also an antilinear bijection, since D is just like that:

$$\hat{D}(aA + bB) = D(aA + bB)D^{-1} = a^*A^* + b^*B^*, \ a, b \in \mathbb{C}.$$

All properties of $\hat{D}$ are inherited from the properties of D.

When we act with A on a ket $|x>$, i.e., $A|x>$, then this is represented in an ON basis $u = \{\bar{u}_1, \ldots, \bar{u}_n\}$ in U_n by the matrix-column $\mathscr{A}\xi$, where the matrix elements of $\mathscr{A} = [a_{ij}]_{n \times n}$ are $a_{ij} = (\bar{u}_i, A\bar{u}_j)$, $i, j = 1, \ldots, n$, and $\xi = [\xi_1 \ldots \xi_n]^T$, where $\xi = (\bar{u}_i, \bar{x})$, $i = 1, \ldots, n$.

The dual vector $D(A|x>) = D(AD^{-1}D|x>) = (DAD^{-1})D|x>$ is represented in the dual basis $\{D\bar{u}_1, \ldots, D\bar{u}_n\}$ in U_n^* by the matrix-column $\mathscr{A}^*\xi^*$, since the dualism D when represented, acts between $\mathbb{C}^n$ and $\mathbb{C}^n$ as the complex conjugation K, so

$$K\mathscr{A}K^{-1} = K\mathscr{A}K = \mathscr{A}^* \text{ and } K\xi = \xi^*.$$

To get the representing row in the ON basis u in U_n, which is one characteristic of Dirac notation, we have only to transpose that column to get

$$\xi^\dagger \mathscr{A}^\dagger.$$

This means that the dual operators A and A^* are represented in the basis u by *adjoint matrices*, but the first one $\mathscr{A}$ acts on matrix-columns from the left as $\mathscr{A}_\rightarrow$, while the second one $\mathscr{A}^\dagger$ acts on matrix-rows from the right as $_\leftarrow\mathscr{A}^\dagger$.

It is quite natural to introduce the same notation for the operators:

$$\hat{D}(A_{\rightarrow}) = {}_{\leftarrow}A^{\dagger} \quad (*)$$

(instead of $\hat{D}(A_{\rightarrow}) = A^*_{\rightarrow}$, which is more natural when we represent vectors and operators in dual bases).

It is immediately clear that the dual of the adjoint of A, i.e., $\hat{D}(\mathscr{A}^{\dagger}_{\rightarrow})$, will be represented in the ON basis u in U_n by the same matrix $\mathscr{A}$ as the matrix of the operator $A_{\rightarrow}$, but the former acts from the right, i.e., ${}_{\leftarrow}\mathscr{A}$.

Therefore, the new rule for writing operators in dual spaces [see $(*)$ above] suggests the following:

$$\hat{D}(A^{\dagger}_{\rightarrow}) = {}_{\leftarrow}A.$$

This indicates that between the operator $A_{\rightarrow}$ from $\hat{L}(U_n, U_n)$ and the operator $\hat{D}(A^{\dagger}_{\rightarrow}) = {}_{\leftarrow}A$ from $\hat{L}(U_n^*, U_n^*)$, there exists some very deep connection. Indeed, this is the realization (for the special case of unitary spaces) of a fundamental connection between two superspaces which exists even in the case of the most general type of vector spaces $V_n(F)$, which we studied in Sect. 4.7, i.e., the invariant isomorphism (isodualism) between $\hat{L}(V_n(F), V_n(F))$ and $\hat{L}(V_n^*(F), V_n^*(F))$ which enables their identification.

7.2.2 Isodualism between Unitary Superspaces

In Sect. 4.7, we defined isodualism between the superspaces $\hat{L}(V_n(F), V_n(F))$ and $\hat{L}(V_n^*(F), V_n^*(F))$ as the *invariant linear bijection* between the pairs of operators from these superspaces that are connected by the fact that they are represented by the same matrix in every basis in $V_n(F)$. The only difference is that the partner from $\hat{L}(V_n^*(F), V_n^*(F))$ is represented by the matrix ${}_{\leftarrow}\mathscr{A}$ which acts on the rows that represent vectors from $V_n^*(F)$, while the partner from $\hat{L}(V_n(F), V_n(F))$ is represented by the matrix $\mathscr{A}_{\rightarrow}$ which acts on the columns that represent vectors from $V_n(F)$.

When we consider special vector spaces, for example, the unitary ones, i.e., U_n and its dual U_n^*, and their superspaces $\hat{L}(U_n, U_n)$ and $\hat{L}(U_n^*, U_n^*)$, then the isodual partners are any $A_{\rightarrow}$ which acts in U_n and $\hat{D}(A^{\dagger}_{\rightarrow}) = {}_{\leftarrow}A$ (the dual of the adjoint of $A_{\rightarrow}$) which acts in U_n^*. They are represented by matrices $\mathscr{A}_{\rightarrow}$ and ${}_{\leftarrow}\mathscr{A}$, respectively, in every ON basis in U_n.

From one ON basis to another related by the unitary replacement matrix $\mathscr{R}$, they both change by the same transformation:

$$\mathscr{R}^* \mathscr{A}_{\rightarrow} \mathscr{R}^T \text{ and } \mathscr{R}^* {}_{\leftarrow}\mathscr{A} \mathscr{R}^T.$$

Proof For the first matrix, we proved this kind of change in Sect. 4.4. For the second one, we know that in the dual ON basis in U_n^* the operator $\hat{D}(A^{\dagger})$ is represented by the matrix $(\mathscr{A}^{\dagger})^* = \mathscr{A}^T_{\rightarrow}$. With the change of ON basis by $\mathscr{R}$ in U_n, the dual (biorthogonal) basis in U_n^* changes by $(\mathscr{R}^{-1})^T = (\mathscr{R}^{\dagger})^T = \mathscr{R}^*$. The representing matrix $\mathscr{A}^T$ changes by the similarity transformation (Sect. 4.3.2) with the matrix contragredient to $\mathscr{R}^*$, i.e., $[(\mathscr{R}^*)^{-1}]^T = [(\mathscr{R}^*)^\dagger]^T = (\mathscr{R}^T)^T = \mathscr{R}$. Thus, the new representing matrix in U_n^* is $\mathscr{R}\mathscr{A}^T\mathscr{R}^{-1} = \mathscr{R}\mathscr{A}^T\mathscr{R}^\dagger$. Transposing this column (to get the representing matrix in U_n), we finally get

$$(\mathscr{R}\mathscr{A}^T_{\rightarrow}\mathscr{R}^\dagger)^T = \mathscr{R}^*_{\leftarrow}\mathscr{A}\mathscr{R}^T. \quad \Delta$$

We shall now give the general definition (Sect. 4.7) of the isodual operator A^T in $V_n^*(F)$

$$[A^T f](\bar{y}) = f(A\bar{y}), \; \forall \bar{y} \in V_n(F),$$

in Dirac notation

$$[<x|_{\leftarrow}A](|y>) = [<x|](A_{\rightarrow}|y>),$$

where we use the square bracket [] to enclose a functional that acts in U_n, while the round bracket ( ) encompasses a ket (a vector from U_n).

However, it is obviously too complicated graphically, since both sides can be written simply (as is done in Dirac notation in Quantum Mechanics) in the form $<x|A|y>$, where it is understood that A can act from the right on the bra $<x|$ as the isodual operator $_{\leftarrow}A$ or from the left on the ket $|y>$ as $A_{\rightarrow}$, in short

$$<x|_{\leftarrow}A|y> = <x|A_{\rightarrow}|y> \Leftrightarrow <x|A|y> .$$

This is one of the basic rules in Dirac notation.

Identification of isodual operators $A_{\rightarrow} \in \hat{L}(U_n, U_n)$ and $_{\leftarrow}A \in \hat{L}(U_n^*, U_n^*)$ in this Quantum Mechanical expression is rarely explained as a consequence of the invariant isomorphism between the superspaces. It is easily understood since such an explanation involves many sophisticated mathematical concepts.

Nevertheless, it should be once properly explained (as well as other features of the relation between U_n and U_n^*), and then use of the rules of Dirac notation implicitly takes care of everything and prevents mistakes in calculations.

We have shown the representation matrices of both isodual partners change as

$$\{\mathscr{R}^*\mathscr{A}\mathscr{R}^T \,|\, \mathscr{R} \in U(n)\},$$

which means that these operators generate mixed second order unitary tensors (once contravariant and once covariant). We can again write this in a short (incorrect) form for operators as $\mathscr{R}^*A\mathscr{R}^T$, or in Dirac notation as $|A|$, where the vertical left and right segments indicate contravariant and covariant change of the representing matrix, respectively, just as with kets and bras. Consequently, we have the following explicit tensor multiplication $<x|\mathscr{R}^T\mathscr{R}^*|A|\mathscr{R}^T\mathscr{R}^*|y>$, which gives a tensor of the fourth order, which with two contractions $\mathscr{R}^T\mathscr{R}^* = (\mathscr{R}^\dagger\mathscr{R})^* = I_n$ becomes finally $<x|A|y> = <x|A|y>$, a tensor of the zeroth order (a scalar) which is the unique number $\xi^\dagger\mathscr{A}\eta$ in any ON basis in U_n, once calculated in the given ON basis $u = \{\bar{u}_1, \ldots, \bar{u}_n\}$.

It is very instructive to compare the above compact definition of isodual operators

$$<x|A|y> \Leftrightarrow <x|_{\leftarrow}A|y> = <x|A_{\rightarrow}|y>$$

with the definition of the adjoint operator (Sects. 4.2 and 7.1)

$$(A^\dagger \bar{x}, \bar{y}) = (\bar{x}, A\bar{y}) \quad \text{for every } \bar{x}, \bar{y} \in U_n.$$

In Dirac notation the left hand side is $(A^\dagger | x >, | y >)$. Using the fundamental formula of dualism (FFD) which is $(| x >, | y >) = [D | x >](| y >)$, this left hand side of the above expression is equal to

$$[D(A^\dagger | x >)](| y >) = [(DA^\dagger D^{-1})D | x >](| y >) =$$
$$= [\hat{D}(A^\dagger)D | x >](| y >) = [< x |_\leftarrow A](| y >) \text{ (Sec. 7.2.1 and 7.2.2)}.$$

The right hand side becomes $(| x >, A | y >) = [D | x >](A | y >) = [< x |](A_\rightarrow | y >)$ (Sect. 7.2.2). In this way, we arrive at $< x |_\leftarrow A | y > = < x | A_\rightarrow | y >$ which is just the definition of isodual operators, and, finally, to its compact form

$$< x | A | y > .$$

Summary The definition of the *adjoint* operator in the space $U_n : (A^\dagger \bar{x}, \bar{y}) = (\bar{x}, A\bar{y})$, $\forall \bar{x}, \bar{y} \in U_n$, becomes one simple expression in Dirac notation $< x | A | y >$, so that the left hand side of the above definition of the adjoint operator $A^\dagger$ is $< x |_\leftarrow A | y >$, and the right hand side is $< x | A_\rightarrow | y >$. From this, it follows again that in a unitary space the isodual operator $_\leftarrow A$ of $A_\rightarrow$ is just the dual of its adjoint:

$$\hat{D}(A^\dagger_\rightarrow) = {}_\leftarrow A$$

or applied to vectors,

$$D(A^\dagger | x >) = < x | A.$$

Therefore, the combination of adjoining as an invariant autodualism in $\hat{L}(U_n, U_n)$ (Sect. 4.2) with an invariant dualism $\hat{D}$ between superspaces (Sect. 7.2.1) produces isodualism as an invariant isomorphism between these superspaces $\hat{L}(U_n, U_n) \overset{inv}{\cong} \hat{L}(U_n^*, U_n^*)$ which enables their identification.

Graphically, this can be presented as:

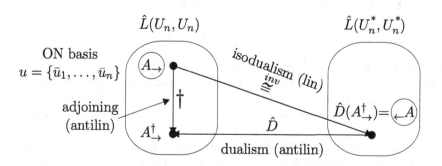

Note that the isodual operator of $\_A$, i.e., $A\_$, is obtained by inverse dualism $\hat{D}^{-1}$ followed by adjoining.

In any ON basis $u = \{\bar{u}_1, \ldots, \bar{u}_n\}$ in U_n, the isodual operators $A\_$ and $\_A$ are represented by the same matrix $\mathscr{A}$, which acts from the left on the representing columns of kets (vectors in U_n) as $\mathscr{A}\_\xi$ and from the right on the representing rows of bras (vectors in U_n^*) as $\varphi\_\mathscr{A}$, respectively.

The basic formula $< x|A|y >$ is the definition of the adjoint operators $A^\dagger$ and A, as well as of isodual operators $< x|\_A|y > = < x|A\_|y >$.

7.3 Superspaces $\hat{L}(U_n, U_n) \overset{inv}{\cong} \hat{L}(U_n^*, U_n^*)$ as the Tensor Product of U_n and U_n^*, i.e., $U_n \otimes U_n^*$

Repetition of the definition. (Sect. 6.2.2):

The tensor product of two unitary spaces U_m and U_n is the third unitary space $U_{m \cdot n}$ (whose dimension is the product of m and n), if we can define a map φ of the Cartesian product $U_m \times U_n$ into U_{mn} which is:

1. Bilinear, and
2. Such that the Cartesian product of two arbitrary ON bases (one in U_m and the other in U_n) is mapped onto an ON basis in U_{mn}. Then, we write

$$U_{m \cdot n} = U_m \otimes U_n.$$

The image of every ordered pair $[\bar{x}, \bar{y}]$, $\bar{x} \in U_m$, $\bar{y} \in U_n$, i.e., $\varphi[\bar{x}, \bar{y}]$, is written as $\bar{x} \otimes \bar{y}$ and it is called an elementary vector in U_{mn}. It is easy to show (Sect. 7.3.4) that the elementary vectors span U_{mn}, i.e., $\text{LIN}[\varphi(U_m \times U_n)] = U_m \otimes U_n$.

7.3.1 The Tensor Product of U_n and U_n^*

This product is very important, especially due to its applications in Quantum Mechanics. By a quite natural choice of the map φ we get

$$U_n \otimes U_n^* = \hat{L}(U_n, U_n) \overset{inv}{\cong} \hat{L}(U_n^*, U_n^*) \text{ (Sect. 7.3.3)},$$

i.e., both superspaces are the tensor product of U_n and U_n^*.

In this case, we shall call elementary vectors by a specific term: *diads*, and write them in Dirac notation as $|x><y|$ instead of $|x> \otimes <y|$, because it is a convention in Dirac notation that $><$ replaces $\otimes$, so $> \otimes <$ would unnecessarily duplicate the same information.

Thus, we map an ordered pair of vectors, consisting of a ket $|x> \in U_n$ and a bra $<y| \in U_n^*$, onto an operator (diad) $|x><y|$ which as isodual acts in both spaces U_n and U_n^*. (Notice that a diad is not a pair of vectors, but an associated operator.)

The action of the operator $|x><y|$ on the right (on kets from U_n) and on the left (on bras from U_n^*) is defined naturally in accordance with the Dirac rule of merging the vertical segments (since it means a contraction $\mathscr{R}^T\mathscr{R}^* = I_n$ in the tensor product).

So, the operator $|x><y|$ acts in U_n on an arbitrary ket $|z>$ as

$$(|x><y|)|z> = |x><y|z> = <y|z> |x>$$

(where $<y|z>$ is a complex number).

Consequently, diad $|x><y|$ is a very specific linear operator which maps the whole space U_n on $L(|x>)$, the one-dimensional subspace in U_n spanned by the ket $|x>$.

The same diad $|x><y|$ acts, as the isodual operator, acting on an arbitrary bra $<z|$ from U_n^* as

$$<z|(|x><y|) = <z|x><y|,$$

i.e., it maps the whole space U_n^* on the line $L(<y|)$.

The diad $|x><y|$ is not necessarily a projector in either case. It will be the projector on $L(|x>)$ if $<y|$ is equal to ort $<x|$ or $<y| = D|x>$, i.e.,

$$|x><x|, \quad \text{with} \quad <x|x> = 1,$$

in the first case, or

$$|y><y| \quad \text{with} \quad <y|y> = 1,$$

in the second case (the projector on $L(<y|)$).

Proof The operator $|x><x|$ is obviously

1. Hermitian: $(|x><x|)^\dagger = |x><x|$, since

$$(|x><y|_\rightarrow)^\dagger = |y><x|_\rightarrow.$$

To verify this last statement, we shall first take the dual of $|x><y|_\rightarrow$, i.e. $D(|x><y|_\rightarrow)D^{-1} =_\leftarrow |y><x|$, and then the isodual of the obtained operator, i.e., $|y><x|_\rightarrow$.

2. Idempotent: $(|x><x|)(|x><x|) = <x|x> (|x><x|) = |x><x|$.
3. Its trace is 1 (so it projects on a one-dimensional subspace): $\text{Tr}\,|x><x| = <x|x> = 1$ (see later Sect. 7.3.3 that $\text{Tr}\,|x><y| = <y|x>$).
4. It projects on $L(|x>)$, since it leaves the vector $|x>$ invariant: $(|x><x|)|x> = <x|x> |x> = |x>$. Δ

7.3.2 Representation and the Tensor Nature of Diads

Diads as operators in both U_n and U_n^* are represented by square $n \times n$ matrices (which act from the left on representing matrix-columns for kets or from the right on matrix-rows that represent bras).

However, the specific property in the representation of diads is that both factors in the tensor product $|x><y|$ are represented in the same ON basis in the first factor space U_n, e.g., in an ON basis of kets $u = \{|u_1>,\ldots,|u_n>\}$. A diad (as the tensor product of a ket $|x>$ and a bra $<y|$) is represented by the Kronecker product of the matrix-column ξ that represents $|x>$, i.e., $\xi = [<u_1|x> \ldots <u_n|x>]^T$, and the matrix-row $\eta^\dagger$ which represents the bra $<y|$, i.e., $\eta^\dagger = [<y|u_1> \ldots <y|u_n>]$ (note that η represents the ket $|y>$). The square matrix $\xi \: \textcircled{K} \: \eta^\dagger$ so obtained is the same as one which is the result of ordinary matrix multiplication of ξ and $\eta^\dagger$:

$$\xi \: \textcircled{K} \: \eta^\dagger = \xi \eta^\dagger \text{ (Sect. 6.1B)}.$$

The (i,j)-th element is obviously

$$<u_i|x><y|u_j>,$$

as it must be, since this result coincides (since the diad is an operator) with the general procedure ($a_{ij} =<u_i|A|u_j>$ for any operator A) for calculating matrix elements of the operator $|x><y|$ in the basis u.

It is useful to calculate the trace of the diad $|x><y|$ (remember that the trace is associated with the operator since it is the same number in all of its representations):

$$\text{Tr}\,|x><y| = \sum_{i=1}^{n} <u_i|x><y|u_i> = \sum_{i=1}^{n} <u_i|y>^*<u_i|x> =$$
$$= \eta^\dagger \xi = (|y>,|x>) =<y|x> \text{ (Sect. 7.1)}.$$

When we change an ON basis in U_n by a unitary replacement matrix $\mathscr{R}$, then the matrix that represents the diad $|x><y|$ becomes $(\mathscr{R}^*\xi)(\eta^\dagger \mathscr{R}^T) = \mathscr{R}^*(\xi\eta^\dagger)\mathscr{R}^T$, which means that this is a mixed unitary tensor of the second order, i.e., it behaves as the matrix of a linear operator, as it should do. This is the tensor product without a contraction when every factor brings its tensor nature, which graphically means that the left and right vertical segment in $|x><y|$ indicate the change $\mathscr{R}^*_\rightarrow$ and $_\leftarrow\mathscr{R}^T$, respectively.

We have already mentioned (Sect. 7.2.2) that with any other linear operator $A_\rightarrow$ in U_n, and for its isodual $_\leftarrow A$ in U_n^*, we can use the same two vertical segments in Dirac notation, i.e., write the operator A as $|A|$ in order to mark its tensor nature.

So, $|A||x>$ and $<x||A|$ are tensors of the third order, which after the contraction (the vertical segments merge) become $|A|x>$ and $<x|A|$, tensors of the first order, i.e., vectors. Similarly, $|A||B|$ is a tensor of the fourth order, which after the contraction $|AB|$ becomes a mixed tensor of second order.

7.3.3 The Proof of Tensor Product Properties

To prove that by the above choice of the map φ of the Cartesian product $U_n \times U_n^*$ into the superspace $\hat{L}(U_n, U_n)$, i.e.,

$$\varphi[|x>,<y|] = |x><y|_\rightarrow,$$

or into the superspace $\hat{L}(U_n^*,U_n^*)$

$$\varphi[|x>,<y|] =_\leftarrow |x><y|,$$

these two identified superspaces are the tensor product of U_n and U_n^*:

$$U_n \otimes U_n^* = \hat{L}(U_n,U_n) \stackrel{inv}{\cong} \hat{L}(U_n^*,U_n^*),$$

we have only to prove that the two axioms of the tensor product are satisfied.

These two axioms are 1) that the map φ is bilinear and 2) that the product of ON bases in factor spaces gives an ON basis in the product space.

Explicitly, the first axiom requires that the tensor product is linear in both factors:

(1) $(a|x_1 > +b|x_2 >) <y| = a|x_1 ><y| + b|x_2 ><y|,$
(2) $|x> (a<y_1| + b<y_2|) = a|x><y_1| + b|x><y_2|.$

These two requirements can be reduced to three more elementary ones:

(1') $(|x_1 > +|x_2 >) <y| = |x_1 ><y| + |x_2 ><y|$ (distributivity of addition in U_n),
(2') $|x> (<y_1| + <y_2|) = |x><y_1| + |x><y_2|$ (distributivity of addition in U_n^*),
(3') $a|x><y| = |x>a<y| = |x><y|a$ (commutation of a scalar with factors).

The procedure for proving these properties consists simply in applying the above operators to an arbitrary $|z>$ for $\hat{L}(U_n,U_n)$ or to an arbitrary $<z|$ for $\hat{L}(U_n^*,U_n^*)$.

To prove the second axiom, we take an arbitrary ON basis $\{|u_1 >,\ldots,|u_n >\}$ in U_n and its dual ON basis $\{<u_1|,\ldots,<u_n|\}$ in U_n^*, and show that by tensor multiplication they produce the ON basis

$$\{|u_i ><u_j| \,|\, i,j = 1,\ldots,n\}$$

in the product space $U_n \otimes U_n^*$. We call these product-vectors $|u_i ><u_j|, i,j = 1,\ldots,n$, *basis diads* in both superspaces $\hat{L}(U_n,U_n)$ and $\hat{L}(U_n^*,U_n^*)$.

Since there are n^2 basis diads, which is the dimension of both superspaces, it will be enough to show that the vectors from that set are orthonormal (which implies their linear independence) to get the desired ON basis.

For this proof, we need two formulas, which we have already demonstrated:

$$(|x><y|)^\dagger = |y><x| \text{ (Sect. 7.4.1) and } \mathrm{Tr}(|x><y|) =<y|x> \text{ (Sect. 7.3.2).}$$

Finally,

$$(|u_i ><u_j|, |u_m ><u_p|) = \mathrm{Tr}(|u_i ><u_j|)^\dagger (|u_m ><u_p|) = \mathrm{Tr}(|u_j ><u_i|)(|u_m ><u_p|)$$
$$= <u_i|u_m > \mathrm{Tr}(|u_j ><u_p|) =<u_i|u_m ><u_p|u_j >= \delta_{im}\delta_{jp}.$$

It is important to note that we would obtain the same result if we took some other ON basis $\{<u_1'|,\ldots,<u_n'|\}$ in U_n^*, different from the dual basis. The set of vectors

$\{|u_i><u'_j| \, | \, i,j = 1,\ldots n\}$ is also an ON basis in the superspaces, since the factor δ_{im} in the above result follows from the orthonormality of the basis in U_n, while the factor δ_{jp} follows from the orthonormality of the basis in U_n^*, regardless of which of them is chosen.

It is worthwhile pointing out that the inner product in $U_n \otimes U_n^*$ is linked with the inner products in U_n and U_n^*:

$$(|x><y|, |u><v|) = \mathrm{Tr}\,(|x><y|)^\dagger (|u><v|) = \mathrm{Tr}\,(|y><x|)(|u><v|) =$$
$$= <x|u> \mathrm{Tr}\,(|y><v|) = (<x|u>)(<v|y>) =$$
$$= (|x>, |u>)(<y|, <v|) \ \ (\text{Sect. 7.1}).$$

This result makes the above proof of orthonormality immediate.

7.3.4 Diad Representations of Operators

From the fact that basis diads make an ON basis in both superspaces $\hat{L}(U_n, U_n)$ and $\hat{L}(U_n^*, U_n^*)$, it follows that the set of all diads (elementary vectors in $U_n \otimes U_n^*$) spans both superspaces, i.e.,

$$\mathrm{LIN}[\varphi(U_n \times U_n^*)] = U_n \otimes U_n^*.$$

Thus, an arbitrary operator A from U_n (or from U_n^*) can be expressed as a sum of diads or as a linear combination of n^2 basis diads. To obtain either of these two expansions, we can use the expression for diad representation of the identity operator (otherwise a very useful formula). To get this representation, we first choose an ON basis in U_n, i.e., $\{|u_1>, \ldots, |u_n>\}$ and its dual basis $\{<u_1| \ldots, <u_n|\}$ in U_n^*. As we have already proved the projectors on the lines determined by these basis vectors are

$$|u_i><u_i|, \ i = 1,\ldots,n$$

$(|u_i><u_i| \rightarrow$ is the projector onto $L(|u_i>)$, while $\leftarrow |u_i><u_i|$ projects onto the line $L(<u_i|))$. Since the unitary space U_n is the orthogonal sum of the orthogonal lines $L(|u_i>)$, i.e.,

$$U_n = \sum_{i=1}^{n} {}^{\oplus} L(|u_i>) \ \ (\text{Sect. 5.5.4}),$$

and analogously for U_n^*, it follows (as we demonstrated in Sect. 5.5.4) that the projector on U_n, i.e., the identity operator I, can be obtained as the sum of projectors on these n orthogonal lines

$$I = \sum_{i=1}^{n} |u_i><u_i|.$$

From this very important and much used diad representation of the identity operator, we immediately get the diad representation of an arbitrary operator A from U_n:

$$A_\rightarrow = AI = \sum_{i=1}^{n} A|u_i\rangle\langle u_i| = \sum_{i=1}^{n} |v_i\rangle\langle u_i|_\rightarrow,$$

where $A|u_i\rangle = |v_i\rangle$ and $D|u_i\rangle = \langle u_i|$, $i = 1, \ldots, n$, i.e., every operator A from U_n can be expressed as the sum of n diads, in which the first vectors are the images by A of the given ON basis in U_n (as we saw previously the set of these images completely determines the operator A), and the second vectors are elements of the dual ON basis.

For the isodual operator $_\leftarrow A$ which acts in U_n^*, we have the analogous expression

$$_\leftarrow A = IA = \sum_{i=1}^{n} |u_i\rangle\langle u_i|A = \sum_{i=1}^{n} {}_\leftarrow|u_i\rangle\langle v_i|.$$

Example Let us consider the diad representation of a real 3×3 matrix $\mathscr{A}$ which is an operator in $\mathbb{R}^3$. The chosen ON basis is the standard basis

$$\{\bar{e}_1, \bar{e}_2, \bar{e}_3\} = \left\{ \begin{bmatrix} 1 \\ 0 \\ 0 \end{bmatrix}, \begin{bmatrix} 0 \\ 1 \\ 0 \end{bmatrix}, \begin{bmatrix} 0 \\ 0 \\ 1 \end{bmatrix} \right\}$$

in $\mathbb{R}^3$:

$$\mathscr{A} = \begin{bmatrix} a_{11} & a_{12} & a_{13} \\ a_{21} & a_{22} & a_{23} \\ a_{31} & a_{32} & a_{33} \end{bmatrix} = \mathscr{A}I_3 = \sum_{i=1}^{3} \mathscr{A}\bar{e}_i\bar{e}_i^T =$$

$$= \mathscr{A}\left(\begin{bmatrix} 1 \\ 0 \\ 0 \end{bmatrix} [1\,0\,0] + \begin{bmatrix} 0 \\ 1 \\ 0 \end{bmatrix} [0\,1\,0] + \begin{bmatrix} 0 \\ 0 \\ 1 \end{bmatrix} [0\,0\,1] \right) =$$

$$= \begin{bmatrix} a_{11} \\ a_{21} \\ a_{31} \end{bmatrix} [1\,0\,0] + \begin{bmatrix} a_{12} \\ a_{22} \\ a_{32} \end{bmatrix} [0\,1\,0] + \begin{bmatrix} a_{13} \\ a_{23} \\ a_{33} \end{bmatrix} [0\,0,1] =$$

$$= \begin{bmatrix} a_{11} & 0 & 0 \\ a_{21} & 0 & 0 \\ a_{31} & 0 & 0 \end{bmatrix} + \begin{bmatrix} 0 & a_{12} & 0 \\ 0 & a_{22} & 0 \\ 0 & a_{32} & 0 \end{bmatrix} + \begin{bmatrix} 0 & 0 & a_{13} \\ 0 & 0 & a_{23} \\ 0 & 0 & a_{33} \end{bmatrix},$$

$$\text{since } I_3 = \sum_{i=1}^{3} \bar{e}_i\bar{e}_i^T. \ \Delta$$

The diad representation of an arbitrary $A \in \hat{L}(U_n, U_n)$ in terms of basis diads can also be obtained easily and naturally:

$$A = IAI = \sum_{i,j=1}^{n} |u_i\rangle\langle u_i|A|u_j\rangle\langle u_j| =$$

$$= \sum_{i,j=1}^{n} \langle u_i|A|u_j\rangle |u_i\rangle\langle u_j| = \sum_{i,j=1}^{n} a_{ij}|u_i\rangle\langle u_j|,$$

where $a_{ij} = <u_i|A|u_j>$ are the elements of the matrix $\mathscr{A} = [a_{ij}]_{n \times n}$ which represents the operator A in the chosen ON basis $\{|u_1>, \ldots, |u_n>\}$ in U_n.

This formula gives an additional importance to the representation matrix $\mathscr{A}$, i.e., its elements are the coefficients in the expansion of A with respect to the ON diad basis and of course can be obtained as Fourier coefficients of A in this basis:

$$(A, |u_i><u_j|) = \mathrm{Tr}\,(|u_i><u_j|^\dagger)A = \mathrm{Tr}\,(|u_j><u_i|)A =$$
$$= \mathrm{Tr}\,(|u_j><u_i|A) = <u_i|A|u_j> . \text{ (Sect. 7.3.1 and 7.3.2)}.$$

The above formula for the expansion of the operator A with respect to the diad basis can be immediately obtained if we use the basic formula for representing the operator A:

$$A|u_j> = \sum_{i=1}^{n} a_{ij}|u_i>,$$

and put it in the diad representation of A:

$$A = \sum_{j=1}^{n} A|u_j><u_j| = \sum_{i,j=1}^{n} a_{ij}|u_i><u_j|.$$

So, these two diad representations are obtainable from each other by making use of the basic formula for the representation of A.

It is also interesting and useful to point out three more applications of the diad representation of the identity operator

$$I = \sum_{i=1}^{n} |u_i><u_i|$$

(here the two representations coincide since I is always represented by the identity matrix $I_n = [\delta_{ij}]_{n \times n}$:

$$I = \sum_{i,j=1}^{n} \delta_{ij}|u_i><u_j| = \sum_{i=1}^{n} |u_i><u_i|).$$

(1) Representation of a vector $|x>$ from U_n in the chosen ON basis (the expansion with Fourier coefficients):

$$|x> = I|x> = \sum_{i=1}^{n} |u_i><u_i|x> = \sum_{i=1}^{n} <u_i|x> |u_i> .$$

An analogous formula is valid for bras. (Sect. 3.2)
(2) Parseval's identity [the inner product of two vectors in the chosen ON basis expressed as the standard inner product of the representing columns (Sect. 3.2)]:

$$<x|y> \; = \; <x|I|y> \; = \sum_{i=1}^{n} <x|u_i><u_i|y> \; =$$

$$= \sum_{i=1}^{n} <u_i|x>^* <u_i|y> \; = \sum_{i=1}^{n} \xi_i^* \eta_i = \xi^\dagger \eta, \quad \text{where } \xi = [\xi_1 \ldots \xi_n]^T.$$

(3) Isomorphism between the algebra of operators $\hat{L}(U_n, U_n)$ and the algebra $\hat{L}(\mathbb{C}^n, \mathbb{C}^n)$ of representing matrices (Sect. 4.1)

(A) The product of operators AB is represented by the product $\mathcal{A}\mathcal{B}$ of the representing matrices:

$$AB = IAIIBI = \sum_{i,p,k,j=1}^{n} |u_i><u_i|A|u_p> \underbrace{<u_p||u_k>}_{\delta_{pk}} <u_k|B|u_j><u_j| =$$

$$= \sum_{i,j,k=1}^{n} |u_i><u_i|A|u_k><u_k|B|u_j><u_j| = \sum_{i,j,k=1}^{n} a_{ik}b_{kj}|u_i><u_j| =$$

$$= \sum_{i,j=1}^{n} \{\mathcal{A}\mathcal{B}\}_{ij}|u_i><u_j|,$$

here we use the alternative notation $a_{ij} = \{\mathcal{A}\}_{ij}$.

(B) The sum of operators $A + B$ is represented by the sum $\mathcal{A} + \mathcal{B}$ of the representing matrices:

$$A + B = IAI + IBI = \sum_{i,j=1}^{n} |u_i><u_i|A|u_j><u_j|$$

$$+ \sum_{i,j=1}^{n} |u_i><u_i|B|u_j><u_j| =$$

$$= \sum_{i,j=1}^{n} a_{ij}|u_i><u_j| + \sum_{i,j=1}^{n} b_{ij}|u_i><u_j| =$$

$$= \sum_{i,j=1}^{n} \{\mathcal{A} + \mathcal{B}\}|u_i><u_j|.$$

(C) The product aA of a complex scalar a and an operator A is represented by the product $a\mathcal{A}$ of a and the representing matrix $\mathcal{A}$.

$$aA = aIAI = a \sum_{i,j=1}^{n} |u_i><u_i|A|u_j><u_j| =$$

$$= a \sum_{i,j=1}^{n} a_{ij}|u_i><u_j| = \sum_{i,j=1}^{n} \{a\mathcal{A}\}_{ij}|u_i><u_j|.$$

Bibliography

1. Birkhoff G. and MacLane S., *A Survey of Modern Algebra*, Macmillan, New York, 1977.
2. Fraleigh J. B. and Beauregard R. A., *Linear Algebra* Addison-Wesley, Reading, 1995.
3. Lipschutz S., *Theory and Problems of Linear Algebra*, Schaum's Outline Series, McGraw-Hill, New York, 1991.
4. Halmos P. R. *Finite Dimensional Vector Spaces*, Springer, Berlin, 1974.
5. Merzbacher E., *Quantum Mechanics*, John Wiley & Sons, New York, 1970.

Index